Introduction to Combinatorial Methods in Geometry

This book offers an introduction to some combinatorial (also, set-theoretical) approaches and methods in geometry of the Euclidean space $\mathbf{R}^m$. The topics discussed in the manuscript are due to the field of combinatorial and convex geometry.

The author's primary intention is to discuss those themes of Euclidean geometry which might be of interest to a sufficiently wide audience of potential readers. Accordingly, the material is explained in a simple and elementary form completely accessible to college and university students. At the same time, the author reveals profound interactions between various facts and statements from different areas of mathematics: the theory of convex sets, finite and infinite combinatorics, graph theory, measure theory, classical number theory, etc.

All chapters (and also the five Appendices) end with a number of exercises. These provide the reader with some additional information about topics considered in the main text of this book. Naturally, the exercises vary in their difficulty. Among them there are almost trivial, standard, nontrivial, rather difficult, and difficult. As a rule, the more difficult exercises are marked by asterisks and are provided with necessary hints.

The material presented is based on the lecture course given by the author. The choice of material serves to demonstrate the unity of mathematics and variety of unexpected interrelations between distinct mathematical branches.

Alexander Kharazishvili is a head of the department of Discrete Mathematics at I. Vekua Institute of Applied Mathematics.

Introduction to Combinatorial Methods in Geometry

Alexander Kharazishvili

CRC Press
Taylor & Francis Group
Boca Raton London New York

CRC Press is an imprint of the
Taylor & Francis Group, an **informa** business
A CHAPMAN & HALL BOOK

First edition published 2024
by CRC Press
2385 Executive Center Drive, Suite 320, Boca Raton, FL 33431

and by CRC Press
4 Park Square, Milton Park, Abingdon, Oxon, OX14 4RN

CRC Press is an imprint of Taylor & Francis Group, LLC

Library of Congress Cataloging-in-Publication Data

Names: Kharazishvili, A. B., author.
Title: Introduction to combinatorial methods in geometry / authored by
Alexander Kharazishvili.
Description: First edition. | Boca Raton, FL : CRC Press , [2024] |
Includes bibliographical references and index.
Identifiers: LCCN 2023049592 | ISBN 9781032594705 (hbk) | ISBN
9781032603483 (pbk) | ISBN 9781003458708 (ebk)
Subjects: LCSH: Combinatorial geometry.
Classification: LCC QA167 .K448 2024 | DDC 516/.13--dc23/eng/20231220
LC record available at https://lccn.loc.gov/2023049592

ISBN: 978-1-032-59470-5 (hbk)
ISBN: 978-1-032-60348-3 (pbk)
ISBN: 978-1-003-45870-8 (ebk)

DOI: 10.1201/9781003458708

Typeset in CMR10 font
by KnowledgeWorks Global Ltd.

Contents

Preface

This manuscript may be treated as an introduction to some combinatorial (also, set-theoretical) approaches and methods in geometry of the Euclidean space $\mathbf{R}^m$. Actually, the topics discussed in the manuscript are due to the field of combinatorial and convex geometry (cf. [36], [76], [123], [132]).

The material presented below is based on the lecture course given by the author on seminars of the Department of Discrete Mathematics at I. Vekua Institute of Applied Mathematics (Tbilisi, Georgia).

Writing this book, we intended to discuss those themes of Euclidean geometry which might be of interest to a sufficiently wide audience of potential readers. Accordingly, we attempted to explain the material in a simple and elementary form completely accessible to college and university students. At the same time, we maximally tried to show profound interactions between various facts and statements from different areas of mathematics: the theory of convex sets, finite and infinite combinatorics, graph theory, measure theory, classical number theory, etc.

The reader of this book will be able to decide himself/herself whether we succeeded (at least, partly) in our choice of the material that serves to demonstrate the unity of mathematics and variety of unexpected interrelations between distinct mathematical branches.

Now, we would like to outline (briefly and schematically) the scope of this manuscript.

In Chapter 1 we introduce and examine the index of an isometric embedding with respect to a given group G of isometries of Euclidean space $\mathbf{R}^m$. The precise definition of the index of an isometric embedding is as follows (see [184]).

For a set $X \subset \mathbf{R}^m$, there exists a least cardinal number κ having the following property:

For every set $Y \subset \mathbf{R}^m$, if any subset Z of Y with cardinality not exceeding κ can be embedded into X by a transformation from G (which, in general, depends on Z), then the entire Y can also be embedded into X by some transformation from G.

It is natural to consider this κ as the index of an isometric embedding into X with respect to the group G, and accordingly it is convenient to use the notation $\kappa = \kappa_G(X)$. Chapter 1 contains some information about the behavior of $\kappa_G(X)$. In particular, the inequalities

$$\kappa_G(X) \leq \mathrm{card}(G), \qquad \kappa_G(X) \leq \mathrm{card}(X) + 1$$

are established. Furthermore, certain connections of the introduced notion with well-known Helly's theorem on intersections of convex sets in $\mathbf{R}^m$ are indicated (cf. Exercises 18 and 23 from Appendix 1). For instance, as readily follows from Helly's theorem, if B is an m-dimensional closed ball in $\mathbf{R}^m$ and G coincides with the group of all isometric transformations of $\mathbf{R}^m$, then

$$\kappa_G(B) = m + 1.$$

In the same chapter a characterization of all discrete groups of isometries of the space $\mathbf{R}^m$ is also presented, exclusively in terms of the index of an isometric embedding (see [184]).

Chapter 2 is devoted to *ot*-subsets of the space $\mathbf{R}^m$, where $m \geq 2$.

By definition, a set X lying in a real pre-Hilbert space H is an *ot*-set in H if each three-element subset of X forms an obtuse-angled triangle.

It should be noticed that *ot*-sets naturally arise in various topics of combinatorial and discrete geometry. For example, the set of all vertices of any so-called Carathéodory–Gale polyhedron produced by the moment curve in $\mathbf{R}^m$, where $m \geq 4$, is an *ot*-subset of $\mathbf{R}^m$.

It is easy to show that every finite *ot*-set in the space $\mathbf{R}^m$ can be expanded to a larger *ot*-set in the same space. So, no finite *ot*-set is maximal with respect to the standard inclusion relation. On the other hand, there are simple examples of maximal *ot*-sets in the plane $\mathbf{R}^2$ which have the cardinality continuum $\mathbf{c}$ (e.g., a semicircle without one of its endpoints).

The question naturally arises whether there exists a countable maximal *ot*-subset of $\mathbf{R}^2$. In Chapter 2 this question is solved in the affirmative (see also [202]). Moreover, it is demonstrated that there exists a maximal discrete *ot*-subset Z of $\mathbf{R}^2$. Clearly, such a Z is always countably infinite.

The analogous result is valid for the m-dimensional Euclidean space $\mathbf{R}^m$, where $m > 2$. Namely, there are maximal discrete *ot*-subsets of $\mathbf{R}^m$. In this context, it worth noticing that a maximal *ot*-set in the plane $\mathbf{R}^2$ is not, in general, a maximal *ot*-subset of the space $\mathbf{R}^3$.

In the same chapter, an *rt*-set Y in a real pre-Hilbert space H is defined as having the property that every three-element subset of Y forms a right-angled triangle. The geometric structure of *rt*-sets is described and, in particular, it is demonstrated that any *rt*-set is necessarily a separable subspace of H (see [177]). This circumstance immediately implies that the cardinality of an arbitrary *rt*-set in H does not exceed the cardinality continuum $\mathbf{c}$.

In Chapter 3 we consider *at*-sets and strong *at*-sets in a finite-dimensional Euclidean space and also in a real infinite-dimensional pre-Hilbert space. The definition of these sets is similar to the definitions of *ot*-sets and of *rt*-sets.

More precisely, a set X in an Euclidean space or in a real pre-Hilbert space is called an *at*-set (respectively, a strong *at*-set) if every three-element subset of X forms either an acute-angled triangle or a right-angled triangle (respectively, forms an acute-angled triangle).

For each natural number $m \geq 2$, denote by $at(m)$ the maximum of the cardinalities of at-sets in the space $\mathbf{R}^m$, and denote by $sat(m)$ the maximum of the cardinalities of strong at-sets in $\mathbf{R}^m$.

Answering the question posed by Erdös in one of his numerous works, Danzer and Grünbaum proved in their joint paper [67] that

$$at(m) = 2^m,$$

and the equality $\text{card}(X) = 2^m$ holds true for an at-set $X \subset \mathbf{R}^m$ if and only if X coincides with the set of all vertices of a right rectangular m-dimensional parallelepiped in $\mathbf{R}^m$.

A much harder problem is to determine the precise value of $sat(m)$. Even the case $m = 3$ is highly nontrivial and, as has been shown by Danzer and Grünbaum, the equality $sat(3) = 5$ holds true. They also conjectured that $sat(m) = 2m - 1$ for all natural numbers $m \geq 2$.

However, in [88] it was unexpectedly established that the growth of $sat(m)$ is exponential with respect to m. The method presented in [88] for estimating $sat(m)$ from below is based on some probabilistic arguments. In Chapter 3 we give a direct deterministic proof of the same fact by computing the total number of right angles in the family of all those triangles whose vertices belong to the set of vertices of a fixed m-dimensional cube in $\mathbf{R}^m$.

We also discuss a function analogous to the function $at(m)$ for a real infinite-dimensional Hilbert space H. Surprisingly, the growth of such a function with respect to the Hilbert dimension of H substantially depends on additional set-theoretical assumptions. For example, even for a Hilbert space H_{ω_1} whose Hilbert dimension is equal to the least uncountable cardinal ω_1, it is impossible to establish within **ZFC** set theory the precise growth of the maximum of all cardinalities of at-sets in H_{ω_1} with respect to ω_1 (cf. [219]).

Chapter 4 is devoted to similar questions concerning it-sets in a finite-dimensional Euclidean space. The definition of an it-set is as follows.

A set $Z \subset \mathbf{R}^m$ is called an it-set in $\mathbf{R}^m$ if all three-element subsets of Z form isosceles triangles.

Denote by $it(m)$ the maximum of the cardinalities of it-subsets of $\mathbf{R}^m$.

Analogously to the topics of Chapters 2 and 3, the problem naturally arises to find the precise values of $it(m)$.

This difficult problem was first formulated by Erdös in [85] and afterwards it was investigated by several authors (cf. [66], [168], [185], [230]). It should be mentioned that even in the case $m = 3$, for obtaining the equality $it(3) = 8$, a rather delicate argument is needed. Keeping in mind the result concerning at-subsets and strong at-subsets of $\mathbf{R}^m$, one might conjecture that the growth of $it(m)$ is also of an exponential character with respect to m. But, fortunately, it turns out that the following upper estimate is valid:

$$it(m) \leq (m + 2)(m + 1)/2.$$

The proof of this inequality is given in Chapter 4, using the method developed in [146] and [147] (see also the references therein). Moreover, we show tight connections of the above problem with 2-distance sets in $\mathbf{R}^m$ (a subset Y of a metric space is called a 2-distance set if the set of all nonzero distances between the points from Y is precisely two-element). Actually, it is demonstrated in Chapter 4 how the problem of finding upper estimates for $it(m)$ can be reduced to the analogous problem for 2-distance sets in $\mathbf{R}^m$ (see [16], [26]).

Speaking of combinatorial methods in mathematics, it is necessary to refer to Ramsey's famous theorem which has numerous applications in various mathematical fields and significantly stimulated further development of combinatorics. As is well known, there are two classical versions of Ramsey's theorem: finite and countably infinite (see [291]). Chapter 5 is devoted to some applications of both these versions to point sets lying in the Euclidean space $\mathbf{R}^m$ or in an infinite-dimensional Hilbert space H over $\mathbf{R}$. In the same chapter we give several examples which show that certain combinatorial geometric facts, obtained with the aid of Ramsey's theorem, concerning finite and countable point sets in $\mathbf{R}^m$ or in H do not admit a natural generalization to uncountable point sets in these spaces (cf. [216], [224]).

In their seminal paper [91], Erdös and Szekeres proved that, for any natural number $n \geq 3$, there exists a smallest natural number $c(n)$ possessing the following property:

Every set consisting of at least $c(n)$ points of the plane $\mathbf{R}^2$, which are in general position, contains a subset of n points which are convexly independent, i.e., they are the vertices of a certain convex polygon (actually, convex n-gon) in $\mathbf{R}^2$.

In connection with this exciting result, Erdös and Szekeres posed the problem to determine the precise value of $c(n)$. Moreover, they conjectured that

$$c(n) = 2^{n-2} + 1.$$

This intriguing problem is typical for combinatorial geometry and, naturally, it was investigated by many mathematicians. However, it is still far from being solved. An extensive survey about this problem and closely related questions of combinatorial geometry can be found in [265] where a long list of references is also presented. It should be noticed that the basic technical tool utilized by Erdös and Szekeres in [91] is the above-mentioned Ramsey combinatorial theorem. By applying this theorem, one easily gets a rough upper bound of the value $c(n)$ (for more details, see again [91]). The same method successfully works for point sets lying in Euclidean spaces of higher dimension.

In Chapter 6 we discuss two infinite variants of the Erdös–Szekeres problem. The main attention in Chapter 6 is paid to those infinite sets of points in $\mathbf{R}^2$, which either are infinitely countable or are of cardinality continuum $\mathbf{c}$. It turns out that every countably infinite set in $\mathbf{R}^2$, whose points are in general position, always contains an infinite convexly independent subset. However, if we are given an uncountable set Z in $\mathbf{R}^2$, whose points are in general position, then it may happen that Z does not contain an uncountable convexly independent subset. To get the latter result, the method of transfinite induction is essentially used.

Chapter 7 is focused on homogeneous coverings of the Euclidean plane $\mathbf{R}^2$, consisting of certain geometric figures (e.g., lines or circles). Here we begin with some comments on a remarkable Mazurkiewicz set in $\mathbf{R}^2$ which meets every straight line of $\mathbf{R}^2$ in exactly two points (see [254]). All known proofs of the existence of such paradoxical sets are based on an uncountable form of the Axiom of Choice (**AC**) and, as in the previous chapter, the method of transfinite induction (transfinite recursion) is utilized in order to construct a Mazurkiewicz set in $\mathbf{R}^2$. Further, we consider the question dual to the question of the existence of Mazurkiewicz type sets, and introduce the notion of a k-homogeneous covering of $\mathbf{R}^2$, where $k \geq 1$ is a natural number. A general result is established in this direction, which yields the existence of such coverings, and some relevant examples are presented.

In Chapter 8 we are dealing with various 3-colorings of the Euclidean plane $\mathbf{R}^2$ and examine the question on the existence of "rainbow" triangles with respect to a given 3-coloring.

By definition, a 3-coloring of $\mathbf{R}^2$ is any surjective function

$$g : \mathbf{R}^2 \to \{1, 2, 3\},$$

where the natural numbers 1, 2 and 3 play the role of colors (say, 1 = green color, 2 = red color, and 3 = blue color).

A triangle $[x, y, z]$ in the plane is called associated with g (or, briefly, g-associated, or rainbow) if the vertices x, y, z of $[x, y, z]$ are colored by pairwise different colors.

The following question naturally arises:

How many g-associated triangles of a prescribed type (for instance, acute-angled, right-angled, obtuse-angled, isosceles, and so forth) can exist for a given 3-coloring g of $\mathbf{R}^2$?

It turns out that there always are continuum many g-associated acute-angled, right-angled, obtuse-angled, and isosceles triangles (see [217]). This fact is established in Chapter 8. It is interesting to notice that in the case of acute-angled triangles the proof of the above-mentioned fact relies on the simple two-dimensional version of Sperner's lemma concerning certain triangulations of simplexes (see, e.g., Exercise 19 from Appendix 1). A radically different situation can be observed when one tries to find g-associated triangles similar to a picked triangle in the plane, for example, similar to a fixed equilateral triangle. In this connection, a 3-coloring g of $\mathbf{R}^2$ can be defined, which does not admit any g-associated equilateral triangle.

The main theme of Chapter 9 is closely connected with the chromatic numbers of those graphs which are canonically associated with certain point systems in the Euclidean plane $\mathbf{R}^2$.

Recall that the chromatic number of a graph (V, E) is the least cardinality of a set of colors which are needed for coloring all vertices of (V, E) so that the endpoints of any edge from E would be of distinct colors.

Equivalently, one may say that the chromatic number of (V, E) is the least cardinal number κ such that there exists a partition $\{V_i : i \in I\}$ of V, where $\mathrm{card}(I) = \kappa$ and the endpoints of any edge from E belong to different members of this partition.

In the case when the set of vertices V is identical with the plane $\mathbf{R}^2$, various graphs over $V = \mathbf{R}^2$ dictated by metrical or other structural properties of $\mathbf{R}^2$, can be considered. For example, the following natural set E of edges is produced by the relation

$$\{u, v\} \in E \Leftrightarrow ||u - v|| = 1 \qquad (u \in V,\ v \in V).$$

The chromatic number of this (V, E) was investigated quite thoroughly, but its precise value t is still unknown. The estimates $4 \leq t \leq 7$ were found many years ago and are not difficult to prove. Recently it was demonstrated, by using an auxiliary computer program, that $t \geq 5$ (see [117]). So, we now have the inequalities $5 \leq t \leq 7$ or, equivalently,

$$t = 5 \ \ \vee \ \ t = 6 \ \ \vee \ \ t = 7.$$

But we do not know which one of these three cases is realizable.

In this connection, Shelah and Soifer pointed out in their joint work [313] that some graph structure over $V = \mathbf{R}^2$ can be effectively (constructively) defined so that it is impossible to determine, within the framework of **ZF** set theory, the precise value of the chromatic number of the obtained graph. For this purpose, they considered in $\mathbf{R}^2$ an appropriate analog of Vitali's classical construction of a Lebesgue nonmeasurable point set in $\mathbf{R}$ (see [313]). In Chapter 9 we follow the argument of [313] and effectively show the existence of a graph $(\mathbf{R}^2, E)$ whose chromatic number turns out to be undetermined within **ZF** & **DC** set theory, where **DC** stands for a certain weak form of the Axiom of Choice (**AC**), which logically implies the countable form of **AC** (see, e.g., [137], [150], [151], [316]).

Chapter 10 may be treated as a variation of those combinatorial themes which are usually devoted to the study of possible mutual arrangements (dispositions) of finite families of points and lines in the Euclidean plane $\mathbf{R}^2$. In this chapter we discuss the famous Szemerédi–Trotter theorem that establishes an asymptotically precise estimate of the incidence number for any two finite systems of points and lines in $\mathbf{R}^2$ (see [336]). Several useful consequences of the above-mentioned theorem are also presented, including Beck's remarkable result [17]. The related extensive work [323] is indicated in which this interesting and important topic is developed for the case of a multi-dimensional Euclidean space.

In Chapter 11, Minkowski's widely known and exceedingly beautiful theorem on a centrally symmetric compact convex body in $\mathbf{R}^m$ is discussed with several standard applications in number theory. Among those applications the most impressive is Lagrange's classical theorem which states that it is possible to express every natural number in the form of a sum of four squares of integers. In the same chapter, we also indicate unexpected interrelations between Minkowski's theorem and the deep question of the existence of a Lebesgue nonmeasurable set in a finite-dimensional Euclidean space (cf. Chapter 9).

In Chapter 12 we consider one interesting problem of combinatorial geometry formulated many years ago by Tarski [338]. As known, Tarski was a famous mathematician and logician, and he worked in different branches of mathematics such as formal logic, axiomatic set theory, model theory, algebra, and geometry. Among his results the most popular are the so-called Banach–Tarski paradox (see [13] and, especially, [355]) and his profound metatheorem on the undefinability of arithmetical truth within Peano arithmetic (see [339]). Also, Tarski was interested in questions concerning the foundations of geometry within the framework of first-order logic (see, for example, his works [340] and [341]). In the area of convex geometry he formulated the following intriguing question:

If a bounded convex set C on the plane $\mathbf{R}^2$ is covered by finitely many strips, can one assert that the minimal width of C does not exceed the sum of the widths of those strips?

For the very special case, when C is a closed disc in $\mathbf{R}^2$, Tarski was able to give an elegant elementary argument which shows that the answer is positive (see [338]). However, the general case remained open for some time and finally was also resolved positively by Bang [14] (cf. [95]). Moreover, afterwards further generalizations and extensions of Bang's result were obtained (see, e.g., [12], [22], [238]). In Chapter 12 we briefly discuss Bang's theorem and some related statements.

Chapter 13 is devoted to Borsuk's conjecture on decompositions of a bounded subset X of $\mathbf{R}^m$ into its parts, all of which have diameters strictly less than the diameter of X. Borsuk proved in [40] that if X is an m-dimensional ball in $\mathbf{R}^m$, where $m \geq 1$, then X admits a decomposition into $m+1$ parts of strictly smaller diameters, but it is impossible to decompose the same X into m parts of strictly smaller diameters. It is remarkable that in [40] the proof of another celebrated theorem was also given, namely, it was demonstrated therein that any continuous mapping f of the Euclidean m-dimensional sphere $\mathbf{S}_m$ into $\mathbf{R}^m$ has the property that there always exist two antipodal points x and y of $\mathbf{S}_m$ such that

$$f(x) = f(y);$$

consequently, f cannot be an injective mapping.

Perhaps, motivated by the above-mentioned results Borsuk asked in [40] whether it is possible to dissect any bounded subset X of $\mathbf{R}^m$, containing at least two distinct points, into $m+1$ subsets of strictly smaller diameter.

This question is now known as Borsuk's problem and, quite often, the affirmative answer to the question is regarded as Borsuk's conjecture (although Borsuk did not conjecture anything in his work [40]).

The history of Borsuk's conjecture is very dramatic. For $m \leq 3$ the conjecture holds true and even the case $m = 3$ needs a rather complicated argument. For $m \geq 4$, several partial results were established providing upper estimates of the number of subsets of X in a required decomposition (see, for instance, [34], [36], [37], [287], [289]). Finally, it was proved in [154] that, for sufficiently large m, the number of such subsets is even non-polynomial with respect to m, so in higher dimensions Borsuk's conjecture automatically fails to be valid. In Chapter 13, closely following the text of [7], we give a detailed proof of the fact that, for large dimensions m, Borsuk's conjecture does not hold (see also [36]).

Chapter 14 is focused on extraordinary combinatorial properties of the so-called Carathéodory–Gale polyhedra (see [104]). These convex polyhedra appear in Euclidean spaces whose dimension is strictly greater than 3. For example, in the space $\mathbf{R}^4$ one can meet a convex four-dimensional polyhedron P which has arbitrarily many vertices but any couple of vertices are the endpoints of an edge of P. This surprising fact is closely connected with various dissections of convex polyhedra into simplices, and it turns out that the total number of dissecting simplices of an m-dimensional Carathéodory–Gale polyhedron Q does not admit a polynomial upper estimate. Speaking more formally, if one wants to approximate a continuous real-valued function of m real variables, given on Q, by continuous piecewise affine functions, then the total number of affine pieces should be taken so large that it cannot be estimated from above by an appropriate polynomial of two variables m and v, where $v = v(Q)$ denotes the number of all vertices of Q.

Chapter 15 is entirely devoted to the following question which again may be considered as a typical problem of elementary (school) geometry:

Given an odd natural number n, is it possible to dissect the unit square $[0, 1]^2$ into n many triangles of equal areas?

This question was raised in [294] and was open for several years. Any approach to solve it within the framework of more or less elementary geometric methods turned out to be successless. At last, Monsky [263] obtained the negative answer to this question by presenting substantially non-elementary argument which involves a Sperner type combinatorial statement and a highly non-trivial extension theorem for non-Archimedean valuations. The latter theorem is essentially based on the Zorn (Kuratowski–Zorn) lemma, so is utterly non-constructive. At this moment, no elementary solution of the above-mentioned problem is known and, therefore, the role of the Axiom of Choice (**AC**) in this situation remains unclear (in this connection, it should be noticed that there are many analogous problems of elementary geometry which cannot be resolved without using **AC** or its logical equivalents).

In Chapter 16 we are concerned with visual realizations of various systems of sets. The widely known Euler–Venn diagrams are usually associated with finite systems of sets and yield geometric interpretations of them (see [124], [226], [227], [243], [333], and also the original manuscript by Venn [352]). Of course, it is reasonable to require that the geometric figures participating in Euler–Venn diagrams would be maximally simple, e.g., would be convex subsets of $\mathbf{R}^2$. But, as turns out, convex figures are not enough to realize any finite family of sets. Moreover, a very particular case of Helly's remarkable theorem shows that even some four-member systems of sets cannot be realized by an Euler–Venn diagram consisting of convex subsets of $\mathbf{R}^2$. So it makes sense to try to find suitable geometric figures for realizations of arbitrary finite systems of sets. The problem becomes much more complicated when we deal with infinite systems of sets. We consider this situation too, and, in particular, show the existence of an independent uncountable family of so-called convex quasi-polygons in $\mathbf{R}^2$. Of course, a delicate non-elementary argument based on the principle of transfinite induction is needed here for obtaining the desired result (let us underline once more that such methods are also utilized in some other chapters of this book).

In the last Chapter 17 we discuss a certain geometric form of the Axiom of Choice, found by Bell and Fremlin [18]. They formulated and proved in [18] an interesting equivalent of **AC** in terms of extreme points of a closed ball in the weak$-^*$ topology.

Actually, they demonstrated that the existence of at least one such point implies **AC**.

We present their argument in Chapter 17 in all details and give a few comments on it. In this context, it must be noticed that **AC** possesses its logical equivalents (within the framework of **ZF** set theory) in many branches of modern mathematics, e.g, in group theory, in general topology, in graph theory, etc (see [25], [137], [150], [165], [241], [316]). Consequently, the formulation of **AC** in terms of extreme points of a convex body seems to be of interest, too.

For the readers' convenience, we added to the main text of this book five Appendixes.

Appendix 1 is devoted to basic properties of convex sets in real vector spaces. It needless to say that the general theory of convex sets and convex functions has a wide range of applications in geometry, in real and complex analysis, in probability and statistics, in optimization theory, in game theory, economical models, etc. Of course, the material given in Appendix 1 is quite fragmentary and does not reflect all geometric aspects of convex sets. For more information about convex sets and functions, we refer the reader to [4], [19], [20], [24], [38], [45], [52], [68], [76], [77], [120], [122], [123], [126], [130], [232], [249], [298], [305], [308], [350], [356], [360]. Naturally, in Appendix 1 we also touch upon some important statements of combinatorial geometry, such as Carathéodory's theorem on convex hulls of subsets of the space $\mathbf{R}^m$ and Helly's theorem on nonempty intersections of convex sets in $\mathbf{R}^m$.

In Appendix 2, which is a natural continuation of Appendix 1, we briefly consider some elementary properties of real-valued convex (and concave) functions defined on convex subsets of real vector spaces. In our presentation we leave aside deep differential properties of such functions and are mainly concentrated on those aspects of convex functions which have a purely geometric or topological character.

There are many purely combinatorial statements which can be successfully applied in various practical situations and in those cases when finite families of objects or structures are given. The so-called Principle of Inclusion and Exclusion should be pointed out at the beginning of the list of such results. This principle is especially useful when the precise arithmetical calculations are needed in the context of problems of finding the total number of appropriate combinations of finitely many objects (see, for instance, [48], [53], [118], [134]). Several very standard examples of such calculations are given in Appendix 3. But the same principle also plays a significant role in those questions which naturally arise in classical Euclidean geometry and, in particular, in the theory of convex polyhedra. We hope that Appendix 3 provides a vivid example of a geometric application of this principle, namely, it is shown therein that a clever analogue of Euler's classical formula is valid for multi-dimensional convex polyhedra. It makes sense to recall that this analogue was first established by Poincaré and Schläfli. As known, further generalizations lead to one of the most important fields of modern algebraic topology - homology theory.

The important role of various kinds of inequalities in different fields of mathematics is absolutely recognized and does not need any comments. In geometry the inequalities are concerned, as a rule, with natural geometric objects: length, area, volume, standard line segments in triangles and tetrahedrons, etc. In this context, typical examples are the so-called isoperimetric inequality and the celebrated Brunn–Minkowski inequality (see, for instance, [4], [20], [52], [107], [123], [130], [249], [308]). Numerous works were devoted to geometric inequalities and their continuous and discrete aspects. In Appendix 4 we are dealing with some of such inequalities. Actually, we are focused on the Erdös–Mordell inequality for a triangle and on its generalization for convex polygons in the plane $\mathbf{R}^2$ (cf. [15], [42], [71], [78], [93], [96], [127], [159], [162], [164], [247], [262], [264], [273], [274], [315]). Exercises of Appendix 4 contain additional material concerning related inequalities for multi-dimensional simplexes.

Appendix 5 serves to recall basic notions and facts from graph theory (see, for example, [91] or [187]). Some of these notions and facts play an auxiliary role and are used in certain sections of the book. The central statement discussed in Appendix 5 is, of course, the famous Euler formula on nonempty connected planar graphs (cf. [4], [5], [20], [33], [80], [123], [124], [136], [249], [268], [312], [330], [364]). It is well known that this formula is a basic tool in diverse topics of combinatorial and convex geometry, as well as in many other branches of mathematics. Moreover, one can say that modern algebraic topology begins with the Euler formula. So, it is natural that rather nontrivial applications of this formula are met in this book. Among them an application to the Sylvester problem on collinear points should be mentioned especially (see [334]).

Finally, we would like to note that all chapters (and also the five Appendixes) end with a number of exercises. The role of those exercises is to provide the reader with some additional information about topics considered in the main text of this book. Naturally, the exercises vary in their difficulty. Among them there are almost trivial, standard, nontrivial, rather difficult, and difficult. As a rule, more or less difficult exercises are marked by asterisks and are provided with necessary hints.

A. Kharazishvili

CHAPTER 1

The index of an isometric embedding

Recall that the symbol $\mathbf{c}$ denotes, as usual, the cardinality of the continuum, i.e., we have the equalities

$$\mathbf{c} = \mathrm{card}(2^{\mathbf{N}}) = \mathrm{card}(\mathbf{N}^{\mathbf{N}}) = \mathrm{card}(\mathbf{R}) = \mathrm{card}([0,1]),$$

where $\mathbf{N} = \{0, 1, 2, ..., n, ...\}$ is the set of all natural numbers and $\mathbf{R}$ is the real line.

For a natural number m, let G be a group of isometric transformations of the Euclidean space $\mathbf{R}^m$ (as a rule, we assume that $m \geq 1$).

Clearly, if X is an arbitrary subset of $\mathbf{R}^m$, then there exists a least cardinal number κ possessing the following property:

For any set $Y \subset \mathbf{R}^m$, if every subset Z of Y with $\mathrm{card}(Z) \leq \kappa$ can be embedded into X by a transformation from G (in general, depending on Z), then the entire Y can also be embedded into X by some transformation from G.

In the sequel, we shall say that this κ is the index of an isometric embedding into X with respect to the given group G (cf. [184]).

For the sake of brevity, in our further considerations we will use the notation $\kappa = \kappa_G(X)$. Obviously, we have the inequality $\kappa_G(X) \leq \mathbf{c}$ and

$$\kappa_G(\emptyset) = 1, \quad \kappa_G(\mathbf{R}^m) = 0.$$

To somehow illustrate the notion introduced above, let us present several typical examples.

Example 1. Take as G the group of all translations of the space $\mathbf{R}^m$, where $m \geq 1$, and let P be an m-dimensional parallelepiped in $\mathbf{R}^m$.

It turns out that $\kappa_G(P) = 2$.

Notice that this equality can easily be deduced from the following fact:

If a family $\{P_i : i \in I\}$ of m-dimensional parallelepipeds in $\mathbf{R}^m$ is such

DOI: 10.1201/9781003458708-1

that the edges of each P_i $(i \in I)$ are respectively parallel to the corresponding vectors of a fixed affine basis of $\mathbf{R}^m$ and the intersection of any two members of this family is nonempty, then $\cap\{P_i : i \in I\} \neq \emptyset$.

For more details, see Exercises 1 and 2 of the present chapter.

Example 2. Let G denote the group of all isometric transformations (i.e., motions) of the space $\mathbf{R}^m$ and let B be an m-dimensional closed ball in $\mathbf{R}^m$.

It can be demonstrated that $\kappa_G(B) = m + 1$.

For a more detailed explanation concerning the above equality, see Exercise 3 of this chapter (cf. also Theorem 5 at the end of the chapter).

Example 3. Let G be the group of all translations of the space $\mathbf{R}^m$ and let X be an arbitrary compact convex set in $\mathbf{R}^m$. Then $\kappa_G(X)$ turns out to be uniformly bounded from above, i.e., bounded from above independently of X. Namely, we have the following simple inequality:

$$\kappa_G(X) \leq m + 1.$$

This inequality can be inferred from Helly's wel-known theorem on intersections of convex sets in the space $\mathbf{R}^m$ (see Exercise 4 for a more detailed explanation). Note that some information about Helly's theorem concerning intersections of convex subsets of $\mathbf{R}^m$ is given in Exercises 18, 22, and 23 of Appendix 1).

In the sequel, we need the following auxiliary statement.

Lemma 1. *Let m be a natural number, let X be an infinite subset of the Euclidean space $\mathbf{R}^m$ (or of the Euclidean sphere $\mathbf{S}_m$), and let*

$$D(X) = \{||x - x'|| : x \in X,\ x' \in X,\ x \neq x'\}.$$

Then the equality $\mathrm{card}(X) = \mathrm{card}(D(X))$ *holds true.*

We omit the proof of Lemma 1 and leave it to the reader (see Exercise 5).

Remark 1. It is essential for the validity of the assertion of Lemma 1 that we are dealing with infinite subsets of a finite-dimensional Euclidean space. Indeed, if we take a real infinite-dimensional separable Hilbert space H instead of $\mathbf{R}^m$, then it becomes clear that there exists an infinite (in fact, countably infinite) set $X \subset H$ such that all distances determined by couples of distinct points from X are equal to $2^{1/2}$ (in other words, every three-element subset of X forms an equilateral triangle). Indeed, we may take as X any orthonormal basis of H.

Theorem 1. *Let G be a subgroup of the group of all isometric transformations of the space $\mathbf{R}^m$, where $m \geq 1$, and let X be an arbitrary subset of $\mathbf{R}^m$.*

Then the inequalities $\kappa_G(X) \leq \mathrm{card}(G)$ *and* $\kappa_G(X) \leq \mathrm{card}(X) + 1$ *hold true.*

Proof. To establish the first of the above-mentioned inequalities, consider the family of sets $\{g^{-1}(X) : g \in G\}$. Let Y be any set in $\mathbf{R}^m$ such that each subset of Y whose cardinality does not exceed $\mathrm{card}(G)$ can be embedded into X by using some element from G. We have to show that Y itself can be embedded into X with the aid of some element from G. In other words, we must demonstrate that

$$(\exists f \in G)(Y \subset f^{-1}(X)).$$

Suppose for a moment otherwise, i.e.,

$$(\forall g \in G)(Y \setminus g^{-1}(X) \neq \emptyset).$$

So we come to the family $\{Y \setminus g^{-1}(X) : g \in G\}$ of nonempty sets and, according to the Axiom of Choice (**AC**), there exists a set $Z \subset Y$ satisfying the relations

$$\mathrm{card}(Z) \leq \mathrm{card}(G),$$

$$(\forall g \in G)(Z \cap (Y \setminus g^{-1}(X)) \neq \emptyset).$$

Further, by virtue of our assumption on Y, for this Z there exists an $f \in G$ such that $f(Z) \subset X$, whence it follows that

$$Z \subset f^{-1}(X), \qquad Z \cap (Y \setminus f^{-1}(X)) = \emptyset,$$

which yields a contradiction. Thus, we must have $\kappa_G(X) \leq \mathrm{card}(G)$.

It should be remarked that the above argument does not use the group structure of G.

In order to demonstrate the second inequality $\kappa_G(X) \leq \mathrm{card}(X) + 1$, consider two possible cases.

1. $\mathrm{card}(X) < \omega$, where ω denotes the least infinite cardinal (ordinal) number which is usually identified with $\mathbf{N}$.

   In this case, the inequality $\kappa_G(X) \leq \mathrm{card}(X) + 1$ follows at once from the definition of the index of an isometric embedding.

2. $\mathrm{card}(X) \geq \omega$.

In this case, $\mathrm{card}(X) + 1 = \mathrm{card}(X)$ and the required inequality is a direct consequence of the fact that the set of all distances determined by couples of points from X has cardinality equal to $\mathrm{card}(X)$ (see Lemma 1).

Theorem 1 has thus been proved. □

As a trivial consequence of the above theorem, we get the following statement.

Theorem 2. *These two assertions are valid:*

(1) *if a set $X \subset \mathbf{R}^m$ is finite, then $\kappa_G(X) < \omega$ for every group G of isometric transformations of $\mathbf{R}^m$;*

(2) *if a group* G *of isometries of* $\mathbf{R}^m$ *is finite, then* $\kappa_G(X) < \omega$ *for every subset* X *of* $\mathbf{R}^m$.

In view of the trivial equality $\kappa_G(\mathbf{R}^m) = 0$ for any group G of isometric transformations of $\mathbf{R}^m$, we can conclude that the assertion converse to (1) of Theorem 2 is not true. At the same time, the next example shows us that the assertion converse to (2) of Theorem 2 is valid.

Example 4. Let G be a subgroup of the group of all isometric transformations of the space $\mathbf{R}^m$. Consider the following equivalence relation $S(x, y)$ in $\mathbf{R}^m$ canonically associated with G:

$$S(x, y) \Leftrightarrow (\exists g \in G)(g(x) = y).$$

As is well known, the equivalence classes with respect to $S(x, y)$ are exactly the G-orbits of points in $\mathbf{R}^m$ (see, e.g., [20]). For any point $x \in \mathbf{R}^m$, the G-orbit containing this point is usually denoted by $G(x)$. Obviously, if X is an arbitrary G-orbit in $\mathbf{R}^m$, then the inequality

$$\text{card}(X) \leq \text{card}(G)$$

holds true, and it can be shown that there always exists a G-orbit Z satisfying the equality

$$\text{card}(Z) = \text{card}(G).$$

For more details about the latter fact, see Exercise 6 of this chapter. Further, choose arbitrarily a point z from Z and define $Z' = Z \setminus \{z\}$. It is easy to verify that each proper subset of Z can be isometrically embedded into Z' by using some transformation from G. However, the entire Z cannot be isometrically embedded into Z' with the aid of a transformation from G. This circumstance implies that

$$\kappa_G(Z') \geq \text{card}(Z') + 1 = \text{card}(Z) = \text{card}(G).$$

Keeping in mind Theorem 1, we get

$$\kappa_G(Z') = \text{card}(Z') + 1 = \text{card}(G).$$

Consequently, if $\kappa_G(X) < \omega$ for every set $X \subset \mathbf{R}^m$, then G is a finite group of isometries of $\mathbf{R}^m$.

We thus see that the inequalities presented in the formulation of Theorem 1 are precise in a certain sense.

Theorem 3. *Let* G *be a closed subgroup of the group of all isometric transformations of the space* $\mathbf{R}^m$ *and let* X *be an arbitrary compact subset of* $\mathbf{R}^m$.

If a set $Y \subset \mathbf{R}^m$ *is such that every finite subset of* Y *can be isometrically embedded into* X *by using an appropriate transformation from* G*, then the entire* Y *can also be isometrically embedded into* X *by some transformation from* G*.*

Proof. Let $Y \subset \mathbf{R}^m$ have the property that every finite subset of Y can be embedded into X by a transformation from G. The assumptions of the theorem immediately imply that Y is bounded in $\mathbf{R}^m$ and the diameter of Y does not exceed the diameter of X. We may suppose, without loss of generality, that both sets X and Y are not empty.

Let $\{y_n : n < \omega\}$ be a sequence of points from Y which is everywhere dense in Y. For each natural number k, let us put

$$Y_k = \{y_n : n \leq k\}$$

and take a transformation $g_k \in G$ such that $g_k(Y_k) \subset X$. Observe that the family of transformations $\{g_k : k < \omega\}$ is relatively compact in the group of all motions of $\mathbf{R}^m$ (cf. Exercise 10). Therefore, in view of the closedness of G, there exists $g \in G$ which is an accumulation point for $\{g_k : k < \omega\}$. Now, for this g, the inclusion $g(Y) \subset X$ holds true. A detailed verification of the last inclusion is not hard and we leave it to the reader. □

Remark 2. It follows from the previous theorem that if X a compact subset of the space $\mathbf{R}^m$ and G is a closed subgroup of the group of all isometric transformations of $\mathbf{R}^m$, then the inequality $\kappa_G(X) \leq \omega$ is always satisfied. Notice also that, for compact metric spaces, there are certain analogues of Theorem 3. One of them is formulated in Exercise 11 of this chapter.

Lemma 2. *Let G be a subset of the group of all isometric transformations of the space $\mathbf{R}^m$, let X be a set in $\mathbf{R}^m$, and let Y be a compact set in $\mathbf{R}^m$. Assume that, for some natural number n, every subset of Y containing at most n points can be isometrically embedded into the interior* int(X) *of X by using a transformation from G.*

Then there exists a real $\varepsilon > 0$ such that the same property holds true for the closed ε-neighborhood $V_\varepsilon(Y)$ of Y, i.e., every subset of $V_\varepsilon(Y)$ containing at most n points can be isometrically embedded into int(X) *by using a transformation from G.*

Proof. Suppose otherwise, i.e., there exists no $\varepsilon > 0$ satisfying the above-mentioned property. Take a decreasing sequence $\{\varepsilon_k : k < \omega\}$ of strictly positive real numbers tending to zero. For any natural number k, there exists a set Y_k such that:

(1) Y_k is a subset of $V_{\varepsilon_k}(Y)$ and $\mathrm{card}(Y_k) \leq n$;

(2) Y_k cannot be isometrically embedded into $\mathrm{int}(X)$ by using a transformation from G.

Consider the sequence of sets $\{Y_k : k < \omega\}$. Since Y is compact, the closed ε_0-neighborhood $V_{\varepsilon_0}(Y)$ of Y is compact, too, and all sets Y_k $(k < \omega)$ are contained in $V_{\varepsilon_0}(Y)$. This implies that there exists a subsequence of $\{Y_k : k < \omega\}$ converging in the Hausdorff metric to some set $Z \subset Y$ with $\mathrm{card}(Z) \leq n$. According to our assumption, Z can be isometrically embedded into $\mathrm{int}(X)$

by using a transformation g from G. It follows from this circumstance that, for some sufficiently large $k < \omega$, the set Y_k can also be embedded into $\mathrm{int}(X)$ by g. But the latter fact contradicts (2).

Lemma 2 has thus been proved. □

Recall that a subset P of the space $\mathbf{R}^m$ is an m-dimensional polyhedron in $\mathbf{R}^m$ if P can be represented as the union of a finite family of closed simplices in $\mathbf{R}^m$, at least one of which has dimension m (cf. Appendix 1).

For such polyhedra, we can formulate and prove the following statement.

Theorem 4. *If G is the group of all isometric transformations of the space $\mathbf{R}^m$, where $m \geq 2$, and P is an arbitrary m-dimensional polyhedron in $\mathbf{R}^m$, then the equality $\kappa_G(P) = \omega$ holds true.*

In particular, $\kappa_G(S) = \omega$ for any m-dimensional closed simplex S in $\mathbf{R}^m$.

Proof. First, notice that the inequality $\kappa_G(P) \leq \omega$ follows immediately from Theorem 3, because P is a compact subset of $\mathbf{R}^m$. So, it remains to show that, for each natural number n, the equality $\kappa_G(P) = n$ is false. Suppose to the contrary that $\kappa_G(P) = n$ and consider a closed ball B in $\mathbf{R}^m$ having maximal radius and entirely contained in P. The existence of such a B is obvious (again, in view of the compactness of P). It is not hard to see that any at most n-element subset of B can be isometrically embedded into the interior of P (it suffices to consider small rotations of $\mathbf{R}^m$ about the center of B). Consequently, by virtue of Lemma 2, there exists a real $\varepsilon > 0$ such that any at most n-element subset of the ball $V_\varepsilon(B)$ can also be isometrically embedded into the interior of P. By definition of the index of an isometric embedding, the latter means that $V_\varepsilon(B)$ can be isometrically embedded into P, which contradicts the maximality of the length of a radius of B.

The obtained contradiction completes the proof of Theorem 4. □

The next statement also substantially relies on Lemma 2 and yields one characterization property of closed discs in the Euclidean plane $\mathbf{R}^2$ (see [297]).

Theorem 5. *Let G be the group of all isometric transformations of the plane $\mathbf{R}^2$ and let D be a compact convex subset of $\mathbf{R}^2$ having nonempty interior and satisfying the relation $\kappa_G(D) = 3$.*

Then the set D is a closed disc in $\mathbf{R}^2$.

Proof. The equality $\kappa_G(D) = 3$ means that if a set $Y \subset \mathbf{R}^2$ is such that every at most 3-element subset of Y can be isometrically embedded into D, then Y itself can be isometrically embedded into D. Suppose to the contrary that D is not a disc and take a closed disc B of maximal radius, which is entirely contained in D (in some sense, B may be treated as an inscribed disc of D; in general, an inscribed disc is not unique). By virtue of our assumption, $D \neq B$, so there exists a point $x \in D \setminus B$. Since D is a convex set, we have the inclusion

$$\mathrm{conv}(B \cup \{x\}) \subset D,$$

where $\mathrm{conv}(B \cup \{x\})$ denotes the convex hull of $B \cup \{x\}$ (see Appendix 1).

Now, it is not difficult to verify that every at most 3-element subset of B can be isometrically embedded into the interior of $\text{conv}(B \cup \{x\})$ (hence can be isometrically embedded into the interior of D). To see this fact more visually, it makes sense to consider separately three possible cases:

(a) a 3-element subset of B forms an acute-angled triangle;

(b) a 3-element subset of B forms a right-angled triangle or an obtuse-angled triangle;

(c) a three-element subset of B is collinear.

All corresponding details connected with the cases (a), (b), and (c) are left to the reader.

According to Lemma 2, there is a strictly positive real ε such that the closed disc $V_\varepsilon(B)$ possesses the same property, i.e., every at most 3-element subset of $V_\varepsilon(B)$ admits an isometric embedding into the interior of $\text{conv}(B \cup \{x\})$ (hence into the interior of D). This circumstance must imply the existence of an isometric embedding of $V_\varepsilon(B)$ into the interior of D, which contradicts the definition of B. The contradiction obtained enables us to conclude that $D = B$, so the original compact set D is a closed disc in $\mathbf{R}^2$.

Theorem 5 has thus been proved. □

EXERCISES

1. Let $m \geq 1$ be a natural number and let $\{P_i : i \in I\}$ be a family of m-dimensional parallelepipeds in the Euclidean space $\mathbf{R}^m$, such that the edges of each P_i $(i \in I)$ are respectively parallel to the corresponding vectors of a fixed affine basis of $\mathbf{R}^m$, and the intersection of any two members of this family is nonempty.

Demonstrate that $\cap\{P_i : i \in I\} \neq \emptyset$.

Argue as follows. First, consider the case $m = 1$ and observe that this case is immediately reduced to Helly's theorem on intersections of convex sets in $\mathbf{R}$ (see Exercise 18 from Appendix 1). Then use induction on m.

2. Let G denote the group of all translations of the space $\mathbf{R}^m$, where $m \geq 1$, and let P be an arbitrary m-dimensional parallelepiped in $\mathbf{R}^m$.

Verify that $\kappa_G(P) = 2$.

For this purpose, take an arbitrary nonempty set $Y \subset \mathbf{R}^m$ having the property that, for any two-element subset Z of Y there is a translate of Z in P, and consider the family of parallelepipeds $\{P - y : y \in Y\}$. Check that

$$(P - y) \cap (P - y') \neq \emptyset \qquad (y \in Y,\ y' \in Y).$$

By using Exercise 1, infer that $\cap\{P - y : y \in Y\} \neq \emptyset$ and take some point

$$z \in \cap\{P - y : y \in Y\}.$$

Finally, conclude that $Y + z \subset P$.

3. Let G denote the group of all isometric transformations of the space $\mathbf{R}^m$ and let B be an m-dimensional closed ball in $\mathbf{R}^m$.

Demonstrated that $\kappa_G(B) = m + 1$.

For this purpose, keep in mind the special case of Helly's theorem, stating that if a family of closed balls in $\mathbf{R}^m$ possesses the property that any $m + 1$ members (or less than $m + 1$ members) of the family have a common point, then the intersection of this family is nonempty.

4. Let G denote the group of all translations of the space $\mathbf{R}^m$ and let C be an arbitrary compact convex set in $\mathbf{R}^m$.

Verify the validity of the inequality $\kappa_G(C) \leq m + 1$.

Argue as follows. Apply again Helly's theorem on intersections of convex sets in $\mathbf{R}^m$. Namely, take an arbitrary set $Y \subset \mathbf{R}^m$ having the property that any at most $(m+1)$-element subset of Y can be translated into C, and consider the family $\{C - y : y \in Y\}$. Check that this family satisfies the assumptions of Helly's theorem, so has a nonempty intersection. Then choose a point

$$z \in \cap\{C - y : y \in Y\}$$

and deduce that $z + Y \subset C$.

5*. Give a detailed proof of Lemma 1.

For this purpose, start with an infinite subset X of $\mathbf{R} = \mathbf{R}^1$ (or with an infinite subset X of $\mathbf{S}_1$) and show that $\text{card}(X) = \text{card}(D(X))$.

Then, for $\mathbf{R}^m$ and for $\mathbf{S}_m$, use the method of induction on m and, simultaneously, the method of transfinite induction on $\text{card}(X)$, where X is an infinite set in $\mathbf{R}^m$ (in $\mathbf{S}_m$). Also, it makes sense to consider separately the following two cases:

(a) $\text{card}(X)$ is a regular cardinal number;
(b) $\text{card}(X)$ is a singular cardinal number.

For necessary information about infinite regular and singular cardinalities, see e.g. [44], [151], [237], [241], [259], [316].

6*. Let G be a subgroup of the group of all isometric (affine) transformations of the space $\mathbf{R}^m$.

Prove that there always exists a point $x \in \mathbf{R}^m$ such that

$$\text{card}(G(x)) = \text{card}(G).$$

For this purpose, consider the two possible cases.

(a) card$(G) < \mathbf{c}$, where $\mathbf{c}$ denotes the cardinality of the continuum.

In this case, show the validity of a more delicate result, according to which there exists a point $x \in \mathbf{R}^m$ having the property that the group G acts freely on the orbit $G(x)$, i.e., for any two points $y \in G(x)$ and $z \in G(x)$, there is a unique $g \in G$ such that $g(y) = z$. Clearly, this circumstance automatically implies the desired equality

$$\text{card}(G(x)) = \text{card}(G).$$

To get the required result, take some set $X \subset \mathbf{R}^m$ of points in general position in $\mathbf{R}^m$ with card$(X) = \mathbf{c}$ and verify that there exists a point $x \in X$ such that G acts freely on the orbit $G(x)$.

(b) card$(G) = \mathbf{c}$.

In this case argue as follows. Suppose on the contrary that all G-orbits have cardinalities strictly less than $\mathbf{c}$ and choose a set $\{x_0, x_1, ..., x_m\} \subset \mathbf{R}^m$ of points in general position in $\mathbf{R}^m$. Consider the orbit $G(x_0)$ and introduce the notation

$$G_0 = \{g \in G : g(x_0) = x_0\}.$$

Check that G_0 is a subgroup of G and card$(G_0) = \mathbf{c}$ (the reader probably knows that this G_0 is usually called the stabilizer of x_0 in G). Afterwards, use induction on $k \leq m$. If the group $G_{k-1} \subset G$ is already defined fulfilling the relations

$$\text{card}(G_{k-1}) = \mathbf{c}, \qquad G_{k-1}(x_{k-1}) = \{x_{k-1}\},$$

then put $G_k = \{g \in G_{k-1} : g(x_k) = x_k\}$.

Proceeding in this manner, come to the group G_m and conclude that, on the one hand, card$(G_m) = \mathbf{c}$ and, on the other hand, G_m cannot contain two distinct isometric (affine) transformations of $\mathbf{R}^m$. The obtained contradiction yields the desired result.

7. Two metric spaces (E_1, d_1) and (E_2, d_2) are called combinatorially isomorphic if there exists a bijective mapping $f : E_1 \to E_2$ such that, for any four points x, y, z, t from E_1, the relation

$$d_1(x, y) = d_1(z, t) \Leftrightarrow d_2(f(x), f(y)) = d_2(f(z), f(t))$$

holds true. The above-mentioned f is called a combinatorial isomorphism between (E_1, d_1) and (E_2, d_2).

Supposing that (E_1, d_1) and (E_2, d_2) are combinatorially isomorphic, check that:

(a) if $\phi : E_1 \to E_1$ is an isometry of (E_1, d_1) onto itself and f is a combinatorial isomorphism between (E_1, d_1) and (E_2, d_2), then the mapping

$$\psi = f \circ \phi \circ f^{-1}$$

is an isometry of (E_2, d_2) onto itself (consequently, the group of all isometries of (E_1, d_1) onto itself turns out to be isomorphic to the group of all isometries of (E_2, d_2) onto itself);

(b) if (E_1, d_1) and (E_2, d_2) are combinatorially isomorphic and the group of all isometries of (E_1, d_1) onto itself acts transitively on E_1, then the group of all isometries of (E_2, d_2) onto itself acts transitively on E_2.

8*. Let m be a nonzero natural number.

Demonstrate that there exists a strictly positive real $\varepsilon = \varepsilon(m) < 1$ having the following property:

For any real numbers $l_1, l_2, \ldots, l_{m(m+1)/2}$ satisfying the relations

$$1 - \varepsilon < l_j < 1 + \varepsilon \qquad (1 \leq j \leq m(m+1)/2),$$

there is an m-dimensional simplex S in the Euclidean space $\mathbf{R}^m$ such that $l_1, l_2, \ldots, l_{m(m+1)/2}$ are the lengths of all edges of S.

For this purpose, use induction on m.

Conclude from the above result that every finite metric space (E, d) with $\text{card}(E) \geq 2$ is combinatorially isomorphic to the set of all vertices of an m-dimensional simplex in the space $\mathbf{R}^m$, where $m = \text{card}(E) - 1$.

9*. Let G be a finite subgroup of the group of all isometric transformations of the space $\mathbf{R}^m$ and let X be a finite subset of $\mathbf{R}^m$. Suppose that, for these G and X, the relation

$$\kappa_G(X) = \text{card}(G) = \text{card}(X) + 1$$

holds true.

Prove that there exists a point $x \in \mathbf{R}^m$ such that $X = G(x) \setminus \{x\}$.

Argue as follows. First, using Exercises 7 and 8, reduce the problem to the situation where X is the set of all vertices of a facet of a certain $\text{card}(X)$-dimensional simplex S in the space $\mathbf{R}^{\text{card}(X)}$, and the facets of S are pairwise congruent.

Then denote by H the group of all those isometries h of $\mathbf{R}^{\text{card}(X)}$, which satisfy $h(S) = S$, and show that H acts transitively on the set of all vertices of S. For this purpose, consider the circumscribed sphere of S and take into account that the center of this sphere is equidistant from all affine hyperplanes carrying the facets of S.

Remark 3. The result of Exercise 9 implies that if all facets of an m-dimensional simplex $S \subset \mathbf{R}^m$ are pairwise congruent, then the isometry group of S acts transitively on the set of all vertices of S (the converse assertion is trivially valid). In this connection, it is natural to formulate the following problem:

Let $S = [x_0, x_1, ..., x_m]$ and $S' = [x'_0, x'_1, ..., x'_m]$ be two m-dimensional simplices in $\mathbf{R}^m$ $(m > 1)$ such that, for any natural index $i \in [0, m]$, the facets S_i and S'_i opposite of the vertices x_i and x'_i, respectively, are congruent. Can one assert that S and S' are also congruent to each other?

As far as we know, the above problem still remains open. A simplified version of this problem, formulated in terms of graph theory, was originally raised by Ulam (see, for instance, [135]). Actually, Ulam's problem is concerned with the special case of S and S', when the lengths of all edges of S (of S') take only two values.

10. Let X and Y be any two nonempty bounded subsets of the space $\mathbf{R}^m$. Consider the family F of all those isometric transformations f of $\mathbf{R}^m$ which have the following property:

There exist some two points $x \in X$ and $y \in Y$ (depending on f) such that $f(x) = y$.

Show that the set F is relatively compact in the group of all isometric transformations of $\mathbf{R}^m$.

11*. Verify that in the formulation of Theorem 3 the closedness of a group G in the group of all isometric transformations of $\mathbf{R}^m$ is essential for the validity of Theorem 3.

Let now (X, d) and (Y, ρ) be two metric spaces such that:

(a) X is compact;

(b) each finite subset of Y can be isometrically embedded into X.

Prove that X contains an isometric copy of Y.

Argue as follows. First, check that Y is a totally bounded space (so, it may be assumed, without loss of generality, that Y is compact as well). Let, as in the proof of Theorem 3, $\{y_k : k < \omega\}$ be a sequence of points from Y, which is everywhere dense in Y. For any natural index n, denote by f_n an isometric embedding of the finite set $Y_n = \{y_k : k \leq n\}$ into X. Further, fix a nontrivial ultrafilter Φ in the power set $\mathcal{P}(\mathbf{N})$ of $\mathbf{N}$ (recall that $\mathbf{N} = \omega$). By virtue of the compactness of X, if $k < \omega$, then there exists a point $x_k \in X$ such that:

For each real $\varepsilon > 0$, the set $\{n \in \mathbf{N} : d(f_n(y_k), x_k) < \varepsilon\}$ belongs to Φ.

Finally, verify that $\rho(y_k, y_{k'}) = d(x_k, x_{k'})$ for any two natural numbers

k and k'. Thus, X contains an isometric copy of the set $\{y_k : k < \omega\}$. Taking into account the everywhere density of $\{y_k : k < \omega\}$ in Y, this circumstance directly implies the desired result.

12. Return to the proof of Theorem 4 and give a more detailed explanation of the fact that any at most n-element subset of the ball B can be isometrically embedded into $\mathrm{int}(P)$ by using some rotation of $\mathbf{R}^m$ about the center of B.

Moreover, show that such rotations constitute an open everywhere dense subset of the group of all rotations of $\mathbf{R}^m$ about the center of B.

13. Complete the proof of Theorem 5 by checking in detail the cases (a), (b), and (c).

14*. Let $X = \{x_i : i \in I\}$ be an arbitrary subset of the Euclidean space $\mathbf{R}^m$ and let $\{\varepsilon_i : i \in I\}$ be a family of strictly positive real numbers.

Show that there exists an injective mapping $f : X \to \mathbf{R}^m$ satisfying the following two relations:

(a) $f(X)$ is a set of points in general position in $\mathbf{R}^m$;

(b) $||f(x_i) - x_i|| < \varepsilon_i$ for each point $x_i \in X$.

In order to establish this fact, use the method of transfinite induction.

Consider the special case when $\varepsilon_i = \varepsilon$ for all indices $i \in I$, and conclude that any set $X \subset \mathbf{R}^m$ can be injectively transformed into a set of points in general position by using some ε-mapping.

15. Let Z be a subset of the space $\mathbf{R}^m$ and let $g : Z \to \mathbf{R}^m$ be an isometric embedding of Z into $\mathbf{R}^m$.

Demonstrate that there exists an isometric transformation $g^* : \mathbf{R}^m \to \mathbf{R}^m$ which extends g.

Find necessary and sufficient conditions on Z, under which g^* turns out to be a unique isometric extension of g.

Finally, consider a real infinite-dimensional separable Hilbert space H and give an example of a set $Y \subset H$ such that there exists an isometric embedding $f : Y \to H$ not admitting an isometric extension $f^* : H \to H$.

16. Check that any at most three-element metric space can be isometrically embedded into the Euclidean plane $\mathbf{R}^2$.

In addition, give an example of a four-element metric space (F, d) such that every at most 3-element subset of F can be isometrically embedded into $\mathbf{R}$, but (F, d) does not admit an isometric embedding into $\mathbf{R}^2$.

Remark 4. A much deeper result providing necessary and sufficient conditions for the existence of an isometric embedding of a metric space (E, ρ) into the Euclidean space $\mathbf{R}^m$ of minimal possible dimension m may be found in the classical monograph by Blumental [27].

17*. Let X be a compact convex body in the Euclidean space $\mathbf{R}^m$, where $m \geq 1$.

Denote by X' the set of all those points $x \in \text{int}(X)$ which have the following property:

Every chord of X passing through x is dissected by x into two line segments with lengths l_1 and l_2, respectively, so that $1/m \leq l_1/l_2 \leq m$.

Demonstrate that:

(a) X' is a nonempty closed convex subset of X;

(b) if X is an m-dimensional simplex in $\mathbf{R}^m$, then $X' = \{x\}$, where x coincides with the geometric barycenter (i.e., centroid) of X.

For proving the relation $X' \neq \emptyset$, argue as follows. Take a point z of the boundary $\text{bd}(X)$ of X and consider the homothety f_z of $\mathbf{R}^m$ whose center is z and whose coefficient is $m/(m+1)$. Check that any $m+1$ members of the produced family of sets $\{f_z(X) : z \in \text{bd}(X)\}$ have a common point. By using Helly's theorem to this family (see Exercise 18 from Appendix 1), infer that

$$\cap\{f_z(X) : z \in \text{bd}(X)\} \neq \emptyset$$

and show that $\cap\{f_z(X) : z \in \text{bd}(X)\} \subset X'$.

The closedness and convexity of X' is not difficult to verify.

18*. Let G be a group of isometric transformations of the space $\mathbf{R}^m$.

Prove that these two assertions are equivalent:

(a) G is a discrete group;

(b) for any bounded set $X \subset \mathbf{R}^m$, one has $\kappa_G(X) < \omega$.

For this purpose, argue step by step as follows.

First, assume (a) and take an arbitrary nonempty bounded subset X of $\mathbf{R}^m$. Consider the family $\{g(X) : g \in G\}$ and check that this family is locally finite in $\mathbf{R}^m$. In particular, for every member $g(X)$ of the family, there are only finitely many members of the same family which have a nonempty intersection with $g(X)$. Denote

$$p(g) = \text{card}(\{f(X) : f \in G \ \& \ f(X) \cap g(X) \neq \emptyset\}).$$

Verify that $p(g) = p(g')$ for any two isometries $g \in G$ and $g' \in G$, so the value $p = p(g)$, where $g \in G$, does not depend on g. Further, put $q = p + 1$ and demonstrate that $\kappa_G(X) \leq q$.

To prove the above inequality, take any set $Y \subset \mathbf{R}^m$ such that every at most q-element subset of Y can be embedded into X by some transformation from G. Since $q \geq 2$, this assumption trivially implies that Y is bounded and

$$Y \subset \cup\{g(X) : g \in G\}.$$

In view of the discreteness of G, there exists a least natural number k satisfying the relation

$$Y \subset f_1(X) \cup f_2(X) \cup ... \cup f_k(X)$$

for certain elements $f_1, f_2, \ldots, f_k$ of G. Suppose for a while that $k > 1$.

Choose a point $y_1 \in Y \setminus f_1(X)$ and put $g_1 = f_1$. So $y_1 \notin g_1(X)$. Observe that there is $g_2 \in G$ such that $y_1 \in g_2(X)$. Also, there is $y_2 \in Y$ such that $y_2 \notin g_2(X)$. Then proceed by induction and show that there are points

$$y_1 \in Y, \quad y_2 \in Y, \quad \ldots, \quad y_q \in Y$$

and the corresponding sets

$$g_1(X), \quad g_2(X), \quad \ldots, \quad g_q(X), \quad g_{q+1}(X)$$

where $\{g_1, g_2, ..., g_q, g_{q+1}\} \subset G$ and

$$\{y_1, y_2, ..., y_i\} \subset g_{i+1}(X) \qquad (i = 1, 2, ..., q),$$

$$y_i \notin g_i(X) \qquad (i = 1, 2, ..., q).$$

Infer from these relations that $g_1(X), g_2(X), \ldots, g_q(X), g_{q+1}(X)$ are pairwise distinct sets and

$$g_{q+1}(X) \cap g_i(X) \neq \emptyset \qquad (i = 2, 3, ..., q).$$

But the latter relation contradicts the definition of p for the set $g_{q+1}(X)$.

Thus, the implication (a) $\Rightarrow$ (b) is established.

Further, suppose that (b) holds true, but (a) is false, i.e., G is not a discrete group of isometries of $\mathbf{R}^m$. Then there exists an injective sequence

$$\{g_n : n \in \mathbf{N}\} \subset G$$

converging to a certain isometry g of $\mathbf{R}^m$ which differs from all g_n ($n \in \mathbf{N}$). Observe that, in general, one cannot assert that g belongs to G.

Keeping in mind Exercise 6, infer that there exists a point x of the space $\mathbf{R}^m$ such that

$$(\forall n \in \mathbf{N})(g_n(x) \neq g(x)).$$

Pick any closed m-dimensional ball $B \subset \mathbf{R}^m$ with center x, take a countable everywhere dense subset Z of $\text{int}(B)$ and put $Y = \{x\} \cup Z$. Finally, consider the bounded set

$$X = g(B) \setminus \{g(x)\}$$

and check that each finite subset of Y can be embedded into X by using a member of $\{g_n : n \in \mathbf{N}\}$, but there is no isometry of $\mathbf{R}^m$ which transforms Y into X. In other words, $\kappa_G(X)$ is not finite, contradicting (b).

The obtained contradiction establishes the converse implication (b) $\Rightarrow$ (a).

Thus, the equivalence (a) $\Leftrightarrow$ (b) is valid.

19. Let (E, ρ) be a metric space, let X be a compact subset of E, and let $f : X \to E$ be an isometric embedding such that $f(X)$ is contained in X.

Show that the equality $f(X) = X$ holds true.

For this purpose, suppose otherwise, i.e., $X \setminus f(X) \neq \emptyset$, and take a point $x \in X \setminus f(X)$. Deduce that $\rho(x, f(X)) = \varepsilon$ for some strictly positive real ε. Then consider the sequence of points $\{f^n(x) : n \in \mathbf{N}\}$ and verify that

$$\rho(f^n(x), f^k(x)) \geq \varepsilon \qquad (n \in \mathbf{N},\ k \in \mathbf{N},\ n \neq k).$$

The last relation contradicts the compactness of X.

CHAPTER 2

Maximal *ot*-subsets of the Euclidean plane

We have already mentioned that many interesting problems and questions in the geometry of Euclidean spaces can be indicated, which are of set-theoretical or combinatorial character and are closely connected with point set theory, with finite or discrete systems of geometric figures, and with the convexity structure (see, for instance, [1], [4], [7], [9], [30], [34], [68], [76], [105], [116], [123], [130], [132], [253], [277], [287], [308], [313], [360]).

A number of problems of such a kind was raised by P. Erdös in the series of his publications (see, e.g., [83], [84], [85], [86]).

Similar questions can be frequently met in algebra, mathematical analysis, classical measure theory, and equidecomposability theory of geometric figures (cf. [31], [47], [130], [198], [235]).

Here we would like to discuss one problem of this type for the Euclidean plane $\mathbf{R}^2$ (see [202]).

We begin with some definitions and introduce the corresponding notation.

Let X be a subset of the Euclidean space $\mathbf{R}^m$, where $m \geq 2$.

We shall say that X is an *ot*-set if every three-element subset of X forms a non-degenerate obtuse-angled triangle.

The following two properties are directly implied by this definition:

(i) any subset of an *ot*-set is also an *ot*-set;

(ii) if $\{X_i : i \in I\}$ is a directed (with respect to the inclusion relation) family of *ot*-sets in the space $\mathbf{R}^m$, then $\cup\{X_i : i \in I\}$ is also an *ot*-set in $\mathbf{R}^m$.

Example 1. In the space $\mathbf{R}^m$, where $m \geq 2$, consider the so-called moment curve given by the mapping

$$t \to (t, t^2, ..., t^m) \qquad (t \in [0, 1]).$$

DOI: 10.1201/9781003458708-2

It is easy to verify that the range of this curve is an *ot*-set, all points of which are in general position (i.e., no $m+1$ of them lie in an affine hyperplane of $\mathbf{R}^m$). The above-mentioned curve plays an important role in the theory of convex polyhedra, because for $m \geq 4$ it provides various examples of so-called Carathéodory–Gale polyhedra (see [104] and Chapter 14 of this book)).

We shall say that an *ot*-set $X \subset \mathbf{R}^m$ is maximal if there is no *ot*-set in $\mathbf{R}^m$ properly containing X.

The following statement is a geometric corollary of the well-known general set-theoretical concepts.

Theorem 1. *Each ot-set in the space* $\mathbf{R}^m$*, where* $m \geq 2$*, is contained in some maximal (with respect to the inclusion relation) ot-subset of* $\mathbf{R}^m$.

Proof. It can readily be seen that the following two relations for a set $Z \subset \mathbf{R}^m$ are equivalent:

(1) Z is an *ot*-set;

(2) every finite subset of Z is an *ot*-set.

In other words, the property to be an *ot*-set is of finite character (see, e.g., [44], [275]).

Thus, the assertion of Theorem 1 trivially follows from the Kuratowski–Zorn lemma or, more precisely, from its consequence concerning the properties of finite character (for more details, see again [44], [275], and Exercise 1 of this chapter). □

As far as we know, the following problem remains unsolved.

Problem. Give a characterization of all maximal *ot*-subsets of $\mathbf{R}^m$.

Here are two simple and effective (i.e., constructed without the aid of the Axiom of Choice) examples of maximal *ot*-sets in the plane $\mathbf{R}^2$.

Example 2. Let Z be a semi-circle on the plane $\mathbf{R}^2$ without one of its end-points. More concretely, take

$$Z = \{(x, y) \in \mathbf{R}^2 : x^2 + y^2 = 1,\ y > 0\} \cup \{(1, 0)\}.$$

It can easily be verified that Z is a maximal *ot*-set in $\mathbf{R}^2$.

Example 3. Let $f : \mathbf{R} \to \mathbf{R}$ be a function, let $\mathrm{Gr}(f)$ denote the graph of f, and suppose that the following three conditions are satisfied:

(1) f is increasing and continuous;

(2) no three points of $\mathrm{Gr}(f)$ belong to a straight line, i.e., all points of $\mathrm{Gr}(f)$ are in general position in the plane $\mathbf{R}^2$;

(3) $\lim_{x \to -\infty} f(x) = -\infty$ and $\lim_{x \to +\infty} f(x) = +\infty$.

Then it is not difficult to show that Gr(f) is a maximal ot-set in $\mathbf{R}^2$ (a slightly more general result is presented in Exercise 3 of this chapter). Notice also that conditions (1) and (2) imply at once that the given function f is strictly increasing.

To present other (less trivial) examples of maximal ot-subsets of $\mathbf{R}^2$, we need several auxiliary notions.

Let x and y be any two distinct points in $\mathbf{R}^2$. As usual, we denote by $l(x, y)$ the straight line passing through x and y.

Let l and l' be any two distinct parallel straight lines on $\mathbf{R}^2$. We recall that all points of $\mathbf{R}^2$ lying between l and l' form the strip (plank) determined by these two lines. This strip will be denoted by $S(l, l')$.

In the sequel, only closed strips are under consideration, i.e., we assume throughout this chapter that $S(l, l')$ contains both lines l and l'.

By definition, the width of $S(l, l')$ is equal to the distance between the straight lines l and l'.

Accordingly, a single straight line may be regarded as a degenerate strip whose width is equal to zero.

Let x and x' be any two distinct points of $\mathbf{R}^2$, let l be the straight line passing through x and perpendicular to the segment $[x, x']$, and let l' be the straight line passing through x' and perpendicular to the same segment.

We shall say that the strip $S(l, l')$ is determined by the pair of points x and x'. In this case we also put $S(x, x') = S(l, l')$.

Let $\{S_i : i \in I\}$ be an injective family of non-degenerate strips in $\mathbf{R}^2$.

We shall say that these strips are in general position if, for any two distinct strips S_i and S_j, where $i \in I$ and $j \in I$, a boundary line of S_i is not parallel to a boundary line of S_j.

It is not hard to show that the whole plane $\mathbf{R}^2$ cannot be covered by finitely many strips. From this circumstance it readily follows that no finite ot-subset of $\mathbf{R}^2$ can be maximal (see Exercise 4).

Notice that the maximal ot-sets which were described in Examples 2 and 3 presented above, are of cardinality continuum ($= \mathbf{c}$). In this connection, it is natural to ask whether there exists a countable maximal ot-set in $\mathbf{R}^2$.

The main goal in this chapter is to demonstrate that there are locally finite (hence countable) maximal ot-subsets of $\mathbf{R}^2$.

In order to establish this fact, we need several additional auxiliary notions and statements.

Let $\{S_i : i \in I\}$ and $\{S_j : j \in J\}$ be two injective families of non-degenerate strips in the plane $\mathbf{R}^2$.

We shall say that these two families are in general position if all strips of the expanded family

$$\{S_k : k \in K\} = \{S_i : i \in I\} \cup \{S_j : j \in J\}$$

are in general position.

Let X be a subset of $\mathbf{R}^2$ and let $\{S_i : i \in I\}$ be an injective family of non-degenerate strips in $\mathbf{R}^2$.

We shall say that X and $\{S_i : i \in I\}$ are in general position if the two families of strips

$$\{S_i : i \in I\}, \quad \{S(x, x') : x \in X,\ x' \in X,\ x \neq x'\}$$

are in general position.

If Z is a subset of $\mathbf{R}^2$ and ε is a strictly positive real number, then the symbol $V(Z, \varepsilon)$ will denote below the ε-neighborhood of Z.

Accordingly, if $Z = \{z\}$ where z is a point of $\mathbf{R}^2$, then we will utilize the notation $V(z, \varepsilon)$ for this z and for any real $\varepsilon > 0$.

Lemma 1. *Let X be a finite set of points in the plane $\mathbf{R}^2$, such that all the strips*

$$S(x, x') \quad (x \in X,\ x' \in X,\ x \neq x')$$

are in general position, and let $C \subset \mathbf{R}^2$ be an open disc containing X.

Then, for every $\varepsilon > 0$, there exists a finite family of strips $\{S_i : i \in I\}$ satisfying the following three relations:

(1) for each $i \in I$, the width of S_i is strictly less than ε;

(2) the set X and the family $\{S_i : i \in I\}$ are in general position;

(3) $X \subset C \setminus \cup\{S_i : i \in I\} \subset V(X, \varepsilon)$.

We omit the easy proof of Lemma 1 and leave it to the reader.

Lemma 2. *Let C be an open disc in the plane $\mathbf{R}^2$, let X be a finite ot-subset of C, and let $\{S_1, S_2, ..., S_n\}$ be an injective finite family of non-degenerate strips in $\mathbf{R}^2$ such that X and $\{S_1, S_2, ..., S_n\}$ are in general position and*

$$X \cap (S_1 \cup S_2 \cup ... \cup S_n) = \emptyset.$$

Then there exist in $\mathbf{R}^2$ some pairwise distinct points

$$y_1,\ z_1,\ y_2,\ z_2,\ \ldots,\ y_n,\ z_n$$

satisfying the following four relations:

(1) $\{y_1, z_1, y_2, z_2, ..., y_n, z_n\} \subset \mathbf{R}^2 \setminus C$;

(2) $X^* = X \cup \{y_1, z_1, y_2, z_2, ..., y_n, z_n\}$ is an ot-subset of $\mathbf{R}^2$;

(3) for each natural number $i \in [1, n]$, one has $S_i = S(y_i, z_i)$;

(4) the family of strips $\{S(x, x') : x \in X^*,\ x' \in X^*,\ x \neq x'\}$ is in general position.

The proof of Lemma 2 can be obtained by induction on n. We omit the corresponding technical details (which are not difficult).

Now, we are able to formulate and prove the main result of the present chapter (see [202]).

Theorem 2. *There exists a locally finite maximal (with respect to the inclusion relation) ot-subset of the plane* $\mathbf{R}^2$.

Proof. Fix a decreasing sequence $\{\varepsilon_k : k \geq 1\}$ of strictly positive real numbers, such that $\lim_{k\to\infty}\varepsilon_k = 0$.

Define, by induction on k, two increasing (with respect to the inclusion relation) families $\{X_k : k \geq 1\}$ and $\{C_k : k \geq 1\}$ of subsets of $\mathbf{R}^2$.

For $k = 1$, take any finite ot-set $X_1 \subset \mathbf{R}^2$ with the property that all strips from the family

$$\{S(x, x') : x \in X_1,\ x' \in X_1,\ x \neq x'\}$$

are in general position, and take also an open disc $C_1 \supset X_1$ with center $(0, 0)$ and radius greater than 1.

Suppose that the sets X_k and C_k have already been determined for a natural number $k \geq 1$.

Applying Lemma 1 to X_k, C_k and ε_k, we can find an injective finite family of strips $\{S_i : i \in I_k\}$ satisfying the following conditions:

(1) for each index $i \in I_k$, the width of S_i is less than ε_k;

(2) the set X_k and the family $\{S_i : i \in I_k\}$ are in general position;

(3) $X_k \subset C_k \setminus \cup\{S_i : i \in I_k\} \subset V(X_k, \varepsilon_k)$.

Now, apply Lemma 2 to the disc C_k, the set X_k, and the family of strips $\{S_i : i \in I_k\}$. According to the above-mentioned lemma, there exists a finite ot-set $X_k^* \subset \mathbf{R}^2$ such that:

(a) $X_k \subset X_k^*$;

(b) $X_k^* \setminus X_k \subset \mathbf{R}^2 \setminus C_k$;

(c) the family of strips $\{S(x, x') : x \in X_k^*,\ x' \in X_k^*,\ x \neq x'\}$ is in general position and contains the family $\{S_i : i \in I_k\}$.

Further, we put $X_{k+1} = X_k^*$ and denote by C_{k+1} an open disc in $\mathbf{R}^2$ which contains the set $X_{k+1} \cup C_k$, whose center coincides with $(0, 0)$ and whose radius is greater than $k + 1$.

Proceeding in this manner, we are able to construct the two desired families $\{X_k : k \geq 1\}$ and $\{C_k : k \geq 1\}$ of subsets of $\mathbf{R}^2$.

Note that, by virtue of our construction, we obviously have the equality

$$\mathbf{R}^2 = \cup\{C_k : k \geq 1\}.$$

Finally, we define

$$X = \cup\{X_k : k \geq 1\}.$$

A straightforward verification shows that X is a locally finite *ot*-subset of the plane $\mathbf{R}^2$, and we are going to demonstrate that X is a maximal *ot*-set in $\mathbf{R}^2$ (with respect to the inclusion relation).

For this purpose, let us take any point $p \in \mathbf{R}^2 \setminus X$ and let us check that $X \cup \{p\}$ cannot be an *ot*-set in $\mathbf{R}^2$. Indeed, since the equality

$$\lim_{k\to\infty}\varepsilon_k = 0$$

holds true, there exists a natural number k_0 such that, for all natural numbers $k > k_0$, we have

$$p \in C_k \setminus \cup\{V(x, \varepsilon_k) : x \in X_k\}$$

and, consequently, by virtue of condition (3),

$$p \in \cup\{S_i : i \in I_k\}.$$

Hence $p \in S_i$ for some index $i \in I_k$. In view of our construction, $S_i = S(y, z)$, where

$$\{y, z\} \subset X_{k+1} \setminus X_k \subset X,$$

and we also remember that the width of S_i is less than ε_k. Now, taking this circumstance into account, it can easily be seen that if k is large enough, then the values of all three internal angles of the triangle $[p, y, z]$ do not exceed $\pi/2$, whence it follows that $X \cup \{p\}$ is not an *ot*-subset of $\mathbf{R}^2$.

Theorem 2 has thus been proved. □

In connection with the above theorem, it is natural to ask whether there exists a countable bounded maximal *ot*-subset of the plane $\mathbf{R}^2$. This question still remains unanswered.

Notice that the statement completely analogous to Theorem 2 can be established for the m-dimensional Euclidean space $\mathbf{R}^m$, where $m > 2$. In other words, we have the following more general result.

Theorem 3. *If $m \geq 2$, then there exists a locally finite maximal (with respect to the inclusion relation) ot-set in the space $\mathbf{R}^m$.*

The proof of this statement is carried out in the same manner as for the Euclidean plane $\mathbf{R}^2$ (some purely technical details occur, but they are not connected with any additional difficulties).

Dealing with maximal *ot*-subsets of Euclidean spaces, the following example seems to be relevant.

Example 4. Consider the three-dimensional Euclidean space $\mathbf{R}^3$ and its two-dimensional vector subspace $\mathbf{R}^2 \times \{0\}$.

Let S be a closed semicircle in $\mathbf{R}^2 \times \{0\}$ with endpoints y and z. Let $l(y,z)$ denote, as usual, the straight line passing through y and z. Take any point x on $l(y,z)$ not belonging to the line segment $[y,z]$ and put

$$X = (S \setminus \{y\}) \cup \{(x,1)\}.$$

It is not hard to verify that X is an *ot*-set in $\mathbf{R}^3$. Also, as we already know, $S \setminus \{y\}$ is a maximal *ot*-subset of the plane $\mathbf{R}^2 \times \{0\}$ (see Example 2). Since X properly contains $S \setminus \{y\}$, we conclude that $S \setminus \{y\}$ is not a maximal *ot*-set in the space $\mathbf{R}^3$. Moreover, we cannot assert that X is a maximal *ot*-subset of $\mathbf{R}^3$. For instance, if the semicircle S is such that

$$(\forall p \in S)(||p - x|| < 1),$$

then the set

$$X' = (S \setminus \{y\}) \cup \{(x,1)\} \cup \{(x,-1)\} = X \cup \{(x,-1)\}$$

turns out to be an *ot*-subset of $\mathbf{R}^3$ which properly contains X.

We thus conclude that a maximal *ot*-subset of the plane $\mathbf{R}^2$ is not, in general, a maximal *ot*-subset of the space $\mathbf{R}^3$.

The notion of an *ot*-set can be introduced for some infinite-dimensional analogues of an Euclidean space, e.g., for real Hilbert spaces. Since the completeness of a ground space is not essential for introducing the notion of an *ot*-set in this space, we may consider real pre-Hilbert spaces instead of real Hilbert spaces.

Let H be a pre-Hilbert space (over the field $\mathbf{R}$ of all real numbers).

As earlier, we shall say that a set $X \subset H$ is an *ot*-set in H if each three-element subset of X forms a non-degenerate obtuse-angled triangle.

We shall say that a set $Y \subset H$ is an *rt*-set in H if each three-element subset of Y forms a right-angled triangle.

In [177] a certain characterization of all *rt*-sets in H was given (for more details, see [185]; cf. also [29]).

In particular, it was demonstrated in [177] that $\mathrm{card}(X) \leq \mathbf{c}$ for any *rt*-set $X \subset H$, where $\mathbf{c}$ denotes the cardinality of the continuum.

This result is of interest, because the Hilbert dimension of the completion of H can be an arbitrarily large cardinal number, so $\mathrm{card}(H)$ can also be arbitrarily large.

Notice also that the analogous result holds true for all *ot*-sets in H. This will be proved later in the book, by using elements of Ramsey theory (see Chapter 5).

Example 5. In the case $H = \mathbf{R}^2$, it can easily be shown that a set $Y \subset H$ is a maximal *rt*-subset of H (with respect to the inclusion relation) if and only if Y coincides with the set of all vertices of a rectangle in H.

Some exercises presented below contain more information about *ot*-sets and *rt*-sets in Euclidean spaces and real pre-Hilbert spaces.

EXERCISES

1. Let ω denote, as usual, the least infinite cardinal (ordinal) number.

A property $R(X)$ of a set X is called a property of finite character if

$$R(X) \Leftrightarrow (\forall Y)((Y \subset X \ \& \ \mathrm{card}(Y) < \omega) \Rightarrow R(Y)).$$

Let E be a ground set and let $R(X)$ be a property of finite character for subsets X of E.

Demonstrate that there exists a set $X_0 \subset E$ such that $R(X_0)$ and X_0 is maximal with respect to inclusion, i.e., X_0 is not properly contained in any set $X \subset E$ satisfying $R(X)$.

Deduce from this general statement that, for every *ot*-subset Z of a real pre-Hilbert space H, there exists a maximal (with respect to inclusion) *ot*-subset Z_0 of H such that $Z \subset Z_0$.

2. Let Z be a half-open semicircle in the Euclidean plane $\mathbf{R}^2$, e.g.,

$$Z = \{(\cos(\phi), \sin(\phi)) : 0 \leq \phi < \pi\}.$$

Verify that Z is a maximal (with respect to inclusion) *ot*-subset of $\mathbf{R}^2$.

3. Let $f : \mathbf{R} \to \mathbf{R}$ be a function possessing the following properties:

(a) f is increasing and continuous;

(b) no three points of the graph of f belong to a straight line, i.e., all points of this graph are in general position in the plane $\mathbf{R}^2$.

Show that in this case the graph of f, i.e., the set

$$\mathrm{Gr}(f) = \{(x, y) \in \mathbf{R}^2 : y = f(x)\}$$

is a maximal (with respect to inclusion) *ot*-subset of $\mathbf{R}^2$.

4. Demonstrate that no finite *ot*-set in the plane $\mathbf{R}^2$ can be maximal with respect to the inclusion relation.

Infer from this fact that, for any finite *ot*-set $X \subset \mathbf{R}^2$, there exists an infinite *ot*-set $X' \subset \mathbf{R}^2$ such that $X \subset X'$.

Formulate and prove the two analogous statements for finite *ot*-sets in the space $\mathbf{R}^m$, where $m \geq 2$.

5. Give a detailed proof of Lemma 1.

For this purpose, use the classical Borel–Lebesgue lemma on open coverings of compact subsets of the plane $\mathbf{R}^2$.

6. By using induction on n, give a proof of Lemma 2.

7*. Following the argument of the proof of Theorem 2, give a detailed proof of Theorem 3.

8. Let X be a subset of the Euclidean plane $\mathbf{R}^2$.

Show that X is a maximal rt-set in $\mathbf{R}^2$ if and only if X coincides with the set of all vertices of some rectangle in $\mathbf{R}^2$.

9*. Let X be a finite rt-subset of the space $\mathbf{R}^m$, where $m \geq 2$, and suppose that this X does not contain the four vertices of a rectangle.

Prove that there exists an enumeration of X in the form

$$X = \{x_1,\ x_2,\ x_3,\ \dots,\ x_q\}$$

satisfying the following condition:

For any three natural indices i, j and k such that $1 \leq i < j < k \leq q$, the triangle $[x_i, x_j, x_k]$ is right-angled and its right angle is at the vertex x_j.

Further, suppose that some distinct enumeration of X in the form

$$X = \{x'_1,\ x'_2,\ x'_3,\ \dots,x'_q\}$$

is given satisfying the analogue of the above-mentioned condition.

Verify that

$$x'_1 = x_q,\ x'_2 = x_{q-1},\ x'_3 = x_{q-2},\ \dots,\ x'_q = x_1.$$

10. Deduce from the result of Exercise 9 that if $X \subset \mathbf{R}^m$ is an rt-set, then either $\mathrm{card}(X) \leq m + 1$ or X coincides with the set of all vertices of some rectangle.

11*. An m-dimensional simplex S in the Euclidean space $\mathbf{R}^m$ $(m \geq 2)$ is called orthogonal if the set of all vertices of S is an rt-subset of $\mathbf{R}^m$.

Show that:

(a) if $m = 2$, then every m-dimensional simplex can be dissected into two orthogonal m-dimensional simplices (i.e., every triangle can be dissected into two right-angled triangles);

(b) if $m = 3$, then every m-dimensional simplex can be dissected into 28 (or less) orthogonal m-dimensional simplices;

(c) if $m = 4$, then every m-dimensional simplex can be dissected into 730 (or less) orthogonal m-dimensional simplices.

Remark 1. In connection with Exercise 11, see [29], [130], [155], [178], [185], [346], [347], where further information on orthogonal simplices may be found. The question of whether there exists a natural number $or(m)$ such that every m-dimensional simplex can be decomposed into $or(m)$ (or less) m-dimensional orthogonal simplices is a long-time standing open problem in Euclidean geometry, which is important for the general theory of polyhedra (cf. [130]).

12. Let H be a real pre-Hilbert space and let X be an rt-subset of H not containing the four vertices of a rectangle.

Demonstrate that there exists a linear ordering $\preceq$ of X having the following property:

For any three points $x \in X$, $y \in X$ and $z \in X$ such that $x \prec y \prec z$, the triangle $[x, y, z]$ is right-angled and its right angle is at the vertex y.

For this purpose, take into account the result of Exercise 9.

Moreover, check that there are exactly two linear orderings with the above-mentioned property (consequently, the second one is reverse to the indicated ordering $\preceq$).

Check that there is a four-point ot-subset X of $\mathbf{R}^2$ having the following property:

There exists no linear ordering $\preceq$ of X such that the equivalence

$$x \prec y \prec z \Leftrightarrow \langle x - y, z - y \rangle < 0$$

holds true for any three distinct points x, y, z from X.

Here $\langle x - y, z - y \rangle$ denotes the inner (scalar) product of the vectors $x - y$ and $z - y$.

Conclude that there is an essential difference between the geometric structures of rt-sets and ot-sets in a real pre-Hilbert space.

13*. Let X be an rt-set in a real pre-Hilbert space H.

Prove that the cardinality of X does not exceed the cardinality of the continuum $\mathbf{c}$.

For this purpose, assuming that X is not empty and does not contain the set of all vertices of a rectangle, equip X with a linear ordering $\preceq$ mentioned in Exercise 12. Further, take a point $x \in X$ and consider the following two linearly ordered subsets of X:

$$(X_1, \preceq) = \{y \in X : y \preceq x\}, \qquad (X_2, \preceq) = \{y \in X : x \preceq y\}.$$

Verify that there exists a monomorphism from $(X_1, \preceq)$ into $(\mathbf{R}, \leq)$ and there exists a monomorphism from $(X_2, \preceq)$ into $(\mathbf{R}, \leq)$.

Infer from the existence of such monomorphisms the desired inequality

$$\mathrm{card}(X) \leq \mathbf{c} = 2^{\omega}.$$

14*. Strengthen the result of Exercise 13 and establish that if X is an arbitrary rt-set in a real pre-Hilbert space H, then X is a separable subset of H (so the cardinality of X automatically does not exceed $\mathbf{c}$).

For this purpose, preserving the notation of Exercise 13, define two functions $f_1 : X_1 \to \mathbf{R}$ and $f_2 : X_2 \to \mathbf{R}$ by the formulas

$$f_1(y) = ||x - y||^2 \quad (y \preceq x), \qquad f_2(y) = ||x - y||^2 \quad (x \preceq y),$$

and check that $|f_i(y) - f_i(y')| = ||y - y'||^2$ for all elements y and y' of X_i, where $i \in \{1, 2\}$. Deduce from this fact that both X_1 and X_2 are separable subsets of H, so $X = X_1 \cup X_2$ is separable, too.

Remark 2. It makes sense to recall the reader one interesting example of an rt-set in a real infinite-dimensional separable Hilbert space H. As is well known, H can be identified with the space $L_2([0,1])$ of all real-valued square integrable (in the Lebesgue sense) functions on $[0,1]$. Considering the characteristic functions $\chi_{[0,t]}$ of the subintervals $[0,t]$ of $[0,1]$, where t ranges over $[0,1]$, we come to an injective continuous mapping

$$W : [0,1] \to L_2([0,1])$$

defined by the formula $W(t) = \chi_{[0,t]}$. A straightforward verification shows that $W([0,1])$ is an rt-subset of $L_2([0,1]])$ and, obviously, this subset has cardinality continuum $\mathbf{c}$.

The mapping W is usually called the Wiener curve in the space $L_2([0,1])$.

CHAPTER 3

The cardinalities of *at*-sets in a real Hilbert space

In Chapter 2 we were concerned with so-called ot-subsets and rt-subsets of a finite-dimensional Euclidean space (or of a real pre-Hilbert space).

Similarly to those subsets, the notion of an at-set in the space $\mathbf{R}^m$ ($m \geq 2$) or, more generally, in a real pre-Hilbert space can be introduced and seems to be natural. In this chapter, we will study the question of finding the maximal cardinality of an at-set lying in $\mathbf{R}^m$ or in a real infinite-dimensional Hilbert space H.

Furthermore, the number of those right angles which are produced by the set of all vertices of the unit cube $[0, 1]^m$ will be calculated. As an application, a simple solution of one well-known combinatorial-geometric problem will be given, concerning the growth of the maximal cardinality of a strong at-set in $\mathbf{R}^m$, when the dimension m tends to infinity.

Let H be a real pre-Hilbert space and let X be a subset of H.

We shall say that X is an at-set (in H) if any three-element subset of X forms either an acute-angled triangle or a right-angled triangle.

We shall say that X is a strong at-set (in H) if any three-element subset of X forms an acute-angled triangle.

It directly follows from the above definitions that each subset of an at-set (respectively, of a strong at-set) is also an at-set (respectively, is a strong at-set).

In this chapter we are planning to discuss the question of finding the maximum value of the cardinalities of all at-subsets (respectively, of all strong at-subsets) of H.

First of all, consider the case when H is finite-dimensional, i.e., H coincides with some Euclidean space $\mathbf{R}^m$, where $m \geq 2$ is a natural number.

DOI: 10.1201/9781003458708-3

Denote by $at(m)$ the maximum of the cardinalities of all at-sets in $\mathbf{R}^m$. Answering two questions posed by P. Erdös and V. L. Klee, it was demonstrated in [67] that the inequality

$$at(m) \leq 2^m$$

always holds true. Moreover, it was also shown in [67] that the equality

$$\mathrm{card}(X) = 2^m$$

for an at-set $X \subset \mathbf{R}^m$ is valid if and only if X coincides with the set of all vertices of some right rectangular m-dimensional parallelepiped in $\mathbf{R}^m$. Thus, one can conclude from these results that $at(m)$ has an exponential growth with respect to m. For more details, see [67], or Chapter 15 in the remarkable book [2], or Exercise 1 below.

Denote by $sat(m)$ the maximum of the cardinalities of all strong at-sets in $\mathbf{R}^m$. It is easy to show that $sat(2) = 3$ and it is also known that $sat(3) = 5$ (see Exercises 6, 7 and 8 of this chapter). It immediately follows from the result of [67] that one of the upper bounds for $sat(m)$ is $2^m - 1$, i.e., we have the inequality

$$sat(m) \leq 2^m - 1.$$

In the general case, the precise value of $sat(m)$ is still unestablished. However, it was proved that $sat(m)$ also has an exponential growth with respect to m (see [2], [88], [290]).

Notice that in [2], [88], and [290] the latter fact is proved with the aid of a probabilistic argument which at first sight looks to be somewhat artificial or irrelevant in this case. Indeed, a simple deterministic proof of the same fact can be presented by using another approach. We give below a different proof of the same fact.

In what follows, the symbol V_n will denote the set of all vertices of the n-dimensional unit cube $C_n = [0, 1]^n$ in $\mathbf{R}^m$, where $0 \leq n \leq m$.

So we have $\mathrm{card}(V_n) = 2^n$ and, in particular, $\mathrm{card}(V_m) = 2^m$.

First of all, we are going to present the precise formula for the total number r_m of those right-angled triangles whose vertices belong to V_m. Clearly, this number coincides with the total number of those right-angled triangles whose vertices belong to the set of vertices of an m-dimensional right rectangular parallelepiped in $\mathbf{R}^m$.

Let t_m stand for the number of all those right-angled triangles in C_m, the right angle of which is at a fixed vertex v from V_m and the other two vertices of which also belong to V_m. Consider some facet C_{m-1} of C_m incident to v. Obviously, we have t_{m-1} right angles with the same vertex v, all of which lie in C_{m-1}. Further, any of the above-mentioned t_{m-1} angles is a projection of exactly two right angles which do not lie in C_{m-1}. Besides, there are precisely $2^{m-1} - 1$ right angles, all of which have a common side, namely, the edge of C_m passing through v and perpendicular to the affine hyperplane in $\mathbf{R}^m$ carrying C_{m-1}.

Thus, we come to the following recursive formula:

$$t_m = 3t_{m-1} + 2^{m-1} - 1.$$

This formula allows us to readily deduce (e.g., by induction on m) that

$$t_m = (3^m + 1)/2 - 2^m.$$

Therefore, ranging v over the whole of V_m, we finally get

$$r_m = 2^m((3^m + 1)/2 - 2^m).$$

As an immediate consequence of the obtained formula, we get that the total number of all those acute-angled triangles whose vertices belong to V_m is equal to

$$\frac{2^m!}{3!(2^m - 3)!} - r_m = \frac{2^m!}{3!(2^m - 3)!} - 2^m((3^m + 1)/2 - 2^m).$$

Now, we would like to give an application of the last formula for evaluating from below the value $k = sat(m)$ (this approach was presented in [214], [219]).

Let $X_1, X_2, ..., X_p$ be an injective enumeration of all $(k+1)$-element subsets of V_m, so

$$p = \frac{2^m!}{(k+1)!(2^m - (k+1))!},$$

and let, for each natural index $i \in [1, p]$, the symbol a_i denote the number of right-angled triangles in X_i. Since no X_i is a strong *at*-set, we may write

$$1 \leq a_i \qquad (1 \leq i \leq p).$$

At the same time, it is clear that

$$a_1 + a_2 + ... + a_p = \frac{(2^m - 3)!}{(k-2)!(2^m - 3 - (k-2))!} \cdot r_m.$$

Since we obviously have

$$\frac{(2^m - 3)!}{(k-2)!(2^m - 3 - (k-2))!} \leq \frac{(2^m)!}{(k-2)!(2^m - (k-2))!},$$

we infer that

$$p \leq a_1 + a_2 + ... + a_p \leq \frac{(2^m)!}{(k-2)!(2^m - (k-2))!} \cdot r_m.$$

Consequently,

$$\frac{2^m!}{(k+1)!(2^m - (k+1))!} \leq \frac{2^m!}{(k-2)!(2^m - (k-2))!} \cdot 2^m((3^m + 1)/2 - 2^m).$$

The above inequality implies

$$(2^m - (k+1))^3 \leq (k+1)^3 \cdot 2^m((3^m+1)/2 - 2^m)$$

or, equivalently,

$$\frac{2^m}{1 + (2^m((3^m+1)/2 - 2^m))^{1/3}} \leq k+1.$$

Further, taking into account the two trivial inequalities

$$1 + (2^m((3^m+1)/2 - 2^m))^{1/3} \leq 2 \cdot (2^m((3^m+1)/2 - 2^m))^{1/3},$$

$$(3^m+1)/2 - 2^m < 3^m,$$

we conclude that

$$\frac{1}{2} \cdot (\frac{2}{6^{1/3}})^m \leq k+1.$$

As an immediate result of the above considerations, we can formulate and prove the following statement.

Theorem 1. *The value $k = sat(m)$ has an exponential growth with respect to m.*

Proof. Since the obvious inequality $2 > 6^{1/3}$ is true and

$$\frac{1}{2} \cdot (\frac{2}{6^{1/3}})^m \leq k+1,$$

the value $k+1$ is of an exponential growth with respect to m. Consequently, $k = sat(m)$ has the same growth with respect to m.

This completes the proof of Theorem 1. □

Remark 1. The argument presented above and the argument given in [2], [88], and [290] are not totally constructive in the sense that they do not allow one to indicate or geometrically describe a concrete strong *at*-subset X of V_m whose cardinality is of an exponential growth with respect to m. In this connection, it would be interesting to have some (as many as possible) concrete examples of such subsets X of V_m and to give their geometric characterization.

Now, let us consider the case of a real infinite-dimensional separable Hilbert space H_ω, where ω denotes the least infinite cardinal (ordinal) number and, simultaneously, is the cardinality of any orthonormal basis of H_ω. Clearly, H_ω can be identified with the standard Hilbert space

$$l_2 = \{\{x_n : n \in \omega\} \in \mathbf{R}^\omega : \sum\{x_n^2 : n \in \omega\} < +\infty\}.$$

Let $\mathbf{c} = 2^\omega$ stand for the cardinality of the continuum. Recall the following nice combinatorial fact first observed by W. Sierpiński:

There exists a family $\{T_i : i \in I\}$ of subsets of ω such that:

(a) $\mathrm{card}(I) = \mathbf{c}$;

(b) $\mathrm{card}(T_i) = \omega$ for each $i \in I$;

(c) $\mathrm{card}(T_i \cap T_j) < \omega$ for any two distinct indices $i \in I$ and $j \in I$.

The conditions (a), (b), and (c) actually state that $\{T_i : i \in I\}$ is an almost disjoint family of infinite subsets of ω and this family has maximal cardinality $\mathbf{c}$ (in this connection, see e.g. [316] or Exercise 9 of the present chapter).

Associate with each index $i \in I$ the element $x_i \in \mathrm{l}_2$ defined as follows:

$$x_i = \{(1/2^n)\chi_i(n) : n \in \omega\},$$

where χ_i denotes the characteristic function of the set $T_i \subset \omega$.

It is not difficult to check that $X = \{x_i : i \in I\}$ is a strong *at*-subset of the space l_2 and, trivially,

$$\mathrm{card}(X) = \mathrm{card}(I) = \mathbf{c} = \mathrm{card}(\mathrm{l}_2).$$

Thus, there exists a bounded *at*-set (even a bounded strong *at*-set) in the space l_2, whose cardinality coincides with the cardinality of this space (in [235] a slightly more complicated proof of the above fact is given).

Below, we will obtain some generalization of the above result.

For this purpose, we need the notation H_α for a real Hilbert space H whose Hilbert dimension (i.e., the cardinality of an orthonormal basis of H) coincides with a given infinite cardinal number α. Clearly, all such Hilbert spaces H are pairwise isomorphic. As is well known, the equality

$$\mathrm{card}(H_\alpha) = \alpha^\omega$$

holds true for any infinite cardinal α.

Theorem 2. *In the space H_α there exists a bounded at-set X whose cardinality is equal to the cardinality of H_α.*

Proof. We argue by transfinite induction on α.

The case $\alpha = \omega$ has already been considered and an *at*-set $X \subset H_\omega$ with $\mathrm{card}(X) = \mathrm{card}(H_\omega)$ was chosen so that it turns out to be bounded in H_ω.

Suppose now that, for every infinite cardinal number $\beta < \alpha$, there exists a bounded *at*-subset X_β of H_β satisfying the relation

$$\mathrm{card}(X_\beta) = \mathrm{card}(H_\beta) = \beta^\omega.$$

Take the space H_α and consider the two possible cases.

1. The cardinal α is cofinal with ω, i.e., there exists a strictly increasing sequence $\{\beta_n : n \in \omega\}$ of cardinal numbers, all of which are strictly less than α and the cardinal sum of all of them is equal to α.

In this case, we may represent H_α as the Hilbert sum of Hilbert spaces H_{β_n} $(n \in \omega)$. In other words, every element $h \in H_\alpha$ admits a unique representation in the form

$$h = h_0 + h_1 + ... + h_n + ...,$$

where $h_n \in H_{\beta_n}$ for each $n \in \omega$, and the vectors h_n and h_m are orthogonal whenever n and m are any two distinct natural numbers.

According to the inductive assumption, for each $n \in \omega$, there exists a bounded *at*-set X_n in H_{β_n} such that

$$\mathrm{card}(X_n) = \beta_n^\omega.$$

Using homotheties of H_α with sufficiently small strictly positive coefficients, we can additionally suppose that

$$(\forall x \in X_n)(||x|| \leq 1/2^n).$$

Now, consider in H_α the vectors of the form

$$x = x_0 + x_1 + ... + x_n + ...,$$

where $x_n \in X_n$ for every $n \in \omega$. The set of all these vectors denote by X. Obviously, X is a bounded subset of H_α. Moreover, a straightforward verification shows that X is an *at*-set in H_α. Finally, we have

$$\mathrm{card}(X) = \prod\{\beta_n^\omega : n \in \omega\} = \alpha^\omega = \mathrm{card}(H_\alpha).$$

For more detailed information concerning the latter equalities, see Exercise 10 of the present chapter.

2. The cardinal α is not cofinal with ω.

In this case we treat α as the set of all those ordinals ξ whose cardinalities are strictly less than α, and we can write

$$\alpha^\omega = \sum\{\xi^\omega : \xi \in \alpha\}.$$

Suppose first that $(\mathrm{card}(\xi))^\omega \leq \alpha$ for each $\xi \in \alpha$. Then we readily come to the relation

$$\mathrm{card}(H_\alpha) = \alpha^\omega = \alpha.$$

If $\{e_\xi : \xi \in \alpha\}$ is an arbitrary orthonormal basis in H_α, then

$$X = \{e_\xi : \xi \in \alpha\}$$

is a bounded *at*-subset of H_α with $\mathrm{card}(X) = \mathrm{card}(H_\alpha)$. Moreover, every three-element subset of this X forms an equilateral triangle.

Further, suppose that there exists $\xi \in \alpha$ for which $(\mathrm{card}(\xi))^\omega > \alpha$. Then it is not difficult to infer that

$$(\mathrm{card}(\xi))^\omega = \alpha^\omega.$$

By virtue of the inductive assumption, there exists a bounded *at*-set X in the space $H_{\mathrm{card}(\xi)}$ such that

$$\mathrm{card}(X) = \mathrm{card}(H_{\mathrm{card}(\xi)}) = (\mathrm{card}(\xi))^{\omega} = \alpha^{\omega}.$$

At the same time, $H_{\mathrm{card}(\xi)}$ can be regarded as a subspace of H_α, so X turns out to be a bounded *at*-set in H_α with $\mathrm{card}(X) = \mathrm{card}(H_\alpha)$.

Theorem 2 has thus been proved. □

Remark 2. A subset Y of a real Hilbert space H will be called an *rot*-set (in H) if any three-element subset of Y forms either a right-angled triangle or an obtuse-angled triangle.

It will be demonstrated later in the book that every *rot*-set Y in H is separable, so $\mathrm{card}(Y) \leq \mathbf{c}$ (see Chapter 5).

In this respect, the geometric-combinatorial structure of *at*-sets substantially differs from the analogous structures of *ot*-sets and *rot*-sets.

A canonical example of a big (in the sense of cardinality) *ot*-set in the standard Hilbert space l_2 is provided by the set Y of all points of the form

$$y(t) = (t, t^2, t^3, ..., t^n, ...) \qquad (0 < t < 1).$$

This Y may be interpreted as an analogue in l_2 of the moment curve in a finite-dimensional Euclidean space. Clearly, the equality $\mathrm{card}(Y) = \mathbf{c}$ holds true for Y and any three-element subset of Y forms an obtuse-angled triangle, so Y is an *ot*-set in l_2.

In this context, the famous Wiener curve $W = \{\chi_{[0,t]} : 0 \leq t \leq 1\}$ in the real Hilbert space $L_2([0, 1])$ should also be mentioned, where $\chi_{[0,t]}$ stands for the characteristic function of $[0, t] \subset [0, 1]$ (see Remark 2 in Chapter 2). It is readily verified that any three-element subset of W forms a right-angled triangle.

Remark 3. There are many uncountable cardinal numbers α satisfying the equality $\alpha^\omega = 2^\alpha$. For instance, such an α can be obtained as the cardinal sum

$$\alpha = \sum \{\alpha_n : n \in \omega\},$$

where $\alpha_0 = \beta \geq \omega$ and $\alpha_{n+1} = 2^{\alpha_n}$ for all $n \in \omega$.

Since the initial cardinal β in this definition can be taken arbitrarily large, all α of the above form constitute a proper class of cardinal numbers. Considering for any such α the Hilbert space H_α and choosing an *at*-subset X of H_α with $\mathrm{card}(X) = \mathrm{card}(H_\alpha)$, we conclude that the cardinality of X has an exponential growth with respect to the Hilbert dimension α.

Remark 4. Take the real Hilbert space H_{ω_1} whose Hilbert dimension is equal to the least uncountable cardinal ω_1. Obviously, we have

$$\mathrm{card}(H_{\omega_1}) = \omega_1^\omega = 2^\omega = \mathbf{c}.$$

One may ask whether the maximal cardinality of an *at*-set in H_{ω_1} is of an exponential growth with respect to ω_1. Surprisingly, the answer to this question depends on additional set-theoretical assumptions. Indeed, under the Continuum Hypothesis (**CH**)

$$2^{\omega} = \omega_1,$$

the answer is negative. On the other hand, if the so-called Second Continuum Hypothesis (**SCH**)

$$2^{\omega} = 2^{\omega_1}$$

holds (first formulated by N. Luzin), then the answer is positive. The proof of the consistency of **SCH** with the standard axioms of set theory can be found in [151] or [237].

EXERCISES

1*. Let X be a subset of the space $\mathbf{R}^m$, where $m \geq 2$.

The line segment $[x, y]$, where $x \in X$ and $y \in X$, is called an affine diameter of X if there exist two distinct parallel affine hyperplanes Γ_1 and Γ_2 in $\mathbf{R}^m$ such that $x \in \Gamma_1$, $y \in \Gamma_2$, and X is entirely contained in the domain (actually, in the multi-dimensional strip) bounded by Γ_1 and Γ_2.

Demonstrate that:

(a) if $[x, y]$ is an affine diameter of X and f is an affine transformation of $\mathbf{R}^m$ onto itself, then $[f(x), f(y)]$ is an affine diameter of $f(X)$;

(b) every diameter of a set $Y \subset \mathbf{R}^m$ is simultaneously an affine diameter of Y (but not conversely);

(c) if Z is an *at*-subset of $\mathbf{R}^m$, then for any two distinct points z and z' from Z the line segment $[z, z']$ is an affine diameter of Z;

(d) if X is a finite subset of $\mathbf{R}^m$ and each point $x \in X$ is such that, for any point $y \in X \setminus \{x\}$, the line segment $[x, y]$ is an affine diameter of X, then the inequality

$$\mathrm{card}(X) \leq 2^m$$

holds true.

For proving (d), assume without loss of generality that X is not contained in an affine hyperplane of $\mathbf{R}^m$. Denote by $h_{x,2}$ the homothety of $\mathbf{R}^m$ with center $x \in X$ and coefficient 2, and put

$$X' = h_{x,2}(\mathrm{conv}(X)),$$

where $\mathrm{conv}(X)$ is the convex hull of X. Further, consider the family

$$\{(y - x) + \mathrm{conv}(X) : y \in X\}$$

of translates of conv(X) and check that any two distinct members of this family can be separated by an affine hyperplane in $\mathbf{R}^m$. Finally, taking into account that

$$\cup\{(y-x)+\mathrm{conv}(X) : y \in X\} \subset X', \qquad \lambda_m(X') = 2^m \lambda_m(\mathrm{conv}(X)),$$

infer the desired result (here λ_m denotes, as usual, the standard m-dimensional Lebesgue measure in $\mathbf{R}^m$).

2. Let X be an *at*-subset of the space $\mathbf{R}^m$, where $m \geq 2$.

Check that:

(a) $\mathrm{card}(X) \leq 2^m$;

(b) $\mathrm{card}(X) = 2^m$ if and only if X coincides with the set of all vertices of an m-dimensional right rectangular parallelepiped in $\mathbf{R}^m$.

For this purpose, keep in mind the fact that, for any two distinct points x and y from X, the line segment $[x, y]$ is an affine diameter of X, and use the result of Exercise 1.

3. Let Z be a finite subset of the Euclidean space $\mathbf{R}^m$ (or a finite subset of a real pre-Hilbert space H).

Show that the following two assertions are equivalent:

(a) Z is a strong *at*-set in $\mathbf{R}^m$ (in H);

(b) there exists a real $\varepsilon > 0$ such that, for any two distinct points x and y from Z, the point $z' = x + \varepsilon(y - x)$ satisfies the relation

$$(\forall z \in Z)(z \neq x \Rightarrow ||z' - z|| < ||x - z||).$$

Is this equivalence valid for all infinite strong *at*-subsets of H?

4. Verify that in the Euclidean space $\mathbf{R}^3$ there exist continuum many mutually non-similar strong *at*-sets Z with $\mathrm{card}(Z) = 5$.

5. Demonstrate that every *at*-set in the space $\mathbf{R}^m$ (in a real pre-Hilbert space H) is convexly independent.

6*. Let $[x_1, x_2, x_3, x_4]$ be a three-dimensional simplex (i.e., tetrahedron) in the space $\mathbf{R}^3$, whose vertices are x_1, x_2, x_3, x_4.

Prove that if there exists a point $x_5 \in \mathbf{R}^3$ such that the vertices

$$x_1, \quad x_2, \quad x_3, \quad x_4, \quad x_5$$

of the convex bi-pyramid (whose opposite vertices are x_4 and x_5) form a strong *at*-set, then the inequality

$$||x_1 - x_2||^2 + ||x_1 - x_3||^2 > ||x_4 - x_2||^2 + ||x_4 - x_3||^2$$

holds true.

For this purpose, argue as follows. Suppose that such a point x_5 does exist and denote

$$y_1 = x_1 - x_5, \quad y_2 = x_2 - x_5, \quad y_3 = x_3 - x_5, \quad y_4 = x_4 - x_5.$$

Check that $y_4 = t_1 y_1 + t_2 y_2 + t_3 y_3$, where

$$0 < t_1, \quad 0 < t_2, \quad 0 < t_3, \quad t_1 + t_2 + t_3 > 1.$$

Moreover, by using the relations

$$\langle y_1, y_2 \rangle > 0, \quad \langle y_2, y_3 \rangle > 0, \quad \langle y_3, y_1 \rangle > 0,$$

$$\langle y_1, y_1 - y_4 \rangle > 0, \quad \langle y_2, y_2 - y_4 \rangle > 0, \quad \langle y_3, y_3 - y_4 \rangle > 0,$$

infer that $t_1 < 1$, $t_2 < 1$, and $t_3 < 1$. Then write

$$\begin{gathered}
\langle y_2 - y_4, y_3 - y_4 \rangle = \langle y_2, y_3 \rangle - \langle y_2 + y_3, y_4 \rangle + ||y_4||^2 = \\
\langle y_2, y_3 \rangle - \langle y_2 + y_3, t_1 y_1 + t_2 y_2 + t_3 y_3 \rangle + \\
\langle y_4, t_1 y_1 + t_2 y_2 + t_3 y_3 \rangle < \\
(t_1 + t_2 + t_3)\langle y_2, y_3 \rangle - \langle y_2 + y_3, t_1 y_1 + t_2 y_2 + t_3 y_3 \rangle + \\
\langle y_4, t_1 y_1 + t_2 y_2 + t_3 y_3 \rangle = \\
t_1(\langle y_2, y_3 \rangle - \langle y_1, y_2 \rangle - \langle y_1, y_3 \rangle + \langle y_1, y_4 \rangle) \\
- t_2(||y_2||^2 - \langle y_2, y_4 \rangle) - t_3(||y_3||^2 - \langle y_3, y_4 \rangle) < \\
t_1(\langle y_2, y_3 \rangle - \langle y_1, y_2 \rangle - \langle y_1, y_3 \rangle + ||y_1||^2) \\
= t_1 \langle y_2 - y_1, y_3 - y_1 \rangle < \langle y_2 - y_1, y_3 - y_1 \rangle.
\end{gathered}$$

Further, taking into account that

$$||x_1 - x_2||^2 + ||x_1 - x_3||^2 = ||y_2 - y_3||^2 + 2\langle (y_2 - y_1), (y_3 - y_1) \rangle,$$

$$||x_4 - x_2||^2 + ||x_4 - x_3||^2 = ||y_2 - y_3||^2 + 2\langle (y_2 - y_4), (y_3 - y_4) \rangle,$$

obtain the inequality

$$||x_1 - x_2||^2 + ||x_1 - x_3||^2 > ||x_4 - x_2||^2 + ||x_4 - x_3||^2,$$

as is required.

Remark 5. It should be especially underlined that the above inequality does not involve a point x_5, the existence of which is supposed.

7*. Let P be a convex 3-dimensional polyhedron in the space $\mathbf{R}^3$, all facets of which are triangles and the total number of vertices of which is equal to 6.

Demonstrate that the combinatorial type of P is one of the following two kinds:

(a) P is combinatorially isomorphic to an octahedron;

(b) P is combinatorially isomorphic to a polyhedron, one vertex of which is incident with exactly 5 edges.

For this purpose, apply to P the classical formula of Euler

$$v + f = e + 2$$

(where v is the number of vertices, f is the number of facets, e is the number of edges) and infer that

$$e(P) = 12, \qquad f(P) = 8.$$

Then denote by v_k the number of all those vertices of P which are incident with exactly k edges, and observe that, for this P, one necessarily has

$$(\forall k \in \mathbf{N})(k \geq 6 \Rightarrow v_k = 0).$$

Consequently,

$$3v_3 + 4v_4 + 5v_5 = 2e = 24.$$

At the same time,

$$v_3 + v_4 + v_5 = 6.$$

Further, consider the following two cases.

1. $v_5 = 0$.

In this case, deduce that

$$v_3 = 0, \qquad v_4 = 6,$$

and the combinatorial type of P coincides with the combinatorial type of an octahedron.

2. $v_5 > 0$.

In this case, obtain $v_5 = 1$.

Summarizing all the above, conclude the validity of the disjunction of (a) and (b).

8*. Prove that in the space $\mathbf{R}^3$ there exists no strong *at*-set whose cardinality is equal to 6, and conclude that $sat(3) = 5$.

For this purpose, suppose on the contrary that there exists a strong *at*-set

$$\{x_1, x_2, x_3, x_4, x_5, x_6\} \subset \mathbf{R}^3$$

containing 6 distinct points. Take the convex polyhedron

$$P = \operatorname{conv}(\{x_1, x_2, x_3, x_4, x_5, x_6\})$$

and consider the two possible cases.

(a) P is combinatorially isomorphic to an octahedron.
In this case, assume without loss of generality that x_5 and x_6 are opposite vertices of P, as well as x_1 and x_3. Then, applying Exercise 6 to the bi-pyramid

$$\operatorname{conv}(\{x_1, x_2, x_4, x_5, x_6\}),$$

get the inequality

$$||x_5 - x_2||^2 + ||x_5 - x_4||^2 < ||x_1 - x_2||^2 + ||x_1 - x_4||^2.$$

On the other hand, applying the same Exercise 6 to the bi-pyramid

$$\operatorname{conv}(\{x_1, x_2, x_3, x_4, x_5\}),$$

get the inequality

$$||x_1 - x_2||^2 + ||x_1 - x_4||^2 < ||x_5 - x_2||^2 + ||x_5 - x_4||^2,$$

which yields a contradiction.

(b) P is combinatorially isomorphic to a polyhedron, one vertex of which is incident with exactly 5 edges.
In this case, assume without loss of generality that the vertex x_1 is incident with 5 edges of P, so these edges are, respectively,

$$[x_1, x_2], \quad [x_1, x_3], \quad [x_1, x_4], \quad [x_1, x_5], \quad [x_1, x_6].$$

Also, assume without loss of generality that the facets of P opposite of x_1 are, respectively,

$$[x_2, x_3, x_4], \quad [x_2, x_4, x_5], \quad [x_2, x_5, x_6].$$

Then, considering the bi-pyramid

$$\operatorname{conv}(\{x_1, x_2, x_4, x_5, x_6\})$$

and applying once again Exercise 6, obtain the inequality

$$||x_4 - x_1||^2 + ||x_4 - x_2||^2 < ||x_5 - x_1||^2 + ||x_5 - x_2||^2.$$

On the other hand, considering the bi-pyramid

$$\mathrm{conv}(\{x_1, x_2, x_3, x_4, x_5\})$$

and utilizing once more Exercise 6, come to the inequality

$$||x_5 - x_1||^2 + ||x_5 - x_2||^2 < ||x_4 - x_1||^2 + ||x_4 - x_2||^2,$$

which also implies a contradiction.

The obtained contradiction yields the desired result.

9*. For any number $x \in [0, 1]$, choose a set Q_x of rational numbers in $[0, 1]$ so that x would be a unique accumulation point of Q_x, and consider the family of sets $\{Q_x : x \in [0, 1]\}$.

Check that:

(1) the set Q_x is infinite for every $x \in [0, 1]$;

(2) the set $Q_x \cap Q_y$ is finite for any two distinct numbers x and y from $[0, 1]$.

By using a bijective correspondence between $\mathbf{N}$ and the set of all rational numbers in $[0, 1]$, establish that there exists a family $\{T_i : i \in I\}$ of subsets of $\mathbf{N}$ such that:

(a) $\mathrm{card}(I) = \mathbf{c}$;

(b) $\mathrm{card}(T_i) = \omega$ for each $i \in I$;

(c) $\mathrm{card}(T_i \cap T_j) < \omega$ for any two distinct indices $i \in I$ and $j \in I$.

10*. Let $\{\mathbf{a}_i : i \in I\}$ and $\{\mathbf{b}_i : i \in I\}$ be two families of cardinal numbers such that $\mathbf{a}_i < \mathbf{b}_i$ for all $i \in I$.

Show that the following inequality of König holds true:

$$\Sigma\{\mathbf{a}_i : i \in I\} < \Pi\{\mathbf{b}_i : i \in I\}.$$

Applying König's inequality, restore the omitted details in the proof of Theorem 2.

Remark 6. Considering in König's inequality the special case when

$$\mathbf{a}_i = 1, \quad \mathbf{b}_i = 2$$

for all $i \in I$, we immediately obtain Cantor's basic inequality

$$\mathrm{card}(I) < \mathrm{card}(\mathcal{P}(I)),$$

where I is an arbitrary set and $\mathcal{P}(I)$ is the family of all subsets of I.

In this context, it should be mentioned that Cantor's inequality is provable effectively, within **ZF** set theory, while König's inequality needs certain forms of the Axiom of Choice. Even the case $\text{card}(I) = \omega$ in König's inequality requires some help of **AC**. Namely, it is not difficult to establish, by using this inequality, that the real line **R** cannot be represented as the union of a countable family of countable sets. On the other hand, at present a model of **ZF** theory is known in which **R** is representable in the form of the union of a countable family of countable sets (see [151]).

11. For the Euclidean space $\mathbf{R}^m$, where $m \geq 2$, give a direct construction of a strong *at*-set $X \subset \mathbf{R}^m$ such that

$$\text{card}(X) = 2m - 1.$$

For this purpose, take any orthonormal basis $\{e_1, e_2, ..., e_m\}$ in $\mathbf{R}^m$, consider the set

$$V = \{e_1, -e_1, e_2, -e_2, ..., e_{m-1}, -e_{m-1}, e_m\},$$

and prove by induction on m that there exists a real $\varepsilon > 0$ such that the required set X is contained in the ε-neighborhood of V.

Moreover, verify that in the same ε-neighborhood of V there are continuum many mutually non-similar strong *at*-sets Z with $\text{card}(Z) = 2m - 1$ (cf. Exercise 4 of the present chapter, where $m = 3$).

12. Show that the algebraic (linear) dimension of the Hilbert space l_2 (over **R**) is equal to **c**.

For this purpose, use the sets of the family $\{T_i : i \in I\}$ described in Exercise 9.

13. Recall that, for an infinite cardinal number α, the symbol H_α denotes a real Hilbert space whose Hilbert dimension (i.e., the cardinality of an orthonormal basis of H_α) is equal to α.

For the space H_α, check the validity of the formula

$$\text{card}(H_\alpha) = \alpha^\omega,$$

which was substantially used in the present chapter.

CHAPTER 4

Isosceles triangles and *it*-sets in Euclidean space

Let (E, ρ) be a metric space and let x, y, z be any three (pairwise distinct) points in E.

The triangle $\{x, y, z\}$ in E is called isosceles if at least two from the distances $\rho(x, y)$, $\rho(y, z)$, $\rho(z, x)$ are equal to each other. Moreover, if one has the equalities

$$\rho(x, y) = \rho(y, z) = \rho(z, x),$$

then the triangle $\{x, y, z\}$ is called equilateral.

A subset X of E is called an it-set if every triangle whose vertices belong to X turns out to be isosceles.

Taking $E = \mathbf{R}^m$ with its standard Euclidean metric, one may consider a typical problem of combinatorial geometry which was posed by Erdös many years ago (in this connection, see e.g. [85], [86]).

For every $m \in \mathbf{N} \setminus \{0\}$, let us denote by $it(m)$ the supremum of the cardinalities of all it-sets in the space $\mathbf{R}^m$.

It is not hard to show that $it(m)$ is finite for any $m \in \mathbf{N} \setminus \{0\}$ (see Exercise 1 of the present chapter).

In his seminal article [85] Erdös asked what are the precise values of $it(m)$ for $m \geq 1$. In other words, he was interested whether it is possible to find an explicit formula for the function it defined on the set $\mathbf{N} \setminus \{0\}$. Up to now, this difficult problem remains open. Even for small values of m such as $m = 2$ and $m = 3$ the formulated problem is far from being trivial. Indeed, it was shown that $it(2) = 6$ and $it(3) = 8$ (see [66], [168], [185], [230]). Also, there were found the values $it(m)$ for some $m > 3$ (cf. [16], [79], [146], [147], [246], [248], [269]). However, even the asymptotic behavior of the function

$$it : \mathbf{N} \setminus \{0\} \to \mathbf{N}$$

DOI: 10.1201/9781003458708-4

was unknown during a long period of time.

One might conjecture that the growth of $it(m)$ is of an exponential order with respect to m (in this context, we suggest the reader to compare the above problem with the material of Chapter 3). Fortunately, it turned out that the growth of the function it is rather moderate. Namely, we will see below that the growth of $it(m)$ is at most $O(m^2)$. The main goal of the present chapter is to demonstrate this interesting fact.

We will begin with several auxiliary statements from the theory of finite graphs (cf. Appendix 5).

Recall that, for every natural number n, the symbol K_n denotes the complete (full) graph with exactly n vertices. Also, recall that Kuratowski's graph K_5 is a minimal (with respect to inclusion) non-planar graph, as well as his another graph $K_{3,3}$ (see Remark 7 of Appendix 5).

As usual, we denote by $V(K_n)$ the set of all vertices of K_n and by $E(K_n)$ the set of all edges of K_n.

For an arbitrary natural number p, we define a p-coloring of all edges of K_n as any surjective function

$$g : E(K_n) \to \{1, 2, ..., p\}.$$

In addition, we shall say that a p-coloring g of the edges of K_n has the connectivity property if, for each color $j \in \{1, 2, ..., p\}$ and for any two distinct vertices x and y of K_n, there exists a chain of edges of K_n, all of which have color j, and x and y are the endpoints of this chain.

Also, given some p-coloring g of all edges of K_n, we shall say that a triangle $\{v_1, v_2, v_3\} \subset V(K_n)$ is isosceles (in the combinatorial sense, with respect to g) if among the three edges

$$\{v_1, v_2\}, \quad \{v_2, v_3\}, \quad \{v_3, v_1\}$$

at least two have the same color.

In the sequel, an important role will be played by a purely combinatorial fact concerning g-colorings of the edges of K_n (cf. [147]).

Lemma 1. *Let g be an arbitrary p-coloring of all edges of K_n without non-isosceles triangles.*

The following assertions (1) and (2) are valid.

(1) Let $j \in \{1, 2, ..., p\}$ be any color and let $(V(K_n), E')$ denote the subgraph of K_n which is obtained from K_n by removing all those edges whose colors differ from j. Let C_1 and C_2 be any two distinct connected components of $(V(K_n), E')$.

Then there exists a color $i \in \{1, 2, ..., p\} \setminus \{j\}$ such that every edge connecting a vertex of C_1 with a vertex of C_2 has color i.

(2) If $p \geq 3$, then g does not possess the connectivity property.

Proof. Let us show the validity of (1). Without loss of generality, we may suppose that one of the components C_1 and C_2 contains at least two distinct vertices. Let C_2 be such a component, let z be an arbitrary vertex from C_1, and let x and y be any two distinct vertices from C_2. It suffices to verify that

$$g(\{z, x\}) = g(\{z, y\}).$$

Indeed, if C_1 also has at least two distinct vertices, then the argument given below is applicable to C_1, and we are done.

Clearly, there exists a chain

$$\{u_1, u_2\}, \quad \{u_2, u_3\}, \quad \ldots, \quad \{u_r, u_{r+1}\}$$

of edges of the graph $(V(K_n), E')$ such that

$$u_1 = x, \qquad u_{r+1} = y,$$

and the equalities

$$g(\{u_1, u_2\}) = g(\{u_2, u_3\}) = \ \ldots \ = g(\{u_r, u_{r+1}\}) = j$$

are fulfilled. Since $z \notin C_2$, all colors of the edges

$$\{z, u_1\}, \quad \{z, u_2, \}, \quad \ldots, \quad \{z, u_r\}, \quad \{z, u_{r+1}\}$$

differ from j. Further, consider the triangles

$$\{z, u_1, u_2\}, \quad \{z, u_2, u_3\}, \quad \ldots, \quad \{z, u_{r-1}, u_r\}, \quad \{z, u_r, u_{r+1}\}.$$

Keeping in mind the fact that all these triangles are isosceles, we easily infer that

$$g(\{z, u_1\}) = g(\{z, u_2, \}) = \ \ldots \ = g(\{z, u_r\}) = g(\{z, u_{r+1}\}).$$

In particular, we have $g(\{z, x\}) = g(\{z, y\})$, and assertion (1) is thus proved .

Now, let us demonstrate the validity of assertion (2). For this purpose, we will argue by induction on n. Assume that $p \geq 3$.

The case $n \leq 3$ is trivial.

Suppose that $n \geq 4$ and that (2) has already been established for K_{n-1}.

Consider any p-coloring g of all edges of K_n with the connectivity property. We must show that there exists a non-isosceles triangle in K_n with respect to g.

Obviously, there are edges of K_n which have, respectively, the colors 1, 2, and 3. Replacing in K_n the colors $4, 5, ..., p$ by color 3, we come to some 3-coloring g' of the edges of K_n. It is not difficult to see that:

(a) the coloring g' also possesses the connectivity property;

(b) if $\{v_1, v_2, v_3\} \subset V(K_n)$ is a non-isosceles triangle with respect to g', then $\{v_1, v_2, v_3\}$ is a non-isosceles triangle with respect to g.

So, it suffices to prove that there is a non-isosceles triangle with respect to the new coloring g'.

Let x be a vertex of K_n and let K_{n-1} denote the complete graph obtained from K_n by removing the vertex x with all its incident edges.

Observe that:

(c) The 3-coloring g' induces some 3-coloring g'' on K_{n-1}.

Actually, (c) readily follows from (a).

Supposing for a moment that all triangles in K_{n-1} are isosceles with respect to g'' and using the inductive assumption, we deduce that g'' does not possess the connectivity property.

So, we may assume that the graph Γ, which is obtained from K_{n-1} by removing all edges of K_{n-1} having colors 2 or 3, is not connected. Denote by

$$C_1, \ C_2, \ \dots, \ C_l$$

all connected components of Γ, where $l \geq 2$. According to (1), if C_q and $C_{q'}$ are any two distinct components, then all edges connecting the points of C_q and $C_{q'}$ are of the same color which belongs to the set $\{2, 3\}$. Further, in view of the connectivity property of g', there are vertices

$$y_1 \in C_1, \ \ y_2 \in C_2, \ \ \dots, \ \ y_l \in C_l$$

such that

$$g'(\{x, y_1\}) = g'(\{x, y_2\}) = ... = g'(\{x, y_l\}) = 1.$$

Also, by virtue of the same connectivity property of g', there are two vertices $u \in V(K_{n-1})$ and $v \in V(K_{n-1})$ such that

$$g'(\{x, u\}) = 2, \qquad g'(\{x, v\}) = 3.$$

Now, only the following two cases are possible in this situation.

1. The vertices u and v belong to one of the connected components $C_1, C_2, ..., C_l$.

   Definitely, we may assume that $\{u, v\} \subset C_1$. By virtue of (1), the edges $\{u, y_2\}$ and $\{v, y_2\}$ have the same color different from 1. Then it is easy to see that one of the triangles $\{x, u, y_2\}$ and $\{x, v, y_2\}$ is not isosceles.

2. The vertices u and v lie in two distinct connected components of Γ.

   We may suppose that $u \in C_1$ and $v \in C_2$. In this case, if all edges connecting the vertices from C_1 with the vertices from C_2 have color 2, then the triangle $\{x, y_1, v\}$ is not isosceles. If all edges connecting the vertices from C_1 with the vertices from C_2 have color 3, then the triangle $\{x, y_2, u\}$ is not isosceles.

The obtained contradiction completes the proof of assertion (2), so Lemma 1 has thus been proved. □

Let $k > 0$ be a natural number and let (E, ρ) be a metric space.

We shall say that $Z \subset E$ is a k-distance set in E if the set of all nonzero distances between the points of Z is precisely k-element.

Accordingly, we shall say that $Z \subset E$ is at most k-distance in E if the set of all nonzero distances between the points of Z is at most k-element.

Example 1. If Z is the set of all vertices of a regular m-simplex in the space $\mathbf{R}^m$, where $m \geq 1$, then Z is a 1-distance set, and it can easily be shown that there exists no 1-distance set in $\mathbf{R}^m$ whose cardinality is strictly greater than $m+1$ (see Exercise 2 of this chapter).

Example 2. There are many 2-distance subsets of $\mathbf{R}^m$ whose cardinalities strictly exceed $m+1$. For instance, there is a 2-distance subset X of the plane $\mathbf{R}^2$ with $\mathrm{card}(X) = 5$ and there is a 2-distance subset Y of the space $\mathbf{R}^3$ with $\mathrm{card}(Y) = 6$ (see Exercise 3).

To find a suitable uniform upper bound for the cardinalities of all 2-distance sets in the Euclidean space $\mathbf{R}^m$, we need several auxiliary statements.

Lemma 2. *Treating a point* $x = (x_1, x_2, ..., x_m)$ *of the space* $\mathbf{R}^m$ *as a family of* m *mutually independent real variables* $x_1, x_2, \ldots, x_m$, *consider the following polynomials of these variables:*

$$||x||^4 = (x_1^2 + x_2^2 + ... + x_m^2)^2,$$

$$x_i ||x||^2 \qquad (1 \leq i \leq m),$$

$$x_i x_j \qquad (1 \leq i \leq j \leq m),$$

$$1, \qquad x_i \qquad (1 \leq i \leq m).$$

The above-mentioned $(m+1)(m+4)/2$ *polynomials are linearly independent over* $\mathbf{R}$.

The easy proof of this lemma is left to the reader (see Exercise 4).

Lemma 3. *If* Z *is an arbitrary* 2*-distance subset of the space* $\mathbf{R}^m$, *then the inequality*

$$\mathrm{card}(Z) \leq (m+1)(m+2)/2$$

holds true.

Proof. Take a 2-distance set $Z \subset \mathbf{R}^m$ and denote by d_1 and d_2 the values of all nonzero distances between the points of Z. For each point $z \in Z$, introduce the real-valued polynomial

$$F_z(x_1, x_2, ..., x_m) = (||x - z||^2 - d_1^2)(||x - z||^2 - d_2^2)$$

defined on the whole space $\mathbf{R}^m$. For our further purposes, it will be convenient to give another representation of the polynomials $F_z(x)$ $(z \in Z)$. Namely, since

$$F_z(x) = (||x||^2 - 2\langle z, x\rangle + ||z||^2 - d_1^2)(||x||^2 - 2\langle z, x\rangle + ||z||^2 - d_2^2),$$

we can write

$$F_z(x) = ||x||^4 - 4 \sum_{1 \le i \le m} z_i(||x||^2 x_i) +$$

$$\sum_{1 \le i \le j \le m} \alpha_{ij} x_i x_j + \sum_{1 \le i \le m} \beta_i x_i + \gamma,$$

where z_i, α_{ij}, β_i, and γ are certain real coefficients.

We thus see that all polynomials $F_z(x)$ $(z \in Z)$ are linear combinations of the linearly independent polynomials described in Lemma 2.

Observe that if u and v are any two distinct points from Z, then

$$F_u(v) = F_v(u) = 0, \qquad F_u(u) = F_v(v) = (d_1 d_2)^2 > 0.$$

It follows from the above relations that the polynomials $F_z(x)$ $(z \in Z)$ are linearly independent over $\mathbf{R}$. Indeed, suppose for a while that

$$\sum_{z \in Z} \tau_z F_z(x) = 0 \qquad (x \in \mathbf{R}^m)$$

for some real numbers τ_z $(z \in Z)$. Taking arbitrarily $z_0 \in Z$, putting in the last equality $x = z_0$ and remembering that $F_z(z_0) = 0$ for any z from $Z \setminus \{z_0\}$, we readily obtain $\tau_{z_0} = 0$, which shows us the required linear independence of the polynomials $F_z(x)$ $(z \in Z)$.

Moreover, it turns out that the polynomials

$$1, \quad x_i \ (1 \le i \le m), \quad F_z(x) \ (z \in Z)$$

are also linearly independent over $\mathbf{R}$. To demonstrate this much deeper fact, suppose that

$$\sum_{z \in Z} \tau_z F_z(x) = t + \sum_{1 \le i \le m} b_i x_i$$

for some real numbers τ_z $(z \in Z)$, t, and b_i $(1 \le i \le m)$. Further, put

$$b = (b_1, b_2, ..., b_m)$$

and rewrite the above equality in the form

$$\sum_{z \in Z} \tau_z F_z(x) = t + \langle b, x\rangle.$$

For each $z \in Z$, we must have

$$\tau_z (d_1 d_2)^2 = t + \langle b, z\rangle.$$

Consequently,

$$\sum_{z \in Z} (t + \langle b, z \rangle) F_z(x) = (d_1 d_2)^2 (t + \langle b, x \rangle).$$

Since both sides of the last formula are linear combinations of the polynomials of Lemma 2, the corresponding real coefficients of those polynomials must be equal. For the polynomial $||x||^4$, we get

$$t\text{card}(Z) + \sum_{z \in Z} \langle b, z \rangle = 0$$

or, equivalently,

$$(1/\text{card}(Z)) \sum_{z \in Z} \langle b, z \rangle = -t \qquad (*).$$

For $i = 1, 2, ..., m$, the equality of the coefficients at $||x||^2 x_i$ gives us

$$\sum_{z \in Z} (t + \langle b, z \rangle) z_i = 0 \qquad (1 \leq i \leq m),$$

whence it follows that

$$\sum_{z \in Z} (t + \langle b, z \rangle) b_i z_i = 0 \qquad (1 \leq i \leq m),$$

$$\sum_{1 \leq i \leq m} \sum_{z \in Z} (t + \langle b, z \rangle) b_i z_i = 0,$$

$$\sum_{z \in Z} \sum_{1 \leq i \leq m} (t + \langle b, z \rangle) b_i z_i = 0,$$

$$\sum_{z \in Z} (t + \langle b, z \rangle) \langle b, z \rangle = 0.$$

Keeping in mind relation (*), we obtain

$$(1/\text{card}(Z)) \sum_{z \in Z} \langle b, z \rangle^2 = t^2 \qquad (**).$$

The formulae (*) and (**) show us that the quadratic mean of the real numbers

$$\langle b, z \rangle \qquad (z \in Z)$$

and the arithmetic mean of the same numbers are equal to each other. Therefore, all these numbers are pairwise equal and so are identical with $-t$. The latter implies that both sides of the equality

$$\sum_{z \in Z} (t + \langle b, z \rangle) F_z(x) = (d_1 d_2)^2 (t + \langle b, x \rangle)$$

are identically zero. In particular, $b = 0$ and $t = 0$, which yields the required result.

Finally, remembering the linear independence of the polynomials of Lemma 2 and taking into account the linear independence of the polynomials

$$1, \quad x_i \ (1 \leq i \leq m), \quad F_z(x) \ (z \in Z),$$

we easily conclude that

$$\mathrm{card}(Z) + (m+1) \leq (m+1)(m+4)/2$$

or, equivalently,

$$\mathrm{card}(Z) \leq (m+1)(m+2)/2,$$

which completes the proof of Lemma 3. □

Lemma 4. *If Z is a 2-distance subset of the space $\mathbf{R}^m$, all points of which belong to some $(m-1)$-dimensional sphere in $\mathbf{R}^m$, then the inequality*

$$\mathrm{card}(Z) \leq (m+1)(m+2)/2 - 1$$

holds true.

The proof of Lemma 4 is completely analogous to the proof of Lemma 3 and is left to the reader (see Exercise 5 of this chapter).

Lemma 5. *Let Z be an it-subset of the space $\mathbf{R}^m$ such that the set of all nonzero distances between the points of Z contains at least three distinct elements.*

Then Z admits a representation in the form $Z = X \cup Y$, where:

(1) $X \cap Y = \emptyset$;

(2) $Y \neq \emptyset$ and $\mathrm{card}(X) \geq 2$;

(3) each point of Y is equidistant from all points of X.

Proof. Denote $n = \mathrm{card}(Z)$ and consider the graph K_n, the set of all vertices of which coincides with Z. Let now

$$\{d_1, \ d_2, \ \ldots, \ d_p\}$$

be the set of all distinct nonzero distances between the points of Z. Clearly, we may treat these distances as different colors of the edges of K_n (cf. Exercise 7 from Chapter 1).

By virtue of our assumption, $p \geq 3$. Applying Lemma 1 to the colored graph K_n, we infer that the subgraph (K_n, E_1) of K_n formed by all edges of color d_1 is not connected. Since K_n contains some edges of color d_1, the graph (K_n, E_1) contains a component C_1 having at least two vertices. Define

$$X = C_1, \quad Y = Z \setminus X.$$

The sets X and Y trivially satisfy the relations (1) and (2). Now, let y be an arbitrary point of Y. Evidently, y belongs to the component of (K_n, E_1) distinct from C_1. According to the same Lemma 1, y must be equidistant from all points of X, i.e., relation (3) is satisfied, too.

Lemma 5 has thus been proved. □

The preceding results enable us to establish the following statement, which is more precise than Lemma 5 (see [147]).

Theorem 1. *Let Z be an it-subset of the space $\mathbf{R}^m$ such that the set of all nonzero distances between the points of Z contains at least three distinct elements.*

Then Z admits a representation in the form $Z = X \cup Y$, where:

(1) $X \cap Y = \emptyset$ and $Y \neq \emptyset$;

(2) X is at most 2-distance in $\mathbf{R}^m$ and $\mathrm{card}(X) \geq 2$;

(3) each point of Y is equidistant from all points of X;

(4) the affine hulls $\mathrm{aff}(X)$ and $\mathrm{aff}(Y)$ of X and Y, respectively, are perpendicular to each other, so the inequality

$$\dim(\mathrm{aff}(X)) + \dim(\mathrm{aff}(Y)) \leq m$$

holds true.

Proof. Consider all those representations of Z which are described in Lemma 5 and choose among them a representation $Z = X \cup Y$ such that $\mathrm{card}(X)$ attains the minimum value. We assert that X is at most 2-distance. Indeed, supposing for a moment otherwise and applying Lemma 5 to X, we get a representation $X = X_1 \cup Y_1$ as in Lemma 5. Now, putting

$$Z = X_1 \cup (Y_1 \cup Y)$$

and taking into account that $\mathrm{card}(X_1) < \mathrm{card}(X)$, we come to a contradiction with the choice of (X, Y). This contradiction shows us that X is at most 2-distance in $\mathbf{R}^m$.

Further, since each point $y \in Y$ is equidistant from all points of X, the orthogonal projection of y to $\mathrm{aff}(X)$ coincides with the circumcenter of X (i.e., the unique point of $\mathrm{aff}(X)$ which is equidistant from all points of X).

Consequently, the affine linear manifolds $\mathrm{aff}(X)$ and $\mathrm{aff}(Y)$ are perpendicular to each other. Therefore,

$$\dim(\mathrm{aff}(X)) + \dim(\mathrm{aff}(Y)) \leq m,$$

which ends the proof of Theorem 1. □

Using Theorem 1, it is not difficult to obtain some more or less suitable upper bound for the number $it(m)$ (see again [147]).

Theorem 2. *If Z is an arbitrary it-set in the space* $\mathbf{R}^m$, *then*

$$\text{card}(Z) \le (m+2)(m+1)/2.$$

Moreover, if the equality

$$\text{card}(Z) = (m+2)(m+1)/2$$

holds true, then either Z is at most 2-distance or

$$Z = X \cup \{y\},$$

where $y \notin X$ and X is at most 2-distance and y is equidistant from all points of X.

Proof. Take any *it*-set Z in $\mathbf{R}^m$ and consider the two possible cases.

1. The set Z is at most 2-distance.

Then Lemma 3 proved above and Exercise 2 below immediately imply that $\text{card}(Z) \le (m+2)(m+1)/2$.

2. The set Z is such that the set of all nonzero distances between the points of Z contains at least three distinct elements.

In this case, we represent Z in the form $Z = X \cup Y$ as in Theorem 1 and use induction on m. Observe that

$$\dim(\text{aff}(X)) \ge 1, \quad \dim(\text{aff}(X)) + \dim(\text{aff}(Y)) \le m,$$

so we have the inequality

$$\dim(\text{aff}(Y)) < m.$$

For the sake of brevity, let us denote

$$\dim(\text{aff}(X)) = r, \qquad \dim(\text{aff}(Y)) = s.$$

The equality $s = 0$ is equivalent to $Y = \{y\}$ for some point $y \in \mathbf{R}^m$. Consequently, if $Z = X \cup \{y\}$, then all points of X are equidistant from y and, in view of Lemma 4,

$$\text{card}(Z) = \text{card}(X) + 1 \le ((m+2)(m+1)/2 - 1) + 1 = (m+2)(m+1)/2.$$

Suppose now that $s \ge 1$. Then, applying the inductive assumption to Y and taking into account that X is contained in some $(r-1)$-dimensional Euclidean sphere, we may write

$$\text{card}(Z) = \text{card}(X) + \text{card}(Y) \le ((r+2)(r+1)/2 - 1) + (s+2)(s+1)/2.$$

Finally, keeping in mind that

$$r \ge 1, \qquad s \ge 1, \qquad r + s \le m,$$

it is not hard to check that

$$(r+2)(r+1)/2 - 1 + (s+2)(s+1)/2 < (m+2)(m+1)/2.$$

Theorem 2 has thus been proved. □

Remark 1. The general results presented in this chapter can successfully be applied for computing $it(2)$ and $it(3)$. Namely, it turns out that

$$it(2) = 6, \quad it(3) = 8.$$

In this connection, see Exercises 7 and 9 below. More detailed information about it-sets in multi-dimensional Euclidean spaces and the values $it(m)$ may be found in [146] and [147] (see also the references therein).

EXERCISES

1. By using the method of induction on m, give a direct simple proof of the fact that $it(m)$ is finite for each $m \in \mathbf{N} \setminus \{0\}$.

2. Verify that:

(a) any 1-distance set X in the space $\mathbf{R}^m$ with $\mathrm{card}(X) = m + 1$ is the set of all vertices of a regular m-simplex in $\mathbf{R}^m$;

(b) there exists no 1-distance set in $\mathbf{R}^m$ whose cardinality is strictly greater than $m + 1$.

3. Check that:

(a) the set of all vertices of a regular pentagon in the plane $\mathbf{R}^2$ is a 2-distance subset of $\mathbf{R}^2$;

(b) the set of all vertices of a regular octahedron in the space $\mathbf{R}^3$ is a 2-distance subset of $\mathbf{R}^3$.

Also, show that there is a convex prism in $\mathbf{R}^3$, all vertices of which form a 2-distance subset of $\mathbf{R}^3$

4. Give a proof of Lemma 2.

For this purpose, compare the growths of all polynomials described in Lemma 2.

5. Give a proof of Lemma 4.

For this purpose, argue as follows. Assume, without loss of generality, that a 2-distance set Z is entirely contained in the sphere $\mathbf{S}_{m-1}$ and consider the polynomials

$$||x||^4 = (x_1^2 + x_2^2 + ... + x_m^2)^2,$$

$$x_i||x||^2 \quad (1 \le i \le m),$$

$$x_i x_j \quad (1 \le i \le j \le m),$$

$$1, \quad x_i \quad (1 \le i \le m)$$

on $\mathbf{S}_{m-1}$, where $x = (x_1, x_2, ..., x_m) \in \mathbf{R}^m$. Exclude from this list the polynomial x_m^2 and check the linear independence on $\mathbf{S}_{m-1}$ of the other polynomials. Then use an argument similar to the proof of Lemma 3.

6. Consider in the space $\mathbf{R}^m$ the m-dimensional analog P of a regular octahedron. More concretely, P is defined as the convex hull of the points

$$(1,0,0,...,0),\ (-1,0,0,...,0),\ (0,1,0,...,0),\ (0,-1,0,...,0),\ \ldots,$$

$$(0,0,0,...,1),\ (0,0,0,...,-1).$$

Check that the set of all vertices of P (i.e., the set of points indicated above) is a 2-distance subset of $\mathbf{R}^m$ with cardinality $2m$.

7. Give a description (up to similarity) of all 2-distance subsets of the plane $\mathbf{R}^2$ and of all *it*-subsets of $\mathbf{R}^2$.

   In particular, verify that the cardinality of any 2-distance set in $\mathbf{R}^2$ does not exceed 5.

   Also, demonstrate that the equality $it(2) = 6$ holds true.

   For this purpose, apply Theorem 2, according to which

$$it(2) \leq (4 \cdot 3)/2 = 6.$$

   Then take into account the fact that the vertices of a regular pentagon with its circumcenter form an *it*-set in $\mathbf{R}^2$ whose cardinality is 6.

8*. Let X be a subset of a real pre-Hilbert space H and let z be a point of $\mathrm{conv}(X) \setminus X$ such that, for any two distinct points x and y from X, the triangle $[x, y, z]$ is isosceles, i.e., at least two sides of $[x, y, z]$ have the same length.

   Show that z is equidistant from all points of X.

   For this purpose, first consider the case when X is a finite subset of H.

   In addition, verify that the assumption $z \in \mathrm{conv}(X)$ is essential for the validity of this result.

9*. Let X be an *it*-set contained in the unit sphere $\mathbf{S}_{m-1} \subset \mathbf{R}^m$ and let Y be an *it*-set contained in the unit sphere $\mathbf{S}_{n-1} \subset \mathbf{R}^n$.

   Denoting by 0 the origin of the product space

$$\mathbf{R}^{m+n} = \mathbf{R}^m \times \mathbf{R}^n,$$

   check that $X \cup Y \cup \{0\}$ is an *it*-set in $\mathbf{R}^{m+n}$.

   Keeping in mind Exercise 7, infer from the above simple fact that there exists an *it*-set Z in the space $\mathbf{R}^3$ with $\mathrm{card}(Z) = 8$.

   In addition, prove that the equality $it(3) = 8$ holds true.

   For this purpose, use the above result and Exercises 7 and 8.

10. Give an example of an *it*-set Y in the space $\mathbf{R}^4$, satisfying the equality $\mathrm{card}(Y) = 11$.

Remark 2. It can be proved that $it(4) = 11$ (in this connection, see e.g. [147] where the values $it(m)$, for some other small dimensions m, are presented; cf. also [26]).

11. Let H be a real separable infinite-dimensional Hilbert space and let Z be an uncountable subset of H.

Under the notation

$$D(Z) = \{||z - z'|| : z \in Z,\ z' \in Z,\ z \neq z'\},$$

demonstrate that $\mathrm{card}(D(Z)) \geq \omega$.

For this purpose, take into account the fact that Z has a condensation point in H which belongs to Z.

Remark 3. It can be shown that in the standard separable infinite-dimensional Hilbert space

$$\mathrm{l}_2 = \{t \in \mathbf{R}^{\mathbf{N}} : \sum\{(t(n))^2 : n \in \mathbf{N}\} < +\infty\}$$

there exists a set X satisfying these two relations:

$$\mathrm{card}(X) = \mathbf{c}, \qquad \mathrm{card}(D(X)) = \omega.$$

For more details about X, see Exercise 9 from Chapter 5.

CHAPTER 5

Some geometric consequences of Ramsey's combinatorial theorem

We have already mentioned that there are interesting geometric applications of the purely combinatorial theorem due to F. Ramsey [291]. Among them, the best known and very impressive application was found many years ago by Erdös and Szekeres. Namely, it was demonstrated in their joint paper that in the Euclidean space $\mathbf{R}^m$ ($m \geq 2$) any finite set, whose cardinality is sufficiently large and all points of which are in general position, contains a prescribed number of points in convex position (see [91]).

An extensive survey about this beautiful result and many closely related statements is given in the work by W. Morris and V. Soltan [265].

In the present chapter we would like to give some other applications of Ramsey's theorem to problems of a geometric flavor.

Actually, the questions considered below are concerned with certain combinatorial properties of point sets lying either in a finite-dimensional Euclidean space or in a real infinite-dimensional pre-Hilbert space.

First, let us recall the fairly standard notation which will be used in our further considerations.

ω denotes, as usual, the least infinite ordinal (cardinal) number.

$\mathbf{c}$ is the cardinality of the continuum, i.e., we have by definition

$$\mathbf{c} = \text{card}(\mathbf{R}) = 2^{\omega}.$$

ω_1 is the least uncountable ordinal (cardinal) number.

DOI: 10.1201/9781003458708-5

If X is an arbitrary set and k is a natural number, then the symbol $[X]^k$ denotes the family of all k-element subsets of X (the same family is also often denoted by the symbol $\mathcal{F}_k(X)$).

Let us formulate the two well-known versions of Ramsey's theorem: finite and countably infinite.

Theorem 1. *Let n, l, and k be three natural numbers such that*

$$1 \leq l, \quad 1 \leq k \leq n.$$

Then there exists a natural number $r = r(n, l, k)$ having the following property:

For any set X with $\text{card}(X) \geq r$ and for any partition $\{\mathcal{A}_1, \mathcal{A}_2, ..., \mathcal{A}_l\}$ of $[X]^k$, there is a subset Y of X with $\text{card}(Y) \geq n$ such that $[Y]^k$ is entirely contained in some member $\mathcal{A}_i$ of this partition.

Theorem 2. *Let $l \geq 1$ and $k \geq 1$ be two natural numbers, X be an infinite set, and let $\{\mathcal{A}_1, \mathcal{A}_2, ..., \mathcal{A}_l\}$ be a partition of $[X]^k$.*

Then there is an infinite subset Y of X such that $[Y]^k$ is entirely contained in some member $\mathcal{A}_i$ of this partition.

For the proofs of Theorems 1 and 2, see e.g. [114], [131], [132], [161], [186], [235], [291] or any other more or less advanced text-book of combinatorics (see also Exercises 1 and 2 of Chapter 6).

Notice that, by using a certain weak form of the Axiom of Choice (**AC**), the finite version of Ramsey's statement (i.e., Theorem 1) can be deduced from the infinite version of Ramsey's statement (i.e., from Theorem 2).

In this context, it should also be mentioned that various uncountable versions of Ramsey's theorem were obtained, too. Among them, the central place is occupied by the celebrated Erdös–Rado theorem (for more details, see [89], [90], [151], [235]).

Now, let us consider several applications of Theorems 1 and 2 to questions of a geometric nature.

We begin with a very simple example concerning mutual positions of straight lines in the standard three-dimensional Euclidean space $\mathbf{R}^3$.

Example 1. For an arbitrary natural number $n \geq 2$, there exists a natural number r having the following property:

If $\mathcal{L}$ is a family of straight lines in $\mathbf{R}^3$ such that $\text{card}(\mathcal{L}) \geq r$ and no two distinct lines from $\mathcal{L}$ are parallel, then there is a subfamily $\mathcal{L}'$ of $\mathcal{L}$ such that $\text{card}(\mathcal{L}') \geq n$ and the disjunction of these two assertions holds true:

(a) all lines from $\mathcal{L}'$ lie in a plane or all of them pass through a point;

(b) no two distinct lines from $\mathcal{L}'$ lie in a plane (or, in other words, any two distinct lines from $\mathcal{L}'$ are skew).

Indeed, take $l = 2$ and $k = 2$. Let $r = r(n, l, k)$ be as in the formulation of Theorem 1. Consider any family $\mathcal{L}$ of straight lines in $\mathbf{R}^3$ such that card$(\mathcal{L}) \geq r$ and no two distinct lines from $\mathcal{L}$ are parallel. Define a partition $\{\mathcal{A}_1, \mathcal{A}_2\}$ of $[\mathcal{L}]^2$ as follows:

If two distinct lines $u \in \mathcal{L}$ and $v \in \mathcal{L}$ have a common point, then put $\{u, v\} \in \mathcal{A}_1$, otherwise put $\{u, v\} \in \mathcal{A}_2$.

According to Theorem 1, $\mathcal{L}$ contains some subfamily $\mathcal{L}'$ with card$(\mathcal{L}') \geq n$ and $[\mathcal{L}']^2$ is entirely contained either in $\mathcal{A}_1$ or in $\mathcal{A}_2$.

Now, it can easily be seen that $\mathcal{L}'$ satisfies the disjunction of assertions (a) and (b).

In a similar manner, utilizing Theorem 2, we obtain that if $\mathcal{M}$ is an infinite family of straight lines in the space $\mathbf{R}^3$ such that no two distinct lines from $\mathcal{M}$ are parallel, then there exists an infinite subfamily $\mathcal{M}'$ of $\mathcal{M}$ satisfying the disjunction of these two assertions:

(c) all lines from $\mathcal{M}'$ lie in a plane or all of them pass through a point;

(d) any two distinct lines from $\mathcal{M}'$ are skew.

Notice that the number r of Example 1 can be roughly estimated from above (in this connection, see Exercise 1).

The next example is very similar to the Erdös–Szekeres result mentioned at the beginning of this chapter.

Example 2. Recall that a point set X in the plane $\mathbf{R}^2$ is in general position if no three distinct points of X are collinear (i.e., no three distinct points of X belong to a straight line). It is easy to check that among any five points in $\mathbf{R}^2$, which are in general position, there always exist three points which form an obtuse-angled triangle. Let us fix natural numbers

$$n \geq 5, \quad l = 2, \quad k = 3$$

and take the number $r = r(n, l, k)$ as in Theorem 1. Further, consider any set $X \subset \mathbf{R}^2$ of points in general position such that card$(X) \geq r$.

Then we can assert that there is a set $Y \subset X$ satisfying the following relations:

(a) card$(Y) \geq n$;

(b) every three-element subset of Y forms an obtuse-angled triangle, i.e., Y is an *ot*-set in $\mathbf{R}^2$.

To see this, define a partition $\{\mathcal{A}_1, \mathcal{A}_2\}$ of $[X]^3$ as follows:

if $V \in [X]^3$ is the set of vertices of an acute-angled triangle or of a right-angled triangle, then put $V \in \mathcal{A}_1$; otherwise put $V \in \mathcal{A}_2$.

According to Theorem 1, there exists a set $Y \subset X$ with card$(Y) \geq n$, all three-element subsets of which lie in exactly one member of $\{\mathcal{A}_1, \mathcal{A}_2\}$. Remembering that $n \geq 5$, we infer that this member cannot be $\mathcal{A}_1$. So we get $[Y]^3 \subset \mathcal{A}_2$, which obviously means that all three-element subsets of Y form obtuse-angled triangles.

Now, let $Z \subset \mathbf{R}^2$ be an infinite set of points in general position.

Then there exists an infinite set $Y \subset Z$ such that every $V \in [Y]^3$ forms an obtuse-angled triangle, i.e., Y is an infinite *ot*-subset of Z.

Indeed, define the partition $\{\mathcal{A}_1, \mathcal{A}_2\}$ of $[Z]^3$ in the same manner as above: if $V \in [Z]^3$ is the set of vertices of an acute-angled triangle or of a right-angled triangle, then put $V \in \mathcal{A}_1$; otherwise put $V \in \mathcal{A}_2$.

According to Theorem 2, there exists an infinite set $Y \subset Z$ such that the disjunction

$$[Y]^3 \subset \mathcal{A}_1 \quad \vee \quad [Y]^3 \subset \mathcal{A}_2$$

holds true. But, as has already been shown, the relation $[Y]^3 \subset \mathcal{A}_1$ is impossible, so $[Y]^3 \subset \mathcal{A}_2$ and we obtain the desired result.

Remark 1. A natural generalization of Example 2 to the case of Euclidean space $\mathbf{R}^m$, where $m \geq 2$, is presented in Exercise 2. In this connection, it should be noticed that an uncountable version of Example 2 fails to be true (cf. Example 10 below).

Example 3. Let H be a Hilbert (more generally, pre-Hilbert) space over $\mathbf{R}$ and let a set $X \subset H$ be such that any three-point subset of X forms either a right-angled triangle or an obtuse-angled triangle.

Then we can assert that X is separable and, consequently, $\mathrm{card}(X) \leq \mathbf{c}$.

Indeed, suppose otherwise, i.e., the given set X is not separable. Then there exists a real $\varepsilon > 0$ such that X contains an uncountable ε-discrete subset Y. This means that $||y - y'|| \geq \varepsilon$ for any two distinct points y and y' from Y. Since Y is uncountable, we may assume without loss of generality that Y is also bounded in H. Furthermore, we may suppose that $\varepsilon = 1$ and the diameter of Y is strictly less than $2^{l/2}$, where l is some sufficiently large natural number.

Further, define a covering $\{\mathcal{A}_1, \mathcal{A}_2, ..., \mathcal{A}_l\}$ of $[Y]^2$ as follows:

$$\{y, y'\} \in \mathcal{A}_i \Leftrightarrow ||y - y'|| \in [2^{(i-1)/2}, 2^{i/2}[,$$

where i ranges over $\{1, 2, ..., l\}$. According to Theorem 2, there are an index $i \in \{1, 2, ..., l\}$ and an infinite set $Y' \subset Y$ such that $[Y']^2 \subset \mathcal{A}_i$. Now, it follows from the definition of $\mathcal{A}_i$ that any three-point subset of Y' forms an acute-angled triangle, which is impossible. The obtained contradiction yields the desired result.

Remark 2. In Chapter 3 a subset X of a real infinite-dimensional separable Hilbert space was indicated which has cardinality $\mathbf{c}$ and all three-element subsets of which form acute-angled triangles. Also, a subset W of a real infinite-dimensional separable Hilbert space was pointed out, which is homeomorphic to the unit segment $[0, 1]$ and all three-element subsets of which form right-angled triangles (see Remark 2 in Chapter 3).

Now, we would like to recall one elementary geometric statement concerning equilateral triangles.

Let $n \geq 3$ be a natural number and let X be a subset of the plane $\mathbf{R}^2$ with $\text{card}(X) = n^2$.

Then there exists a set $Y \subset X$ with $\text{card}(Y) \geq n$ such that no three distinct points of Y form an equilateral triangle.

To prove this statement, consider a maximal (with respect to inclusion) subset Y of X no three points of which form an equilateral triangle, and denote $k = \text{card}(Y)$. According to the definition of Y, for any point $x \in X \setminus Y$, there are two points y and z in Y such that the triangle $[x, y, z]$ is equilateral. Since the number of all possible line segments $[y, z]$ does not exceed $k(k-1)/2$ and there are at most two equilateral triangles, for which $[y, z]$ is one of their sides, we easily infer the inequality

$$n^2 - k \leq 2(k(k-1)/2) = k^2 - k,$$

whence the relation $n \leq k$ immediately follows, showing that Y is as required.

However, this very simple argument does not work in the case where X is a subset of $\mathbf{R}^3$. Therefore, the argument must be changed by another reasoning based on Theorem 1.

Example 4. Let $\triangle$ be a fixed non-degenerate triangle in the space $\mathbf{R}^3$ and let $n \geq 5$ be a natural number.

There exists a natural number r satisfying the following condition:

For any set $X \subset \mathbf{R}^3$ with $\text{card}(X) \geq r$, there is a set $Y \subset X$ such that $\text{card}(Y) \geq n$ and no three-element subset of Y forms a triangle similar to $\triangle$.

To demonstrate this fact, first observe that there is no five-point set in $\mathbf{R}^3$, all three-element subsets of which form triangles similar to $\triangle$. Keeping in mind the above-mentioned circumstance, take as earlier

$$l = 2, \quad k = 3, \quad r = r(n, l, k)$$

and produce a partition $\{\mathcal{A}_1, \mathcal{A}_2\}$ of $[X]^3$ as follows:

$V \in \mathcal{A}_1$ if V is the set of vertices of a triangle similar to $\triangle$, otherwise $V \in \mathcal{A}_2$.

Applying Theorem 1 to this partition, we readily get the desired result.

An analogous application of Theorem 2 gives us that if a set $X \subset \mathbf{R}^3$ is infinite, then there exists an infinite set $Y \subset X$, no three-element subset of which forms a triangle similar to $\triangle$.

In connection with Example 4, see also Exercise 5.

Example 5. It is easy to check that the vertices of a regular pentagon in the plane $\mathbf{R}^2$ with its center constitute a six-point set, all three-element subsets of which form isosceles triangles. Also, it can be shown that there exists no seven-point set in $\mathbf{R}^2$, all three-element subsets of which form isosceles triangles (see Chapter 4 of this book). Now, let $n \geq 7$ be a natural number.

One may assert that there exists a natural number r satisfying the following condition:

For any set $X \subset \mathbf{R}^2$ with $\mathrm{card}(X) \geq r$, there is a set $Y \subset X$ such that $\mathrm{card}(Y) \geq n$ and no three-element subset of Y forms an isosceles triangle.

To demonstrate this, take again

$$l = 2, \quad k = 3, \quad r = r(n, l, k)$$

and produce a partition $\{\mathcal{A}_1, \mathcal{A}_2\}$ of $[X]^3$ as follows:

$V \in \mathcal{A}_1$ if V is the set of vertices of an isosceles triangle, otherwise $V \in \mathcal{A}_2$.

Applying Theorem 1 to this partition, we come to the required result.

In a similar manner, using Theorem 2, we infer that if a set $X \subset \mathbf{R}^2$ is infinite, then there exists an infinite set $Y \subset X$ such that no three-element subset of Y forms an isosceles triangle.

Now, for any set Z lying in $\mathbf{R}^m$ (or in a real pre-Hilbert space H), recall the notation

$$D(Z) = \{||z - z'|| : z \in Z, z' \in Z, z \neq z'\}.$$

Utilizing this notation, let us consider the next example.

Example 6. If $n \geq 6$ is a natural number, then there exists a natural number r satisfying the following condition:

For any set $X \subset \mathbf{R}^2$ with $\mathrm{card}(X) \geq r$, one has $\mathrm{card}(D(X)) \geq n$.

To see this circumstance, take

$$n' = n + 1, \quad l = 2, \quad k = 3$$

and denote $r = r(n', l, k)$. By virtue of Example 5, if $X \subset \mathbf{R}^2$ and $\mathrm{card}(X) \geq r$, then there exists a set $Y \subset X$ with $\mathrm{card}(Y) \geq n'$ such that no three-element subset of Y forms an isosceles triangle. Fix a point $y_0 \in Y$ and consider the real numbers

$$||y - y_0|| \quad (y \in Y \setminus \{y_0\}).$$

Clearly, all these numbers are pairwise different, so

$$\mathrm{card}(D(Y)) \geq n' - 1 = n,$$

which yields the desired result.

Both Examples 5 and 6 can be generalized to the case of $\mathbf{R}^m$, where $m \geq 2$ (see Exercise 6 of this chapter).

Example 7. Let X be an infinite subset of the plane $\mathbf{R}^2$. It follows from Example 6 that the set $D(X)$ is infinite. Moreover, one may assert that if X is uncountable, then $D(X)$ is also uncountable. Indeed, assume on the contrary that $\mathrm{card}(X) > \omega$ but $\mathrm{card}(D(X)) = \omega$. Fix a point $x_0 \in X$. Then all other points of X belong to the union of countably many circles. Consequently, one of those circles, say C, contains uncountably many points of X. But it is easy to see that the set $D(C \cap X)$ is uncountable, which contradicts our assumption. A slightly more complicated argument enables one to conclude that the equality

$$\mathrm{card}(D(X)) = \mathrm{card}(X)$$

holds true (see Lemma 1 and Exercise 5 from Chapter 1).

Note that in Exercises 7 and 8 of the present chapter the corresponding analogs of Examples 6 and 7 for the space $\mathbf{R}^m$ are considered. On the other hand, in Exercise 9 a subset X of a real infinite-dimensional separable Hilbert space H is indicated, such that $\mathrm{card}(X) = \mathbf{c}$ and $\mathrm{card}(D(X)) = \omega$.

Example 8. Let $\mathcal{D}$ be an infinite (respectively, uncountable) family of line segments on $\mathbf{R}$. We do not exclude the situation when some of these segments are degenerate, i.e., are singletons.

One can assert that there exists an infinite (respectively, uncountable) family $\mathcal{D}' \subset \mathcal{D}$ satisfying the disjunction of the following two assertions:

(a) all segments from $\mathcal{D}'$ are pairwise disjoint;

(b) all segments from $\mathcal{D}'$ have a common point.

To see this, first consider the case when the family $\mathcal{D}$ is countably infinite. In this case, introduce a partition $\{\mathcal{A}_1, \mathcal{A}_2\}$ of $[\mathcal{D}]^2$ as follows:

$\{x, y\} \in \mathcal{A}_1$ if $x \cap y = \emptyset$, otherwise $\{x, y\} \in \mathcal{A}_2$.

According to Theorem 2, there exists an infinite family $\mathcal{D}' \subset \mathcal{D}$ such that either $[\mathcal{D}']^2 \subset \mathcal{A}_1$ or $[\mathcal{D}']^2 \subset \mathcal{A}_2$. If $[\mathcal{D}']^2 \subset \mathcal{A}_1$, then $\mathcal{D}'$ satisfies (a). If $[\mathcal{D}']^2 \subset \mathcal{A}_2$, then any two segments from $\mathcal{D}'$ have a nonempty intersection. This circumstance implies that any finite subfamily of $\mathcal{D}'$ has a nonempty intersection. Taking into account the compactness of the segments, we get (b).

Now, consider the case when the family $\mathcal{D}$ is uncountable. We may assume, without loss of generality, that all segments from $\mathcal{D}$ are non-degenerate and, moreover, we may assume that the length of each segment from $\mathcal{D}$ is strictly greater than $\delta > 0$, where δ is some real number. Let us put

$$T = \{n\delta : n \in \mathbf{Z}\},$$

where $\mathbf{Z}$ denotes the set of all integers. Obviously, the set T is countable and any member of $\mathcal{D}$ has a common point with T. So there exists a point $t \in T$ belonging to uncountably many segments from $\mathcal{D}$. Denoting by $\mathcal{D}'$ the family of all those segments from $\mathcal{D}$ which contain t, we conclude that $\mathcal{D}'$ satisfies (b).

Example 9. Recall that a linearly ordered set $(E, \leq)$ is Dedekind complete if, for every nonempty and bounded from above subset X of E, there exists $\sup(X)$. In such a set $(E, \leq)$ consider any infinite family $\mathcal{D}$ consisting of subsegments of E. Then, similarly to the previous example, one can deduce from Theorem 2 that there exists an infinite family $\mathcal{D}' \subset \mathcal{D}$ satisfying the disjunction of the following two assertions:

(a) all segments from $\mathcal{D}'$ are pairwise disjoint;

(b) all segments from $\mathcal{D}'$ have a common point.

Remark 3. Unfortunately, an uncountable variant of the previous example is not provable for $(E, \leq)$ in contemporary set theory. Indeed, it is consistent with the standard axioms of **ZFC** theory that there exists a linearly ordered set $(S, \leq)$ such that:

(i) $(S, \leq)$ is Dedekind complete, dense, nonseparable in its order topology, and has neither least nor greatest element;

(ii) S satisfies the countable chain condition, i.e., any disjoint family of nonempty open subintervals of S is at most countable.

Actually, S is the so-called Suslin line (see, e.g., [82], [151], [237]). Recall that the existence of S is valid in Gödel's Universe $\mathbf{L}$, where the Continuum Hypothesis $\mathbf{c} = \omega_1$ holds true (see [106], [177]). Also, observe that in the same model $\mathbf{L}$ the equality

$$\mathrm{card}(S) = \omega_1$$

is fulfilled, so S can be represented as an ω_1-sequence of points $\{s_\xi : \xi < \omega_1\}$. Now, by using the method of transfinite recursion up to ω_1, one can readily construct a family

$$\mathcal{D} = \{d_\xi : \xi < \omega_1\}$$

of non-degenerate segments in S such that, for any ordinal $\xi < \omega_1$, the segment d_ξ does not intersect the closure of $\{s_\zeta : \zeta < \xi\}$ (it suffices to use the fact that S itself is nonseparable, while the closure of $\{s_\zeta : \zeta < \xi\}$ is separable). It directly follows from the construction of $\mathcal{D}$ that every point of S belongs to at most countably many segments from $\mathcal{D}$. Thus, we may conclude that $\mathcal{D}$ does not contain an uncountable subfamily $\mathcal{D}'$ satisfying the disjunction of the assertions (a) and (b) of Example 9.

Example 10. In 1924, by assuming the Continuum Hypothesis, Sierpiński constructed an uncountable set in $\mathbf{R}$ whose intersection with every Lebesgue measure zero subset of $\mathbf{R}$ is at most countable. Sierpiński's construction is considered in detail in the well-known monograph by Kuratowski [240], where some applications of Sierpiński's sets to certain questions of general topology are also touched upon. By utilizing an argument similar to Sierpiński's construction, it is possible to show that, under the same Continuum Hypothesis, there exists an uncountable set $S \subset \mathbf{R}^2$ of points in general position, such that every λ_2-measure zero subset of $\mathbf{R}^2$ has at most countably many common points with S (here λ_2 denotes the standard two-dimensional Lebesgue measure on $\mathbf{R}^2$). These properties of S allow one to infer that if S' is any uncountable subset of S, then:

(a) there are three points in S' which form an acute-angled triangle;

(b) there are three points in S' which form an obtuse-angled triangle.

We thus deduce that, within **ZFC** set theory, there is no uncountable analogue of Example 2.

Additional information around this topic is given in exercises below which provide some generalizations of the presented examples.

In this book we do not touch upon numerous applications of Ramsey's theorem in other branches of contemporary mathematics (number theory, functional analysis, general topology, etc.).

EXERCISES

1. Show that if a family $\mathcal{L}$ of straight lines in the space $\mathbf{R}^3$ is such that any two lines of $\mathcal{L}$ have a common point, then either all lines from the family pass through a point or all of them lie in some affine plane of $\mathbf{R}^3$.

Also, demonstrate that in Example 1 the number r can be taken to be equal to $(2(n-1))!/((n-1)!)^2$.

In addition, prove that, for every natural number n, there exists a natural number $u(n)$ such that if $\mathcal{L}$ is an arbitrary family of straight lines in $\mathbf{R}^3$ with $\mathrm{card}(\mathcal{L}) \geq u(n)$, then the disjunction of these four assertions holds true:

(a) there exists a subfamily $\mathcal{L}_1$ of $\mathcal{L}$ with $\mathrm{card}(\mathcal{L}_1) \geq n$, the members of which are pairwise parallel;

(b) there exists a subfamily $\mathcal{L}_2$ of $\mathcal{L}$ with $\mathrm{card}(\mathcal{L}_2) \geq n$, all members of which lie in a certain affine plane of $\mathbf{R}^3$;

(c) there exists a subfamily $\mathcal{L}_3$ of $\mathcal{L}$ with $\mathrm{card}(\mathcal{L}_3) \geq n$, all members of which pass through a certain point of $\mathbf{R}^3$;

(d) there exists a subfamily $\mathcal{L}_4$ of $\mathcal{L}$ with $\mathrm{card}(\mathcal{L}_4) \geq n$, the members of which are pairwise skew.

2. Let Z be a subset of the space $\mathbf{R}^m$ such that any three points from Z form either an acute-angled triangle or a right-angled triangle. Then, as shown in Exercise 1 from Chapter 3,

$$\mathrm{card}(Z) \leq 2^m.$$

Starting with this fact and applying Theorem 1 of the present chapter, demonstrate that, for any natural numbers $m \geq 2$ and n, there exists a natural number $r = r(m, n)$ having the following property:

If $X \subset \mathbf{R}^m$ and $\text{card}(X) \geq r$, then a set $Y \subset X$ can be found such that $\text{card}(Y) \geq n$ and all three-element subsets of Y form obtuse-angled triangles (in other words, Y is an *ot*-subset of X).

This result generalizes Example 2 (here, for the sake of convenience, any angle whose measure is equal to π is assumed to be obtuse).

In a similar way, apply Theorem 2 of the present chapter and demonstrate that if $X \subset \mathbf{R}^m$ is an infinite set, then there exists an infinite *ot*-subset Y of X.

3. Take the closed interval $[0, 1/2]$ and consider the mapping

$$f : [0, 1/2] \to \text{l}_2$$

defined by the formula

$$f(t) = (t, t^2, t^3, ..., t^n, ...) \qquad (t \in [0, 1/2]).$$

Verify that:

(a) this f is injective and continuous;

(b) the set $X = f([0, 1/2])$ is homeomorphic to $[0, 1]$;

(c) all three-element subsets of X form obtuse-angled triangles;

(d) all points of X are in general position.

4*. Let (E, ρ) be a metric space, let ε be a strictly positive real number, and let $\{x, y, z\} \subset E$ be a triangle in E, the side lengths of which are a_1, a_2, a_3.

It is said that $\{x, y, z\}$ is an equilateral triangle with exactness to ε if the inequalities

$$1 - \varepsilon < a_i/a_j < 1 + \varepsilon$$

are valid for all $i \in \{1, 2, 3\}$ and for all $j \in \{1, 2, 3\}$.

Show that if (E, ρ) is a nonseparable space, then, for every $\varepsilon > 0$, there exists an infinite set $Z \subset E$ such that all three-element subsets of Z form equilateral triangles with exactness to ε.

For this purpose, keep in mind the argument of Example 3.

Remark 4. Obviously, the result of Exercise 4 generalizes Example 3. It is also clear that, for sufficiently small real $\varepsilon > 0$, all triangles in the space $\mathbf{R}^m$ which are equilateral with exactness to ε are simultaneously acute-angled. In this context, it makes sense to recall that $Z \subset \mathbf{R}^m$ is a strong *at*-set if all three-element subsets of Z form acute-angled triangles (cf. Chapter 3). If X is a strong *at*-set in $\mathbf{R}^3$, then $\text{card}(X) \leq 5$ and there exists a strong *at*-set $Y \subset \mathbf{R}^3$ such that $\text{card}(Y) = 5$. For sufficiently large natural numbers m, it is possible to indicate a strong *at*-set $Z \subset \mathbf{R}^m$ whose cardinality has an exponential growth with respect to m, i.e.,

$$\mathrm{card}(Z) \geq \alpha^m,$$

where $\alpha > 1$ is some real constant not depending on m. A purely combinatorial proof of the this fact was given in Chapter 3.

5. Let $\triangle$ be a non-degenerate triangle in the Euclidean space $\mathbf{R}^m$, where $m \geq 2$, and let n be a natural number.

Show that there exists a natural number r satisfying the following condition:

For any set $X \subset \mathbf{R}^m$ with $\mathrm{card}(X) \geq r$, there is a set $Y \subset X$ such that $\mathrm{card}(Y) \geq n$ and no three-element subset of Y forms a triangle similar to $\triangle$.

To demonstrate the existence of r, take into account the following two circumstances:

(a) if $\triangle$ is not an equilateral triangle and a set $Z \subset \mathbf{R}^m$ is such that all three-element subsets of Z form triangles similar to $\triangle$, then $\mathrm{card}(Z) \leq 4$;

(b) if $\triangle$ is an equilateral triangle and a set $Z \subset \mathbf{R}^m$ is such that all three-element subsets of Z form triangles similar to $\triangle$, then $\mathrm{card}(Z) \leq m + 1$;

Also, applying Theorem 2 show that if a set $X \subset \mathbf{R}^m$ is infinite, then there exists an infinite set $Y \subset X$ such that no three-element subset of Y forms a triangle similar to $\triangle$.

6. Give a simple inductive proof of the fact that, for every natural number $m \geq 2$, there exists a minimal natural number $it(m)$ having the following property:

If a set X in the space $\mathbf{R}^m$ is such that all three-element subsets of X form isosceles triangles, then $\mathrm{card}(X) \leq it(m)$.

For this purpose, denote by $h(m)$ the maximum number of points in the unit sphere $\mathbf{S}_{m-1} \subset \mathbf{R}^m$, all nonzero distances between which are greater than or equal to 1. By using induction on m, verify that

$$it(m+1) \leq it(m) + 2h(m) + 2 \quad (m \geq 2).$$

Keeping in mind the existence of $it(m)$, generalize Examples 5 and 6 to the case of $\mathbf{R}^m$.

Remark 5. We would like to recall the reader that in Chapter 4 it was proved, by using a much more delicate argument, that for the value $it(m)$ the inequality

$$it(m) \leq (m+2)(m+1)/2$$

is valid. In particular, this upper estimate of $it(m)$ is of a polynomial growth with respect to m. However, even the latter estimate does not give us the precise values of $it(m)$. Recall also that in the ordinary space $\mathbf{R}^3$ there exists a set X with $\text{card}(X) = 8$, all three-element subsets of which form isosceles triangles, and 8 is the maximal cardinality of a set in $\mathbf{R}^3$ with the above-mentioned property (see, e.g., [66], [168], [185], and Chapter 4 of this book). In $\mathbf{R}^4$ it is not hard to indicate a point set of cardinality 11, having the same property, and it can be proved that 11 is the maximum value for this property (see [147]).

7*. A four-element subset $\{x, y, z, t\}$ of the Euclidean space $\mathbf{R}^m$ $(m \geq 1)$ is called admissible if at least two numbers from the set

$$\{||x - y||, ||x - z||, ||x - t||, ||y - z||, ||y - t||, ||z - t||\}$$

are equal to each other.

(a) Show that there exists a natural number $p(m)$ having the following property:
If $Z \subset \mathbf{R}^m$ is such that all four-element subsets of Z are admissible, then $\text{card}(Z) \leq p(m)$.

(b) Prove that, for an arbitrary natural number n, there exists a natural number r possessing the following property:

If $X \subset \mathbf{R}^m$ and $\text{card}(X) \geq r$, then there exists a set $Y \subset X$ such that $\text{card}(Y) \geq n$ and all nonzero distances determined by the couples of points from Y are pairwise different.

For (a), keep in mind the result of Exercise 6 and use induction on m.

For (b), assume without loss of generality that $n > p(m)$ and put

$$l = 2, \qquad k = 4, \qquad r = r(n, l, k).$$

Let $X \subset \mathbf{R}^m$ and $\text{card}(X) \geq r$. Consider the partition $\{\mathcal{A}_1, \mathcal{A}_2\}$ of $[X]^4$ defined as follows:

$V \in [X]^4$ belongs to $\mathcal{A}_1$ if V is admissible, otherwise $V \in [X]^4$ belongs to $\mathcal{A}_2$.

Apply to this partition Theorem 1 of the present chapter and the fact indicated in (a).

8. Let X be a set in the space $\mathbf{R}^m$ (or in the sphere $\mathbf{S}_m$).

Prove that:

(a) if X is infinite, then there exists an infinite set $Y \subset X$ such that all nonzero distances determined by the couples of points from Y are pairwise different;

(b) if X is uncountable, then there exists an uncountable set $Y \subset X$ such that all nonzero distances determined by the couples of points from Y are pairwise different.

For this purpose, use induction on m.

Moreover, generalize (a) and (b) by showing that if card(X) is an infinite regular cardinal number, then there exists a set $Y \subset X$ such that

$$\mathrm{card}(Y) = \mathrm{card}(X)$$

and all nonzero distances determined by the couples of points from Y are pairwise different.

9*. Consider the standard real Hilbert space

$$\mathrm{l}_2 = \{t \in \mathbf{R}^{\mathbf{N}} : \sum\{(t(n))^2 : n \in \mathbf{N}\} < +\infty\}$$

and let $\{N_j : j \in J\}$ be an almost disjoint family of infinite subsets of $\mathbf{N}$ such that card$(J) = \mathbf{c}$ (see Exercise 9 from Chapter 3).

For any index $j \in J$, let x_j be an element of l_2 satisfying the following three conditions:

(a) the l_2-norm of x_j is equal to 1;

(b) $x_j(n) \in \mathbf{Q}$ for all $n \in \mathbf{N}$;

(c) if $n \in \mathbf{N} \setminus N_j$, then $x_j(n) = 0$.

Putting $X = \{x_j : j \in J\}$, verify that the equalities

$$\mathrm{card}(X) = \mathbf{c}, \qquad \mathrm{card}(D(X)) = \omega$$

are valid, where $D(X) = \{||x - x'|| : x \in X,\ x' \in X,\ x \neq x'\}$.

10. Let m and n be two natural numbers and let $\mathcal{D}$ be a family of segments in $\mathbf{R}$ such that card$(\mathcal{D}) = mn + 1$.

Show that the disjunction of these two assertions holds true:

(a) there exist $m + 1$ segments from $\mathcal{D}$ which are pairwise disjoint;

(b) there exist $n + 1$ segments from $\mathcal{D}$ which have a common point.

For this purpose, use induction on m (or on n).

11*. Let $(E, \leq)$ be a Dedekind complete linearly ordered set and let $\mathcal{L}$ be an uncountable family of segments in E.

Demonstrate that the disjunction of these two statements is valid:

(a) there exists an uncountable subfamily of $\mathcal{D}$, all segments of which are pairwise disjoint;

(b) there exists an infinite subfamily of $\mathcal{D}$, all segments of which have a common point.

Also, demonstrate that the disjunction of these two statements is valid:

(a′) there exists an uncountable subfamily of $\mathcal{D}$, all segments of which have a common point;

(b′) there exists an infinite subfamily of $\mathcal{D}$, all segments of which are pairwise disjoint.

For this purpose, utilize a Ramsey type theorem due to Dushnik and Miller (see [74]).

12* Let $[x, y, z]$ be a triangle in the plane $\mathbf{R}^2$, the lengths of the sides of which are, respectively, a_1, a_2, a_3, and let ε be a strictly positive real number.

Recall that $[x, y, z]$ is an equilateral triangle with exactness to ε (in short, an ε-equilateral triangle) if the inequalities

$$1 - \varepsilon < a_i/a_j < 1 + \varepsilon \qquad (i \in \{1, 2, 3\}, \ j \in \{1, 2, 3\})$$

are fulfilled for this triangle (see Exercise 4).

Let S be a Sierpiński set in the plane $\mathbf{R}^2$, all points of which are in general position (see Example 10).

Verify that:

(a) any uncountable subset S' of S is also a Sierpiński set in $\mathbf{R}^2$;

(b) for every real $\varepsilon > 0$, there exist three distinct points x, y and z in S' such that the triangle $[x, y, z]$ turns out to be ε-equilateral.

To show the validity of (b), consider a λ_2-measurable hull Q of S' and observe that

$$\lambda_2(Q) > 0.$$

According to the classical theorem of Lebesgue, there is a density point w of Q (see, for instance, [49]).

Demonstrate that the desired ε-equilateral triangle $[x, y, z]$ with vertices in S' can be found in an appropriate neighborhood of w.

Deduce from these facts that there are uncountably many acute-angled triangles whose vertices belong to S' and conclude that, in general, the uncountable version of Example 2 fails to be true.

13*. Let $\kappa \le \mathbf{c}$ be an uncountable regular cardinal and let $\mathcal{L}$ be an injective family of straight lines in the Euclidean space $\mathbf{R}^3$ such that $\mathrm{card}(\mathcal{L}) = \kappa$.

Show that the disjunction of these four assertions holds true:

(i) $\mathcal{L}$ contains a subfamily $\mathcal{L}_1$ with $\mathrm{card}(\mathcal{L}_1) = \kappa$, all members of which are pairwise parallel;

(ii) $\mathcal{L}$ contains a subfamily $\mathcal{L}_2$ with $\text{card}(\mathcal{L}_2) = \kappa$, all members of which lie in some affine hyperplane of $\mathbf{R}^3$;

(iii) $\mathcal{L}$ contains a subfamily $\mathcal{L}_3$ with $\text{card}(\mathcal{L}_3) = \kappa$, all members of which pass through a certain point of $\mathbf{R}^3$;

(iv) $\mathcal{L}$ contains a subfamily $\mathcal{L}_4$ with $\text{card}(\mathcal{L}_4) = \kappa$, any two distinct members of which are skew.

For this purpose, suppose that the assertions (i), (ii), (iii) do not hold for $\mathcal{L}$ and construct by transfinite recursion a family $\mathcal{L}_4 \subset \mathcal{L}$ satisfying (iv).

Compare the obtained result with Example 1 of this chapter.

In addition, for any infinite singular cardinal number $\kappa \leq \mathbf{c}$, give an example of an injective family $\mathcal{L}$ of straight lines in $\mathbf{R}^3$ such that $\text{card}(\mathcal{L}) = \kappa$ and the assertions (i), (ii), (iii), (iv) are all false for $\mathcal{L}$.

Remark 6. As is known (see, e.g., [151], [237]), each of the following two statements concerning the cardinality continuum $\mathbf{c}$ is consistent with **ZFC** set theory:

(1) $\mathbf{c}$ is a regular cardinal number;

(2) $\mathbf{c}$ is a singular cardinal number.

Therefore, according to Exercise 13, if one has an injective family $\mathcal{L}$ of straight lines in the space $\mathbf{R}^3$ such that

$$\text{card}(\mathcal{L}) = \kappa = \mathbf{c},$$

then, in general, one cannot prove the disjunction of the assertions (i), (ii), (iii), and (iv).

CHAPTER 6

Convexly independent subsets of infinite sets of points

The present chapter is dedicated to an infinite version of a concrete problem of combinatorial geometry and to some set-theoretical phenomena emerging in connection with this version.

Let E be a vector space (over the field $(\mathbf{R}, +, \cdot)$ of all real numbers) and let X be a subset of E.

We recall that X is said to be convexly independent in E if, for each point $x \in X$, the relation

$$x \notin \mathrm{conv}(X \setminus \{x\})$$

holds true, where $\mathrm{conv}(\cdot)$ denotes, as usual, the operation of taking the convex hull of a subset of E.

Recall once more that Erdös and Szekeres proved in their seminal paper [91] that, for any natural number $n \geq 3$, there exists a smallest natural number $c(n)$ possessing the following property:

Every set consisting of at least $c(n)$ points of the plane $\mathbf{R}^2$, which are in general position, contains a subset of n points which are convexly independent, i.e., they are the vertices of a certain convex polygon (actually, the vertices of a convex n-gon) in $\mathbf{R}^2$.

In addition to this result, Erdös and Szekeres posed the problem to determine the precise value of $c(n)$. Moreover, in connection with their problem, they conjectured that

$$c(n) = 2^{n-2} + 1 \qquad (n \geq 3).$$

The above-mentioned intriguing problem was then investigated by many mathematicians. However, it is still far from being solved. An extensive survey about this problem and closely connected questions of combinatorial geometry

DOI: 10.1201/9781003458708-6

can be found in [265] where a long list of references to related works is also presented.

Here we would like to discuss two infinite variants of the Erdös–Szekeres problem. The main attention will be paid to those infinite sets of points in $\mathbf{R}^2$, which either are countable (i.e., having cardinality ω) or are of cardinality continuum ($= \mathbf{c}$).

It should be underlined that the basic technical tool utilized by Erdös and Szekeres in [91] is the classical combinatorial theorem of Ramsey [291]. By applying this theorem, one easily gets some upper bound for $c(n)$. The completely analogous method successfully works for point sets lying in Euclidean space of higher dimension (see Exercise 5 for this chapter).

In our further considerations we need a certain infinite version of the Ramsey theorem, which is formulated as follows (cf. Chapter 5).

Ramsey Theorem. *Let i be a nonzero natural number, A be an arbitrary infinite set, let $\mathcal{F}_i(A)$ $(= [A]^i)$ denote the family of all i-element subsets of A, and let $\{\mathcal{F}_1, \mathcal{F}_2, \dots, \mathcal{F}_k\}$ be a finite covering of $\mathcal{F}_i(A)$.*

Then there exist a natural number $r \in [1, k]$ and an infinite subset A' of A such that $\mathcal{F}_i(A') \subset \mathcal{F}_r$.

A particular (but very important) case of the above infinite version of the Ramsey theorem is usually stated in terms of graph theory.

Namely, if (V, E) is an arbitrary infinite graph, then the disjunction of these two assertions holds true:

(i) the graph (V, E) contains an infinite full subgraph;

(ii) the complementary graph $(V, \mathcal{F}_2(V) \setminus E)$ contains an infinite full subgraph.

This special case of Ramsey's theorem is popular in the mathematical literature and has a lot of applications. Moreover, it substantially stimulated the development of infinite combinatorics and the theory of so-called large cardinal numbers (for more details, see [151], [158], [237], [241]).

It is well known that the presented infinite (in fact, countable) version of Ramsey's theorem implies its finite version (see, e.g., [131], [132]).

By applying the countable version of Ramsey's theorem, the following result can also be obtained.

Theorem 1. *Let $m \geq 2$ be a natural number and let A be an infinite subset of the Euclidean space $\mathbf{R}^m$, such that $\mathrm{card}(A \cap l) < \omega$ for all straight lines l lying in $\mathbf{R}^m$.*

Then there exists an infinite convexly independent set $B \subset A$.

Consequently, if $\mathrm{card}(A) = \omega$, then for the above-mentioned set B the relation

$$\mathrm{card}(B) = \mathrm{card}(A) = \omega$$

holds true.

Proof. The assumption that A is infinite directly implies that there are affine linear manifolds L in $\mathbf{R}^m$ satisfying the relation

$$\mathrm{card}(A \cap L) \geq \omega.$$

For example, $L = \mathbf{R}^m$ satisfies this relation. Let L' be an affine linear manifold in $\mathbf{R}^m$ such that

$$\mathrm{card}(A \cap L') \geq \omega$$

and the dimension $\dim(L')$ takes the minimum value. Observe that $\dim(L') > 1$ and every affine hyperplane H in L' has the property that

$$\mathrm{card}(H \cap A) < \omega.$$

Then, by using an easy recursion, one can define an infinite set

$$A' \subset A \cap L'$$

of points in general position in L'. Finally, applying to A' the infinite version of Ramsey's theorem, namely, putting

$$i = \dim(L') + 2, \quad k = 2,$$

we come to the required infinite convexly independent set $B \subset A' \subset A$ (cf. Exercise 3).

This ends the proof of Theorem 1. □

Remark 1. It should be noticed that the use of Ramsey's theorem for obtaining the assertion of Theorem 1 is not necessary. An absolutely different argument also yields the validity of the same assertion. Indeed, we first consider the case of the Euclidean plane $\mathbf{R}^2$. Suppose that A is an infinite subset of $\mathbf{R}^2$ such that

$$\mathrm{card}(A \cap l) < \omega$$

for any straight line l lying in $\mathbf{R}^2$. Then A contains a countably infinite subset A' of points in general position. Here only three cases are possible:

(1) there exist a ray $p \subset \mathbf{R}^2$ and an infinite set $\{a_n : n < \omega\} \subset A'$, such that

$$\{a_n : n < \omega\} \cap p = \emptyset,$$

the sequence of points $\{a_n : n < \omega\}$ converges to the endpoint o of p and the rays oa_n $(n < \omega)$ converge to p;

(2) there exist a ray $p \subset \mathbf{R}^2$ and an infinite set $\{a_n : n < \omega\} \subset A'$, such that

$$\{a_n : n < \omega\} \cap p = \emptyset, \quad \lim_{n \to \infty} ||a_n|| = +\infty, \quad \lim_{n \to \infty} \mathrm{dist}(a_n, p) = 0;$$

(3) there exist a ray $p \subset \mathbf{R}^2$ and an infinite set $\{a_n : n < \omega\} \subset A'$, such that

$$\{a_n : n < \omega\} \cap p = \emptyset, \quad \lim_{n\to\infty}||a_n|| = +\infty, \quad \lim_{n\to\infty}\mathrm{dist}(a_n, p) = +\infty,$$

and the rays oa_n $(n < \omega)$ converge to p, where o is again the endpoint of p.

In the cases (2) and (3) the symbol $\mathrm{dist}(a_n, p)$ denotes, as usual, the distance between a_n and p.

In each of these three cases, it is not difficult to define (by recursion) a polygonal convex curve P with infinitely many sides, all vertices of which belong to the set $\{a_n : n < \omega\}$ (hence to the set A, as well). In the first case, P can be constructed so that the lengths of its sides would tend to zero. In the second and third cases, P can be constructed so that the lengths of its sides would tend to infinity. Thus, the case of infinite point sets lying in the plane $\mathbf{R}^2$ is completely considered.

Suppose now that A is an infinite subset of the Euclidean space $\mathbf{R}^m$, where $m > 2$, and suppose that

$$\mathrm{card}(A \cap l) < \omega$$

for all straight lines l lying in $\mathbf{R}^m$. Again, there exists a countably infinite set $A' \subset A$ of points in general position in some affine linear submanifold of $\mathbf{R}^m$. Without loss of generality, we may assume that this submanifold coincides with $\mathbf{R}^m$. Further, since A' is countable, we can find a plane L in $\mathbf{R}^m$ such that the orthogonal projection of $\mathbf{R}^m$ to L, restricted to A', is injective and transforms A' to a set of points in general position in L. As we already know, the latter set contains an infinite convexly independent subset. Now, one can readily verify that the pre-image of this subset, with respect to the above-mentioned restricted projection, turns out to be a convexly independent subset of A' (hence of A, as well).

Remark 2. It would be interesting to consider the question of whether it is possible to derive the Erdös–Szekeres result on finite point sets in $\mathbf{R}^2$ directly from Theorem 1 (without using the finite version of Ramsey's theorem).

Let us turn our attention to the cases where various uncountable point sets in the plane $\mathbf{R}^2$ are given and let us examine an analogue of the Erdös–Szekeres problem for those sets.

Let A be an uncountable subset of $\mathbf{R}^2$. Assuming that the points of A are in general position, the following natural question arises:

Does there exist an uncountable convexly independent subset of A?

We shall demonstrate below that, in general, the existence of such a subset of A cannot be guaranteed. For this purpose, we need some delicate set-theoretical techniques based on an uncountable form of the Axiom of Choice (**AC**). Namely, in what follows the possibility of endowing the continuum with a well-ordering relation plays a key role (cf. also the next chapter where

Mazurkiewicz type subsets of the plane $\mathbf{R}^2$ are discussed and similar transfinite machinery is exploited).

Let α denote the least ordinal number of cardinality continuum (in fact, we may identify α with $\mathbf{c} = 2^\omega$). By definition, α has the property that

$$\mathrm{card}(\xi) < \mathrm{card}(\alpha) = \mathbf{c}$$

for every ordinal number $\xi < \alpha$.

In the sequel, we shall say that C is a convex curve in the plane $\mathbf{R}^2$ if C coincides with the boundary of a closed convex subset B of $\mathbf{R}^2$ such that B differs from $\mathbf{R}^2$.

Consider the family of all convex curves in $\mathbf{R}^2$. Obviously, we can represent this family in the form of a transfinite sequence $\{C_\xi : \xi < \alpha\}$. By utilizing the method of transfinite recursion, let us construct a family $\{a_\xi : \xi < \omega\}$ of points in general position in $\mathbf{R}^2$. Suppose that, for an ordinal $\xi < \alpha$, the partial family of points $\{a_\zeta : \zeta < \xi\}$ has already been defined. Take the ξ-sequence $\{C_\zeta : \zeta < \xi\}$. Since every C_ζ is convex, the family of all those straight lines whose intersections with C_ζ contain more than two elements, is at most countable (here we use the simple fact that any disjoint family of non-degenerate line intervals contained in C_ζ is at most countable). Hence there exists a straight line $l_\xi \subset \mathbf{R}^2$ such that

$$\mathrm{card}(l_\xi \cap C_\zeta) \leq 2$$

for all ordinal numbers $\zeta < \xi$. Clearly, we can find a point $a_\xi \in l_\xi$ not belonging to the set

$$\cup\{C_\zeta : \zeta < \xi\} \cup \{a_\zeta : \zeta < \xi\}.$$

Moreover, slightly modifying the above argument, we can choose a point a_ξ so that the set of points $\{a_\zeta : \zeta \leq \xi\}$ would be in general position in $\mathbf{R}^2$. The details of such a modification are left to the reader.

Proceeding in this manner, we are able to construct the required family of points $\{a_\xi : \xi < \alpha\}$. Finally, we put

$$A = \{a_\xi : \xi < \alpha\}.$$

Then the following statement holds true for the just constructed A.

Theorem 2. *The set A possesses these three properties:*

(1) A is in general position in $\mathbf{R}^2$;

(2) A is of cardinality continuum;

(3) no subset of A of cardinality continuum is convexly independent.

Proof. The properties (1) and (2) follow immediately from our construction.

Let us check the validity of (3). Suppose to the contrary that some set $B \subset A$ of cardinality continuum is convexly independent and put

$$B' = \mathrm{cl}(\mathrm{conv}(B)),$$

$$C' = \mathrm{bd}(B'),$$

where, as usual, the symbol $\mathrm{cl}(\cdot)$ denotes the operation of taking the closure of a subset of $\mathbf{R}^2$ and the symbol $\mathrm{bd}(\cdot)$ stands for the operation of taking the boundary of a subset of $\mathbf{R}^2$. Then B' is a closed convex set in $\mathbf{R}^2$ with nonempty interior and C', being the boundary of B', is a convex curve.

Notice now that all points of B must belong to C'. This fact can easily be deduced, e.g., from the well-known Steinitz theorem which says that a point x belongs to the interior of the convex hull of a subset X of $\mathbf{R}^m$ $(m \geq 1)$ if and only if there exists a set $Y \subset X$ with $\mathrm{card}(Y) \leq 2m$, such that x belongs to the interior of $\mathrm{conv}(Y)$ (see, e.g., [68] or Exercise 16 from Appendix 1). Consequently, we must have

$$\mathrm{card}(A \cap C') \geq \mathrm{card}(B) = \mathrm{card}(\alpha) = \mathbf{c} = 2^{\omega}$$

and, therefore,

$$\mathrm{card}(A \cap C') = \mathbf{c} = 2^{\omega}.$$

On the other hand, there exists an ordinal $\xi < \alpha$ such that $C_\xi = C'$. Taking into account the definition of A, we may write the inclusion

$$A \cap C_\xi \subset \{a_\zeta : \zeta \leq \xi\},$$

from which we readily infer that

$$\mathrm{card}(A \cap C') = \mathrm{card}(A \cap C_\xi) \leq \mathrm{card}(\xi) + \omega < \mathbf{c}.$$

Thus, we come to a contradiction with the above equality

$$\mathrm{card}(A \cap C') = \mathbf{c}.$$

The contradiction obtained completes the proof of Theorem 2. □

Remark 3. It should be observed that we were able to construct the desired set A of Theorem 2, because of the sufficiently nice structure of convex curves lying in $\mathbf{R}^2$. Speaking more precisely, all convex curves in $\mathbf{R}^2$ turn out to be so thin that any family $\{C_i : i \in I\}$ of such curves, with

$$\mathrm{card}(I) < \mathbf{c} = 2^{\omega},$$

does not form a covering of $\mathbf{R}^2$. Indeed, this moment is crucial for carrying out our transfinite construction. Notice that for more general classes of curves (lying in the plane) the situation can be significantly different. Namely, there is a model of set theory in which there exists a family $\{L_i : i \in I\}$ of curves satisfying the following three relations:

(1) $\mathrm{card}(I) < \mathbf{c}$;

(2) each curve L_i $(i \in I)$ is homeomorphic to the unit circle $\mathbf{S}_1 \subset \mathbf{R}^2$;

(3) $\cup\{L_i : i \in I\} = \mathbf{R}^2$.

For more details, see [195], [198], and Exercise 9 of the present chapter. It is clear that, in such a situation, no construction analogous to the given above can be carried out for general classes of curves in $\mathbf{R}^2$.

Remark 4. Theorem 2 admits a direct extension to the case of the Euclidean space $\mathbf{R}^m$ where $m > 2$. In other words, there exists a set $A \subset \mathbf{R}^m$ of cardinality continuum, whose points are in general position in $\mathbf{R}^m$, but no subset of A of the same cardinality is convexly independent. The proof of this result is analogous to the argument presented in the proof of Theorem 2 (however, some technical details occur in the case of $\mathbf{R}^m$).

Remark 5. In various questions of measure theory and set-theoretic topology similar constructions are well known for obtaining a subset of an infinite measure space E (or of an infinite topological space E) which almost avoids the members of a given family $\mathcal{F}$ of sets in E. In most situations, the family $\mathcal{F}$ forms a σ-ideal of subsets of an original space E (for example, the σ-ideal of all measure zero sets in a nonzero measure space E or the σ-ideal of all first category sets in a nonempty Baire topological space E). As a rule, the corresponding transfinite constructions need some additional set-theoretical axioms, e.g., the Continuum Hypothesis (**CH**) or much weaker Martin's Axiom (**MA**). In this connection, see [151] and [240] where, assuming the Continuum Hypothesis, two classical constructions of the so-called Luzin sets and Sierpiński sets are discussed in detail.

EXERCISES

1*. Prove the countable version of Ramsey's theorem (formulated in this chapter) by using induction on i.

For this purpose, argue as follows. Notice that the case $i = 1$ is trivial, because of $k < \omega$ and $\mathrm{card}(A) \geq \omega$, where $\omega = \mathrm{card}(\mathbf{N})$.

Suppose that the statement is true for i and let

$$[A]^{i+1} = \mathcal{F}_1 \cup \mathcal{F}_2 \cup \ \ldots \ \cup \mathcal{F}_k.$$

Define by recursion a sequence of pairs $\{(A_n, a_n) : n < \omega\}$, where A_n is an infinite subset of A and $a_n \in A_n$.

First of all, put $A_0 = A$ and pick an element $a_0 \in A$.

Assume that the pair (A_n, a_n) has already been defined for a natural index n, and consider the family $[A_n \setminus \{a_n\}]^i$. Obviously,

$$[A_n \setminus \{a_n\}]^i = \mathcal{F}'_1 \cup \mathcal{F}'_2 \cup \ \ldots \ \cup \mathcal{F}'_k,$$

where each family $\mathcal{F}'_j$ $(1 \le j \le k)$ is introduced by the formula

$$Z \in \mathcal{F}'_j \Leftrightarrow Z \cup \{a_n\} \in \mathcal{F}_j \qquad (Z \in [A_n \setminus \{a_n\}]^i).$$

By the inductive assumption, there exists an infinite set $T \subset A_n \setminus \{a_n\}$ such that all i-element subsets of T belong to some $\mathcal{F}'_{j(n)}$, where $1 \le j(n) \le k$.

Put $A_{n+1} = T$ and pick an element $a_{n+1} \in T$.

Proceeding in this manner, obtain the desired sequence of pairs

$$\{(A_n, a_n) : n < \omega\}.$$

Further, consider the corresponding sequence $\{j(n) : n < \omega\}$ of natural numbers, all of which belong to the interval $[1, k]$. Observe that there are a natural index $j \in [1, k]$ and an infinite set $N \subset \mathbf{N}$ satisfying the relation

$$(\forall n \in N)(j(n) = j),$$

and deduce from this fact that the infinite set

$$A' = \{a_n : n \in N\}$$

is as required, i.e., all $(i+1)$-element subsets of A' belong to $\mathcal{F}_j$.

2*. The general finite version of Ramsey's theorem is as follows (cf. Chapter 5).

Let i, k, h_1, h_2, . . . , h_k be natural numbers such that

$$i \ge 1, \quad k \ge 1, \quad h_1 \ge i, \quad h_2 \ge i, \quad . . . , \quad h_k \ge i.$$

Then there exists a natural number $m_i(h_1, h_2, ..., h_k)$ having the following property:

For any set E with $\mathrm{card}(E) \ge m_i(h_1, h_2, ..., h_k)$ and for any covering

$$\{\mathcal{B}_1, \mathcal{B}_2, ..., \mathcal{B}_k\}$$

of the set $[E]^i$, there exist an index $j \in \{1, 2, ..., k\}$ and a set $X_j \subset E$ such that $[X_j]^i \subset \mathcal{B}_j$ and $\mathrm{card}(X_j) = h_j$.

Prove this finite version of Ramsey's theorem.

For this purpose, argue by the method of mathematical induction on $i + k$.

First of all, observe that if $i = 1$ or $k = 1$, then everything is clear, because one may put, respectively,

$$m_1(h_1, h_2, ..., h_k) = h_1 + h_2 + ... + h_k - (k-1),$$

$$m_i(h_1) = h_1.$$

Let us consider the situation when $i > 1$ and $k > 1$. In this case, assume that Ramsey's assertion holds true for $i-1$ and all $(h_1, h_2, ..., h_k)$, for i and all $(h_1, h_2, ..., h_{k-1})$, and for all those $(k+1)$-tuples $(i, h'_1, h'_2, ..., h'_k)$ which satisfy

$$h'_1 + h'_2 + ... + h'_k < h_1 + h_2 + ... + h_k.$$

Verify that Ramsey's assertion is also valid for $(i, h_1, h_2, ..., h_k)$.

Note that if there is some h_j equal to i, then the situation is reduced to i and $k-1$, and in this case the inductive assumption works.

So one may restrict further considerations to the case where

$$h_1 > i, \quad h_2 > i, \quad \dots, \quad h_k > i.$$

For each natural number $j \in [1, k]$, introduce the notation

$$h'_j = m_i(h_1, h_2, ..., h_{j-1}, h_j - 1, h_{j+1}, ..., h_k),$$

that is correct, because of

$$h_1 + ... + (h_j - 1) + ... + h_k < h_1 + ... + h_j + ... + h_k,$$

and stipulate

$$m_i(h_1, h_2, ..., h_k) = m_{i-1}(h'_1, h'_2, ..., h'_k) + 1.$$

It remains to demonstrate that the number $m_i(h_1, h_2, ..., h_k)$ is as required.

For this purpose, take a set E with

$$\text{card}(E) \geq m_i(h_1, h_2, ..., h_k)$$

and consider any covering $\{\mathcal{B}_1, \mathcal{B}_2, ..., \mathcal{B}_k\}$ of $[E]^i$. Choose an element $e \in E$ and put

$$E^* = E \setminus \{e\}.$$

Further, define a covering $\{\mathcal{B}^*_1, \mathcal{B}^*_2, ..., \mathcal{B}^*_k\}$ of $[E^*]^{i-1}$ by the formula

$$\{e_1, e_2, ..., e_{i-1}\} \in \mathcal{B}^*_j \Leftrightarrow \{e_1, e_2, ..., e_{i-1}, e\} \in \mathcal{B}_j,$$

where $j \in \{1, 2, ..., k\}$.

By the inductive assumption applied to E^* and to $\{\mathcal{B}^*_1, \mathcal{B}^*_2, ..., \mathcal{B}^*_k\}$, there exist a natural index $p \in [1, k]$ and a set $Y_p \subset E^*$ such that

$$[Y_p]^{i-1} \subset \mathcal{B}^*_p, \quad \text{card}(Y_p) = h'_p.$$

Analogously, remembering that

$$h'_p = m_i(h_1, h_2, ..., h_{p-1}, h_p - 1, h_{p+1}, ..., h_k),$$

applying the inductive assumption to Y_p and to the covering $\{\mathcal{B}_1, \mathcal{B}_2, ..., \mathcal{B}_k\}$ of $[Y_p]^i$, deduce that there exist a natural index $q \in [1, k]$ and a set $Z_q \subset Y_p$ such that:

(a) $[Z_q]^i \subset \mathcal{B}_q$;

(b) if $q \neq p$, then $\mathrm{card}(Z_q) = h_q$, and if $q = p$, then $\mathrm{card}(Z_q) = h_p - 1$.

If $q \neq p$, then define $j = q$ and $X_j = Z_q$.

If $q = p$, then define $j = q$ and $X_j = Z_q \cup \{e\}$.

In both above cases, it is easy to check that

$$X_j \subset E, \quad \mathrm{card}(X_j) = h_j, \quad [X_j]^i \subset \mathcal{B}_j,$$

which completes the proof by induction.

3. Let Z be a point set in the plane $\mathbf{R}^2$ with $\mathrm{card}(Z) \geq 4$.

Show that if every four-element subset of Z is convexly independent, then Z is also convexly independent.

More generally, let Z be a point set in the Euclidean space $\mathbf{R}^m$, where $m \geq 2$ and $\mathrm{card}(Z) \geq m + 2$.

Show that if every $(m+2)$-element subset of Z is convexly independent, then Z is convexly independent.

For this purpose, use Carathéodory's theorem (see Appendix 1).

4. Let $Z \subset \mathbf{R}^m$ be a set of points in general position and suppose that $\mathrm{card}(Z) \geq m + 3$.

Demonstrate that there exists a convexly independent subset Z' of Z with $\mathrm{card}(Z') = m + 2$.

5*. Following the argument of Erdös and Szekeres, establish the existence of $c(n)$ for every natural number $n \geq 3$.

For this purpose, fix an arbitrarily $n \in \mathbf{N} \setminus \{0, 1, 2\}$ and consider the Ramsey number

$$d = m_4(5, n)$$

defined in Exercise 2. Further, by starting with the results of Exercises 3 and 4, deduce that if $Z \subset \mathbf{R}^2$ is a point set in general position with $\mathrm{card}(Z) \geq d$, then Z contains a convexly independent subset Z' with $\mathrm{card}(Z') = n$.

Also, keeping in mind the result of Exercise 4, verify that the analogous method works for finite point sets in general position, which lie in the Euclidean space $\mathbf{R}^m$, where $m \geq 2$.

6*. Demonstrate that the countable version of Ramsey's theorem (presented in this chapter) implies its finite version.

Do it in the following three different ways:

(a) by using the existence of a nontrivial ultrafilter on the set $\mathbf{N}$ $(= \omega)$ of all natural numbers;

(b) by applying the König lemma from graph theory (see Appendix 5);

(c) by using Gödel's fundamental theorem on the existence of a model for any consistent first-order mathematical theory (see, e.g., [259], [296]).

7. Check in detail that, in Remark 1 of this chapter, only three cases (1), (2) and (3) are possible.

Keeping in mind this circumstance, in each of the above-mentioned cases construct recursively a convex polygonal curve with infinitely many sides.

8. Give a simple direct proof of a very special case of the Steinitz theorem, namely, demonstrate that, for a point $x \in \mathbf{R}^2$ and a set $X \subset \mathbf{R}^2$, the following two assertions are equivalent:

(a) x belongs to the interior of the convex hull of X;

(b) there exists a set $Y \subset X$ such that $\mathrm{card}(Y) \leq 4$ and $x \in \mathrm{int}(\mathrm{conv}(Y))$.

Remark 6. The full version of the Steinitz theorem can easily be deduced from Carathéodory's theorem (cf. Exercise 16 of Appendix 1).

9*. There is a model $\mathcal{M}$ of set theory in which the real line $\mathbf{R}$ contains a subset whose cardinality is strictly less than the cardinality continuum $\mathbf{c}$ and which is nonmeasurable in the Lebesgue sense (see [151]).

Show that, in the same model $\mathcal{M}$, the real line can be represented in the form

$$\mathbf{R} = \cup\{T_i : i \in I\},$$

where $\mathrm{card}(I) < \mathbf{c}$ and all T_i $(i \in I)$ are compact nowhere dense subsets of $\mathbf{R}$.

Further, observe that each set T_i $(i \in I)$ is zero-dimensional, so all topological products

$$T_i \times T_j \quad (i \in I,\ j \in I)$$

are zero-dimensional compact subsets of the plane $\mathbf{R}^2$.

Infer from the fact stated above the existence of a family $\{L_i : i \in I\}$ of homeomorphic images of the unit circle $\mathbf{S}_1 \subset \mathbf{R}^2$, satisfying the relations

$$\mathrm{card}(I) < \mathbf{c}, \qquad \cup\{L_i : i \in I\} = \mathbf{R}^2.$$

For this purpose, keep in mind a well-known topological result, according to which every zero-dimensional compact subset of $\mathbf{R}^2$ is contained in a plane curve homeomorphic to $\mathbf{S}_1$ (see, e.g., [240]).

10*. Generalize Theorem 2 to the case of a multi-dimensional Euclidean space.

More precisely, demonstrate that if $m \geq 2$, then there exists a set $A \subset \mathbf{R}^m$ of cardinality continuum $\mathbf{c}$, whose points are in general position in $\mathbf{R}^m$, but no subset of A of the same cardinality is convexly independent.

11. Let $(E, \preceq)$ and $(F, \leq)$ be two linearly ordered sets and let $g : E \to F$ be a mapping. Suppose that E is infinite.

Using the countable version of Ramsey's theorem, verify that there exists an infinite set $E' \subset E$ for which the restriction $g|E'$ is a monotone function (in general, one cannot assert that $g|E'$ is a strictly monotone function).

12*. Assuming the Continuum Hypothesis (**CH**) and applying the method of transfinite induction, prove that there exists a function $h : \mathbf{R} \to \mathbf{R}$ such that, for every uncountable subset X of $\mathbf{R}$, the restricted function $h|X$ is not continuous on X (any function acting from $\mathbf{R}$ into $\mathbf{R}$ and having this strong discontinuity property is usually called a Sierpiński–Zygmund function).

Conclude that no restriction of h to an uncountable subset of $\mathbf{R}$ can be monotone (compare this fact with the result of Exercise 11).

13*. Show that there exists a graph (V, E) satisfying the following three relations:

(a) $\mathrm{card}(V) = 2^\omega$;

(b) (V, E) does not contain an uncountable full subgraph;

(c) the complementary graph $(V, \mathcal{F}_2(V) \setminus E)$ also does not contain an uncountable full subgraph.

For this purpose, put $V = \mathbf{R}$ and take into account the fact that $\mathbf{R}$ equipped with its standard linear ordering $\leq$ does not contain strictly increasing or strictly decreasing ω_1-sequences of points (here ω_1 denotes, as usual, the least uncountable ordinal number). Consider also some well-ordering $\preceq$ of $\mathbf{R}$ and define the required set E of edges by using the two orderings $\leq$ and $\preceq$.

Conclude from the above result that the direct analogue of Ramsey's theorem for graphs of cardinality ω_1 fails to be true (observe that, assuming **CH**, the same conclusion implicitly follows from Exercise 12).

Remark 7. The properties (1)–(3) of the set A of Theorem 2 also imply that the direct analogue of Ramsey's theorem fails to hold for some uncountable sets. However, there are certain weakened versions of Ramsey's theorem in the case of uncountable sets, which turn out to be valid and very useful in modern set theory (see especially the original work [90] by Erdös and Rado). Quite many papers and monographs concerning those versions were written. The theory developed in this direction is now called the Partition Calculus. It found many applications in various branches of contemporary mathematics, primarily, in infinite combinatorics, model theory, universal algebra, general topology, functional analysis, and measure theory (see, for instance, [82], [89], [151], [186]).

CHAPTER 7

Homogeneous coverings of the Euclidean plane

At present, a lot of examples are known of subsets of the Euclidean plane $\mathbf{R}^2$ which have strange and, sometimes, very paradoxical geometric properties. Usually, such sets in $\mathbf{R}^2$ are obtained by utilizing rather delicate set-theoretical techniques. One of the earliest examples of this kind is due to Mazurkiewicz [254]. In 1914 he proved the following intriguing statement.

Theorem 1. *There exists a set $X \subset \mathbf{R}^2$ such that every straight line in $\mathbf{R}^2$ meets X in exactly two points.*

Any set $X \subset \mathbf{R}^2$ with the extraordinary property described in the above theorem is called a Mazurkiewicz subset of $\mathbf{R}^2$.

Actually, the construction of X is essentially non-elementary, because it is heavily based on an uncountable form of the Axiom of Choice (one of the proofs of Theorem 1 is sketched later in this chapter).

Sierpiński obtained a certain generalization of the Mazurkiewicz result. In particular, according to Sierpiński's theorem, for any natural number $k \geq 2$, there exists a set $Y \subset \mathbf{R}^2$ which meets every line of $\mathbf{R}^2$ in exactly k points. The proof of this generalized statement does not need any new ideas and is carried out similarly to the original argument of Mazurkiewicz (cf. [316]).

Moreover, replacing the family of all lines in $\mathbf{R}^2$ by the family of all circles in $\mathbf{R}^2$, one can establish an analogous result on the existence of a Mazurkiewicz type subset of $\mathbf{R}^2$ for the latter family. A very similar result can also be obtained for the family of all circles lying on the two-dimensional unit sphere $\mathbf{S}_2 \subset \mathbf{R}^3$.

Briefly speaking, one has the following statement.

DOI: 10.1201/9781003458708-7

Theorem 2. *These three assertions are valid:*

(1) *for every natural number $k \geq 3$, there exists a set $Z \subset \mathbf{R}^2$ such that any circle lying in $\mathbf{R}^2$ meets Z in precisely k points;*

(2) *for every natural number $k \geq 3$, there exists a set $Z' \subset \mathbf{S}_2$ not containing any pair of antipodal points of $\mathbf{S}_2$ and such that each circle lying on $\mathbf{S}_2$ meets Z' in precisely k points;*

(3) *for every natural number $k \geq 2$, there exists a set $X \subset \mathbf{S}_2$ not containing any pair of antipodal points of $\mathbf{S}_2$ and such that each great circle of $\mathbf{S}_2$ meets X in precisely k points.*

Notice that the proof of Theorem 2 can be done by using an argument analogous to the proof of Theorem 1.

However, in this chapter we would like to focus our attention on some "dual" versions of the two results presented above. Here the duality is meant in the sense of projective geometry, namely, we require that such a duality should preserve the incidence relation between appropriate geometric objects in a given ground (base) space E.

In particular, if $E = \mathbf{R}^2$, then instead of looking for a set of points such that every line (or circle) meets this set in precisely k points, for some fixed natural number k, we will be looking for a family of lines (or circles) such that every point of E belongs to precisely k members of the family.

For our further purposes, it will be convenient to introduce the general notion of a homogeneous covering of a ground space E.

Let $\{X_i : i \in I\}$ be a family of pairwise distinct nonempty subsets of E and let $k > 0$ be a natural number (or, more generally, cardinal number).

We shall say that $\{X_i : i \in I\}$ is a k-homogeneous family (in fact, a k-homogeneous covering of E) if, for any point $x \in E$, we have

$$\mathrm{card}(\{i \in I : x \in X_i\}) = k.$$

If $k = 1$, then the k-homogeneity of a family $\{X_i : i \in I\}$ simply means that $\{X_i : i \in I\}$ is a partition of the initial space E.

More generally, let $\{\mathcal{P}_t : t \in T\}$ be an arbitrary disjoint family of partitions of E and let

$$k = \mathrm{card}(T),$$

$$\mathcal{P} = \cup\{\mathcal{P}_t : t \in T\}.$$

It can easily be seen that $\mathcal{P}$ is a k-homogeneous covering of E.

In this case, we say that $\mathcal{P}$ is a decomposable k-homogeneous covering of E and the family $\{\mathcal{P}_t : t \in T\}$ is a decomposition of $\mathcal{P}$ into the partitions $\mathcal{P}_t$ $(t \in T)$.

Observe, by the way, that there are many kinds of homogeneous coverings which are not decomposable in the above-mentioned sense (see, for instance, Exercise 2 of this chapter).

The family of all horizontal and vertical lines in the plane $\mathbf{R}^2$ is clearly a 2-homogeneous covering of $\mathbf{R}^2$. Similarly, for any natural number $k > 2$, it is easy to construct a k-homogeneous family of lines in $\mathbf{R}^2$. Indeed, pick k distinct straight lines $\{l_1, l_2, ..., l_k\}$ in $\mathbf{R}^2$ passing through the origin of $\mathbf{R}^2$ and define

$$\mathcal{L} = \{l \subset \mathbf{R}^2 \ : \ l \text{ is a line parallel to some } l_i, \text{ where } 1 \le i \le k\}.$$

A direct verification shows that $\mathcal{L}$ is the required family. Moreover, the description of $\mathcal{L}$ is effective, i.e., it does not appeal to the Axiom of Choice.

Recall that a mathematical argument (or construction) is effective if it does not rely on the Axiom of Choice, i.e., it can be carried out within **ZF** set theory (cf. [137], [150], [240], [241], [316]).

In this context, the following example seems to be of some interest.

Example 1. Consider the question of whether there exists a family of great circles on the sphere $\mathbf{S}_2$ such that every point of $\mathbf{S}_2$ belongs to exactly 2 circles of the family. In order to answer this question, take the set $X \subset \mathbf{S}^2$ described in assertion (3) of Theorem 2, with $k = 2$. For each $x \in X$, denote by $C(x)$ the great circle of $\mathbf{S}_2$ consisting of all points $y \in \mathbf{S}_2$ that are orthogonal to x (naturally, here we think of each point $x \in \mathbf{S}_2$ as representing the vector in $\mathbf{R}^3$ directed from the center of $\mathbf{S}_2$ to x). Notice that since X does not contain any pair of antipodal points, the circles $C(x)$ for $x \in X$ are distinct. We claim that $\{C(x) : x \in X\}$ is a 2-homogeneous covering of $\mathbf{S}_2$. To see why, consider an arbitrary point $y \in \mathbf{S}_2$. Clearly, for any $x \in X$, we have:

$$y \in C(x) \text{ if and only if } x \in C(y).$$

But, by the definition of X, there are precisely two points $x \in X$ such that $x \in C(y)$, so there are precisely two points $x \in X$ such that $y \in C(x)$.

Evidently, a similar construction can be done for any natural number $k \ge 2$.

The argument just presented makes use of Theorem 2, whose proof requires the Axiom of Choice (**AC**), and no effective construction is known of a 2-homogeneous family of great circles in $\mathbf{S}_2$. It would be interesting to investigate the problem of whether such a construction does exist. Notice that in the case of the existence of such a construction, we directly obtain (by using the same duality) an effective example of a Mazurkiewicz type set on $\mathbf{S}_2$.

In this context, it should be noticed that the analogous question for $\mathbf{R}^2$ and for the circles in $\mathbf{R}^2$ having one and the same radius can be readily solved.

The following simple example illustrates this circumstance.

Example 2. Denote by $\mathbf{Z}$ the set of all integers and consider the countable and discrete family

$$\{\mathbf{R} \times \{n\} : n \in \mathbf{Z}\}$$

of horizontal lines in the plane $\mathbf{R}^2$. We say that a circle $C \subset \mathbf{R}^2$ is admissible if there exists an integer n such that C is tangent to both of the lines $\mathbf{R} \times \{n\}$ and $\mathbf{R} \times \{n+1\}$. It is readily verified that all the admissible circles have diameter 1 and every point of $\mathbf{R}^2$ belongs to exactly two admissible circles. Thus, denoting the family of all admissible circles by

$$\mathcal{C} = \{C_i : i \in I\},$$

we conclude that $\mathcal{C}$ is a 2-homogeneous covering of $\mathbf{R}^2$.

More generally, take a natural number $k \geq 2$ and consider in $\mathbf{R}^2$ the finite sequence of vectors

$$e_1 = (0, 1/2), \quad e_2 = (0, 1/2^2), \quad \ldots, \quad e_k = (0, 1/2^k).$$

Then the family of pairwise congruent circles

$$(\mathcal{C} + e_1) \cup (\mathcal{C} + e_2) \cup \cdots \cup (\mathcal{C} + e_k)$$

has the property that every point of $\mathbf{R}^2$ belongs to precisely $2k$ circles of the family. Clearly, this construction is completely effective and, moreover, is done within the framework of elementary geometry.

Now, consider the question whether there exists a family of pairwise congruent circles in the plane $\mathbf{R}^2$, such that every point of $\mathbf{R}^2$ belongs to exactly 3 circles of the family. The trick of Example 1, modified in an appropriate way, works here, too. So we are going to resolve the problem just formulated by using some kind of duality between the points of $\mathbf{R}^2$ and the circles lying in $\mathbf{R}^2$ and having a fixed radius.

For this purpose, let us introduce one auxiliary notion.

Let $k \geq 2$ be a natural number and let $r > 0$ be a real number.

We shall say that a subset Z of the plane $\mathbf{R}^2$ is a (k, r)-Sierpiński set if every circle in $\mathbf{R}^2$ of diameter r has exactly k common points with Z.

It is not difficult to show that the following two statements are effectively equivalent:

(*) there exists a k-homogeneous covering of $\mathbf{R}^2$ with circles of diameter 1;

(**) there exists a $(k, 1)$-Sierpiński subset of $\mathbf{R}^2$.

To check this equivalence, suppose that (*) holds true, i.e, there exists a k-homogeneous covering $\{C_i : i \in I\}$ of $\mathbf{R}^2$ such that all C_i are circles of radius $1/2$. Let x_i denote the center of C_i for each $i \in I$. We assert that the set

$$X = \{x_i : i \in I\}$$

is a $(k, 1)$-Sierpiński set. To see why, let $C \subset \mathbf{R}^2$ be any circle of diameter 1 and let y be its center. Then, for any $i \in I$, we have:

$$x_i \in C \text{ if and only if } y \in C_i,$$

and since $\{C_i : i \in I\}$ is a k-homogeneous covering of $\mathbf{R}^2$, there are exactly k values of i for which $x_i \in C$ is true. We thus conclude that (**) is valid.

Conversely, suppose that (**) holds true and let X be a $(k, 1)$-Sierpiński subset of $\mathbf{R}^2$. For each point $x \in X$, denote by $C(x)$ the circle of radius $1/2$ whose center coincides with x. An argument similar to the previous one shows that the family of circles $\{C(x) : x \in X\}$ forms a k-homogeneous covering of $\mathbf{R}^2$. We thus conclude that (*) is valid.

The effective equivalence (*) $\Leftrightarrow$ (**) directly indicates that there is a close relationship (of duality type) between the k-homogeneous coverings of $\mathbf{R}^2$ with circles of diameter 1 and the $(k, 1)$-Sierpiński sets. By virtue of Example 2 considered earlier, this circumstance allows us to give an effective construction of a $(2k, 1)$-Sierpiński set (actually, this set coincides with the union of some countable family of pairwise parallel straight lines in $\mathbf{R}^2$).

In connection with the statement above, let us remark that it is unknown whether there exists an effective example of a $(k, 1)$-Sierpiński set (or, equivalently, of a k-homogeneous covering of $\mathbf{R}^2$ with circles of diameter 1) for an odd natural number $k \geq 3$.

On the other hand, it was already indicated that, by means of the method of transfinite induction, a subset Z of $\mathbf{R}^2$ can be constructed, which meets every circle in $\mathbf{R}^2$ in precisely $k \geq 3$ points (recall that assertion (1) of Theorem 2 guarantees this). Clearly, such a Z is simultaneously a $(k, 1)$-Sierpiński set.

By the way, there are $(k, 1)$-Sierpiński subsets of $\mathbf{R}^2$ which have no common points with some fixed circle in $\mathbf{R}^2$ (see Exercise 8).

It is easy to show the validity of the following statement.

Theorem 3. *If $k \geq 2$ is a natural number, then there exists a k-homogeneous covering of $\mathbf{R}^2$ with circles of diameter 1.*

Proof. The main difficulty here is to establish this statement for odd $k \geq 3$, because, as has already been mentioned, for even $k \geq 2$ the corresponding result is effective and elementary. Take an odd $k \geq 3$ and consider any $(k, 1)$-Sierpiński set $X \subset \mathbf{R}^2$. Keeping in mind the equivalence between the statements (*) and (**), we obtain at once the required result.

This completes the proof of Theorem 3. □

Another proof of the same theorem can be done directly, by using some transfinite construction of the desired k-homogeneous covering of $\mathbf{R}^2$ with circles of diameter 1 (constructions of this sort are typical in point set theory; for instance, compare Theorem 5 below).

It would be interesting to study in more details the problem of whether there exists an effective method of constructing a 3-homogeneous family of pairwise congruent circles in $\mathbf{R}^2$. We have already demonstrated that a concrete 2-homogeneous covering of $\mathbf{R}^2$ with circles of diameter 1 enables us to present an effective example of a $2k$-homogeneous covering of $\mathbf{R}^2$ with circles of the same diameter (see Example 2). So, the natural question arises whether the existence of a 3-homogeneous covering of $\mathbf{R}^2$ by pairwise congruent circles effectively implies the existence of a $(2k+1)$-homogeneous covering of $\mathbf{R}^2$ by circles of the same radius. This question remains open (compare with Exercise 11 at the end of the present chapter).

It is widely known that the first more or less rigorous system of axioms for elementary geometry was formulated by D. Hilbert in his classical work "Foundations of Geometry" published in 1899 (cf. [143] containing many additional valuable comments). It should be mentioned that the preliminary version of his system had a defect, namely, the so-called "completeness axiom" was not included in the list of axioms, and only later on (after some critical remarks made by Poincaré), was the above-mentioned axiom added to the other ones. But it is also well known that most of elementary geometry can be developed without appealing to the completeness axiom which is not expressible within the framework of first-order logic. In particular, when we deal with various geometric constructions, e.g., using only a compass and a straight-edge, then we do not need this axiom at all, because geometric constructions touch upon a finite (or at most countable) system of geometric objects in $\mathbf{R}^2$. In any case, from the position of elementary geometry, one may suppose that the Euclidean plane is not complete and, for instance, contains only those points of $\mathbf{R}^2$ both Cartesian coordinates of which are algebraic real numbers.

Let us denote this impoverished plane by the symbol $\mathbf{A}^2$, where $\mathbf{A}$ is the field of all algebraic real numbers. According to celebrated Cantor's theorem, $\mathbf{A}^2$ is effectively countably infinite, so its cardinality is ω (= the smallest infinite cardinal number). The circles in $\mathbf{A}^2$ are defined in a natural way: their centers must belong to $\mathbf{A}^2$ and their radii must be algebraic (and, of course, strictly positive) numbers. It is not difficult to verify that every circle in $\mathbf{A}^2$ is also countably infinite. Actually, in $\mathbf{A}^2$ we have the following situation.

Example 3. For any natural number $k \geq 2$, there exists a k-homogeneous covering of $\mathbf{A}^2$ with pairwise congruent circles. The existence of such a covering holds true without using the Axiom of Choice. Indeed, this covering can be constructed effectively, by applying the method of ordinary recursion, instead of the method of transfinite recursion. To show this, denote by

$$\{x_n \ : \ n = 0, \ 1, \ 2, \ \ldots\}$$

a sequence consisting of all points of $\mathbf{A}^2$. Evidently, we may assume without loss of generality that every point of $\mathbf{A}^2$ occurs in this sequence at least k times. Now, we recursively define a family

$$\{C_n \ : \ n = 0, \ 1, \ 2, \ \ldots\}$$

of pairwise distinct circles in $\mathbf{A}^2$ satisfying the following three conditions:

(i) all circles C_n $(n \in \mathbf{N})$ have diameter 1;

(ii) any point of $\mathbf{A}^2$ belongs to at most k members from the family of circles $\{C_n : n = 0, 1, 2, ...\}$;

(iii) for every natural number m, if the point x_m belongs to fewer than k members from the family of circles $\{C_n : n < m\}$, then $x_m \in C_m$.

Suppose that, for a natural number m, the partial family $\{C_n : n < m\}$ of circles satisfying the above-mentioned conditions has already been determined. For each $n < m$, denote by y_n the center of C_n and put

$$Z = \{z \in \mathbf{A}^2 : z \in C_i \cap C_j \text{ for some distinct indices } i < m \text{ and } j < m\}.$$

Observe that Z is a finite set. Now, consider the point x_m. Only two cases are possible.

Case 1. There are pairwise distinct natural numbers

$$n_1 < m, \quad n_2 < m, \quad \ldots, \quad n_k < m$$

such that

$$x_m \in C_{n_1} \cap C_{n_2} \cap \cdots \cap C_{n_k}.$$

In this case, let y be a point of $\mathbf{A}^2$ satisfying the relations

$$y \neq y_n \qquad (n = 0, 1, ..., m-1),$$

$$||y - z|| \neq 1/2 \qquad (z \in Z).$$

The existence of y is obvious. Define C_m as the circle with center y and radius $1/2$.

Case 2. The point x_m belongs to fewer than k members from the constructed family $\{C_n : n < m\}$.

In this case, let y be a point of $\mathbf{A}^2$ such that

$$y \neq y_n \qquad (n = 0, 1, ..., m-1),$$

$$||y - z|| \neq 1/2 \qquad (z \in Z \setminus \{x_m\}),$$

$$||y - x_m|| = 1/2.$$

The existence of y follows from the facts that the circle in $\mathbf{A}^2$ centered at x_m with radius $1/2$ is countably infinite and the first two requirements concerning y rule out only finitely many points on this circle, so there are infinitely many y satisfying all three requirements. Again, we define C_m as the circle with center y and radius $1/2$.

Notice that in both cases the point y can be picked effectively, e.g., as having the least index in the sequence of points

$$\{x_n \ : \ n = 0, \ 1, \ 2, \ \ldots\}$$

and, by virtue of the definition of y, we readily deduce that the conditions (i), (ii), and (iii) remain true for the extended partial family of circles $\{C_n : n \leq m\}$.

Proceeding in this manner, we come to the desired k-homogeneous covering $\{C_n : n = 0, 1, 2, ...\}$ of $\mathbf{A}^2$.

In particular, taking $k = 3$, we may conclude that:

(a) the problem of the existence of a 3-homogeneous covering of the Euclidean plane with pairwise congruent circles is formulated within the framework of elementary geometry;

(b) this problem admits an effective solution if the plane is identified with $\mathbf{A}^2$ (which is sufficient for many purposes of elementary geometry);

(c) for the usual Euclidean plane $\mathbf{R}^2$, the above-mentioned problem can be resolved with the aid of the Axiom of Choice (and, so far, it is unclear whether the use of this axiom is necessary for solving the problem).

Example 4. The well-known Hopf fibration gives us an effective and nice example of a partition of the three-dimensional unit sphere $\mathbf{S}_3 \subset \mathbf{R}^4$ into great circles. More precisely, the Hopf fibration is the mapping

$$\phi = (\phi_1, \phi_2, \phi_3) : \mathbf{S}_3 \to \mathbf{S}_2$$

defined as follows:

$$\phi_1(x_1, x_2, x_3, x_4) = x_1^2 + x_2^2 - x_3^2 - x_4^2,$$

$$\phi_2(x_1, x_2, x_3, x_4) = 2(x_1 x_4 + x_2 x_3),$$

$$\phi_3(x_1, x_2, x_3, x_4) = 2(x_2 x_4 - x_1 x_3).$$

Here (x_1, x_2, x_3, x_4) is any point of the unit sphere $\mathbf{S}_3 \subset \mathbf{R}^4$, so

$$x_1^2 + x_2^2 + x_3^2 + x_4^2 = 1.$$

It can be checked that ϕ maps $\mathbf{S}_3$ onto $\mathbf{S}_2$ and the pre-images of the points of $\mathbf{S}_2$ form a partition of $\mathbf{S}_3$ into great circles.

At the same time, by applying the method of transfinite induction, one can construct many 1-homogeneous coverings of $\mathbf{S}_3$ with great circles. Of course, those coverings are essentially different from the above-mentioned Hopf fibration and, moreover, they are bad from the analytical point of view.

It needless to say that the Hopf fibration plays an outstanding role in topology and geometry (see, e.g., [20], [250], [268]).

Our next example is again concerned with Mazurkiewicz subsets of $\mathbf{R}^2$.

Example 5. It has been proved that there exists a Mazurkiewicz subset X of the plane $\mathbf{R}^2$, which is nowhere dense and has λ_2-measure zero, where λ_2 denotes the standard two-dimensional Lebesgue measure on $\mathbf{R}^2$ (in this connection, see [108]).

Moreover, it is easy to see that if a Mazurkiewicz set in $\mathbf{R}^2$ is λ_2-measurable, then its measure is necessarily equal to zero.

On the other hand, it was also established that there exists a Mazurkiewicz set $Y \subset \mathbf{R}^2$ which is thick with respect to λ_2, i.e., we have $Y \cap B \neq \emptyset$ whenever

$B \subset \mathbf{R}^2$ is a Borel set with $\lambda_2(B) > 0$ (see again [108]). Obviously, such a Y cannot be measurable with respect to λ_2. Indeed, supposing otherwise, we first derive that Y is not of λ_2-measure zero, because it meets every Borel subset of $\mathbf{R}^2$ with strictly positive λ_2-measure. But applying to the same Y the classical Fubini theorem, we readily conclude that Y must be of λ_2-measure zero, which leads to a contradiction.

The existence of X and Y directly indicates that the descriptive structure of a Mazurkiewicz set can be substantially different in various situations. Let us remark, in this context, that it is unknown whether there exists a Mazurkiewicz set which is simultaneously a Borel subset of $\mathbf{R}^2$. At the same time, the structure of a Mazurkiewicz set cannot be rather simple. For instance, the following statement due to Larman is valid (see [245]).

Theorem 4. *No Mazurkiewicz set admits a representation in the form of a union of countably many closed subsets of* $\mathbf{R}^2$*, i.e., no Mazurkiewicz set is of type* F_σ *in* $\mathbf{R}^2$*.*

Proof. Let X be a Mazurkiewicz set in $\mathbf{R}^2$. First, let us show that X does not contain a simple arc (i.e., X does not contain a homeomorphic image of the closed unit interval $[0, 1]$). Suppose otherwise and consider a simple arc $L \subset X$ with endpoints a and b. Let $l(a, b)$ denote the straight line passing through a and b, and let l be a line in $\mathbf{R}^2$ parallel to $l(a, b)$, having nonempty intersection with L and such that the distance between l and $l(a, b)$ attains its maximum value. Obviously, we can write

$$l \cap X = \{a', b'\} \quad (a' \neq b').$$

Also, it is clear that either $a' \in L$ or $b' \in L$ (or both). We may assume, without loss of generality, that $a' \in L$. Then it is not difficult to verify that there exists a line l' passing through b' and having at least three common points with X, which contradicts the definition of X. Thus, X cannot contain a simple arc.

Now, fix a point $y \in X$ and a circle $C(y) \subset \mathbf{R}^2$ whose center coincides with y. For any point $x \in X \setminus \{y\}$, consider the ray $p(y, x)$ which passes through x and whose endpoint is y. Let $g(x)$ denote the point in which $p(y, x)$ meets $C(y)$. A straightforward verification shows that the mapping

$$g : X \setminus \{y\} \to C(y)$$

is a continuous injection and

$$g(X \setminus \{y\}) \cup (s \circ g)(X \setminus \{y\}) = C(y),$$

where s denotes the central symmetry of $\mathbf{R}^2$ with respect to y.

Suppose now that X admits a representation in the form of a union of countably many closed subsets of $\mathbf{R}^2$. Then the same must be true for $X \setminus \{y\}$. Consequently, we come to the equality

$$X \setminus \{y\} = \cup\{F_n : n < \omega\},$$

where all sets F_n are closed and bounded (i.e., compact) in $\mathbf{R}^2$. This implies that

$$C(y) = (\cup\{g(F_n) : n < \omega\}) \cup (\cup\{(s \circ g)(F_n) : n < \omega\}),$$

where all sets

$$g(F_n), \quad (s \circ g)(F_n) \qquad (n < \omega)$$

are closed in $C(y)$. By virtue of the classical Baire theorem (see, e.g., [82], [240]), at least one set $g(F_n)$ has nonempty interior in $C(y)$, which means that $g(F_n)$ contains some arc $L' \subset C(y)$. Keeping in mind the compactness of F_n and the injectivity of g, we deduce that g^{-1} is continuous on L'. Therefore, the set $X \setminus \{y\}$ contains the simple arc $g^{-1}(L')$, which yields a contradiction with the fact established above.

This contradiction ends the proof of Theorem 4 (for a more general result than Theorem 4, see [43]). □

Now, for the sake of completeness, let us sketch the proof of Theorem 1.

We have already mentioned that the construction of a Mazurkiewicz set $X \subset \mathbf{R}^2$ is essentially non-elementary: it is heavily based on Zermelo's well-ordering theorem, which is logically equivalent to the Axiom of Choice, and on the method of transfinite induction (transfinite recursion). More precisely, in his argument Mazurkiewicz exploits the fact that any set of cardinality continuum $\mathbf{c}$ can be well-ordered. He starts with an α-sequence $\{l_\xi : \xi < \alpha\}$ which consists of all distinct lines in $\mathbf{R}^2$, where α denotes the smallest ordinal number of cardinality continuum (this simply means that $\mathrm{card}(\alpha) = \mathbf{c}$ and $\mathrm{card}(\xi) < \mathbf{c}$ for each ordinal $\xi < \alpha$).

Then Mazurkiewicz demonstrates that it is possible to construct a family $\{X_\xi : \xi < \alpha\}$ of subsets of $\mathbf{R}^2$ satisfying the following four relations:

(1) $X_\zeta \subset X_\xi$ if $\zeta < \xi < \alpha$;

(2) $\mathrm{card}(X_\xi) \le \mathrm{card}(\xi) + \omega$;

(3) the points of X_ξ are in general position (i.e., no three of them belong to a straight line);

(4) $\mathrm{card}(X_\xi \cap l_\xi) = 2$.

Indeed, at the first step put $X_0 = \{x, y\}$, where x and y are some distinct points of the line l_0.

Further, suppose that, for a nonzero ordinal number $\xi < \alpha$, the partial family of sets $\{X_\zeta : \zeta < \xi\}$ satisfying the above-mentioned relations (1)–(4) has already been defined. Consider the set

$$D = \cup\{X_\zeta : \zeta < \xi\}.$$

The cardinality of D does not exceed $\mathrm{card}(\xi) + \omega$, so is strictly less than $\mathbf{c}$. As usual, for any two distinct points $x \in D$ and $y \in D$, denote by $l(x, y)$ the straight line passing through x and y. The set of lines

$$\mathcal{L} = \{l(x, y) : x \in D, y \in D, x \neq y\}$$

also has cardinality strictly less than $\mathbf{c}$. If $l_\xi \in \mathcal{L}$, then we put $X_\xi = D$. If $l_\xi \notin \mathcal{L}$, then, taking into account that

$$\mathrm{card}(D \cap l_\xi) < 2,$$

we can define a set $D' \subset l_\xi \setminus \cup\mathcal{L}$ such that

$$\mathrm{card}((D \cup D') \cap l_\xi) = 2.$$

Then we put

$$X_\xi = D \cup D'.$$

It is not difficult to see that in both cases the relations (1)–(4) remain valid.

Proceeding in this manner, we are able to construct the desired family of sets $\{X_\xi : \xi < \alpha\}$. Finally, denoting

$$X = \cup\{X_\xi : \xi < \alpha\},$$

we readily conclude that the set X meets every straight line in $\mathbf{R}^2$ in precisely two points.

Theorem 1 has thus been proved.

Obviously, the argument just presented is of set-theoretical flavor. So, one may anticipate that an analogous approach based on the method of transfinite induction (transfinite recursion) works in many other situations where much more general geometric figures are taken instead of straight lines or circles in the plane $\mathbf{R}^2$.

Actually, the following purely set-theoretical statement (see [204]) is a generalization of some results discussed earlier in this chapter.

Theorem 5. *Let k be a nonzero natural number, E be an infinite ground set, $\mathcal{T}$ be a family of subsets of E, each of which contains at least two elements, and suppose that $\mathcal{T}$ satisfies these two conditions:*

(1) there exists a cardinal number $\tau < \mathrm{card}(E)$ such that

$$\mathrm{card}(X_1 \cap X_2 \cap \ldots \cap X_k) \leq \tau$$

for any pairwise distinct sets $X_1 \in \mathcal{T}$, $X_2 \in \mathcal{T}$, ..., $X_k \in \mathcal{T}$;

(2) for every set $Y \subset E$ with $\mathrm{card}(Y) < \mathrm{card}(E)$ and for every element x from $E \setminus Y$, there exists a set $X \in \mathcal{T}$ such that $x \in X$ and $X \cap Y = \emptyset$.

Then $\mathcal{T}$ contains a k-homogeneous covering of E.

Proof. For each element $z \in E$ and for any family $\mathcal{T}' \subset \mathcal{T}$, let us denote

$$i(z, \mathcal{T}') = \mathrm{card}(\{Z \in \mathcal{T}' : z \in Z\}).$$

Also, denote by α the least ordinal number satisfying the equality

$$\mathrm{card}(\alpha) = \mathrm{card}(E),$$

and fix an enumeration $\{x_\xi : \xi < \alpha\}$ of all elements of the ground set E.

We are going to construct, by the method of transfinite recursion, a certain injective family of sets $\{X_\xi : \xi < \alpha\} \subset \mathcal{T}$.

Suppose that, for an ordinal $\beta < \alpha$, the partial injective family

$$\mathcal{T}^* = \{X_\xi : \xi < \beta\} \subset \mathcal{T}$$

has already been defined so that the inequality $i(z, \mathcal{T}^*) \leq k$ holds true for all elements $z \in E$. Put

$$Y = \cup\{X_{\xi_1} \cap X_{\xi_2} \cap ... \cap X_{\xi_k} : \xi_1 < \xi_2 < ... < \xi_k < \beta\}.$$

If β is a finite ordinal, then in view of condition (1), we readily obtain that

$$\mathrm{card}(Y) \leq C^k_\beta \cdot \tau < \mathrm{card}(E),$$

where C^k_β is a binomial coefficient, i.e., $C^k_\beta = \beta!/(k!(\beta - k)!)$.

If β is an infinite ordinal, then by virtue of the same condition (1), we get

$$\mathrm{card}(Y) \leq \mathrm{card}(\beta) \cdot \tau < \mathrm{card}(E).$$

Hence, in both cases above, we have the relation $\mathrm{card}(Y) < \mathrm{card}(E)$.

Now, let $\zeta < \alpha$ be the smallest ordinal number such that $i(x_\zeta, \mathcal{T}^*) < k$ (if ζ with this property does not exist, then $\mathcal{T}^*$ is a k-homogeneous covering of E, and we are done).

Observe that x_ζ does not belong to the set Y.

According to condition (2), there exists a set $X \in \mathcal{T}$ for which $x_\zeta \in X$ and $X \cap Y = \emptyset$. We may assume, without loss of generality, that $X \neq X_\xi$ for all $\xi < \beta$ (because every member of $\mathcal{T}$ contains at least two distinct elements). Keeping in mind this assumption, we then define $X_\beta = X$.

Notice that the extended partial family

$$\mathcal{T}^{**} = \{X_\xi : \xi \leq \beta\}$$

is such that the relation $i(z, \mathcal{T}^{**}) \leq k$ holds true for all elements $z \in E$.

Proceeding in this way, we are able to construct the required family of sets

$$\{X_\xi : \xi < \alpha\} \subset \mathcal{T}.$$

Denote it by $\mathcal{T}_0$. It turns out that $\mathcal{T}_0$ is a k-homogeneous covering of E. Indeed, it directly follows from our construction that $i(x, \mathcal{T}_0) \leq k$ for each element $x \in E$. Moreover, we can even assert that $i(x, \mathcal{T}_0) = k$. Namely, the last equality is valid because $x = x_\xi$ for some ordinal number $\xi < \alpha$ and

$$k\text{card}(\{x_\zeta : \zeta < \xi\}) \leq \text{card}(\xi) + \omega < \text{card}(\alpha) = \text{card}(E).$$

This completes the proof of Theorem 5. □

In a similar manner, the next statement can be established.

Theorem 6. *Let k be a nonzero natural number, E be an infinite ground set, and let $\mathcal{T}$ be a family of subsets of E each of which contains at least two elements. Suppose that the following three conditions are satisfied:*

(1) $\text{card}(E)$ *is a regular cardinal number;*

(2) the inequality $\text{card}(X_1 \cap X_2 \cap ... \cap X_k) < \text{card}(E)$ *holds true for all pairwise distinct sets* $X_1 \in \mathcal{T}$, $X_2 \in \mathcal{T}$, ..., $X_k \in \mathcal{T}$;

(3) for every set $Y \subset E$ *with* $\text{card}(Y) < \text{card}(E)$ *and for every element* x *from* $E \setminus Y$, *there exists a set* $X \in \mathcal{T}$ *such that* $x \in X$ *and* $X \cap Y = \emptyset$.

Then $\mathcal{T}$ contains a k-homogeneous covering of E.

The proof of this statement (by the same scheme as for Theorem 5) is left to the reader.

Now, let us give several geometric applications of Theorem 5.

Actually, the first of them has already been discussed earlier in this chapter.

Fix a natural number $k \geq 2$, put $E = \mathbf{R}^2$ and take as $\mathcal{T}$ the family of all those circles in E which are congruent to $\mathbf{S}_1$. It is not difficult to verify that both conditions (1) and (2) of Theorem 5 are valid in this situation. Consequently, we again come to the fact that there exists a k-homogeneous covering of $\mathbf{R}^2$, all members of which are pairwise congruent circles. Repeat once more that no such covering is decomposable.

Some more complicated examples are presented below.

Example 6. Fix a natural number $k \geq 3$, put $E = \mathbf{R}^3$, and take as $\mathcal{T}$ the family of all spheres in E which are congruent to the two-dimensional unit sphere $\mathbf{S}_2 \subset \mathbf{R}^3$. Again, it is easy to verify that the conditions (1) and (2) of Theorem 5 are satisfied in this situation. Consequently, there exists a k-homogeneous covering of $\mathbf{R}^3$ consisting of pairwise congruent two-dimensional spheres.

Example 7. If $E = \mathbf{R}^4$, then it is not difficult to show that there are four pairwise congruent three-dimensional spheres X_1, X_2, X_3, X_4 in E such that

$$X_1 \cap X_2 \cap X_3 \cap X_4$$

is a circle and, hence, is of cardinality continuum $\mathbf{c}$. Therefore, a straightforward application of Theorem 5 is impossible in this case. However, a slight modification of the argument used in the proof of Theorem 5 yields the corresponding result for $\mathbf{R}^4$ as well. Namely, one can assert that, for any natural number $k \geq 4$, there exists a k-homogeneous covering of $\mathbf{R}^4$ consisting of pairwise congruent three-dimensional spheres. An analogous method works for the Euclidean space $\mathbf{R}^m$, where $m > 4$, and we obtain that, for any natural number $k \geq m$, there exists a k-homogeneous covering of $\mathbf{R}^m$, all members of which are pairwise congruent $(m-1)$-dimensional spheres.

Example 8. Let $E = \mathbf{R}^m$, where $m \geq 3$. If L is an affine hyperplane in E, then the symbol $e(L)$ will denote the exterior normal vector of L.

Let $\mathcal{L}$ be a family of affine hyperplanes in E satisfying the following two conditions:

(i) for any pairwise distinct hyperplanes

$$L_1 \in \mathcal{L}, \quad L_2 \in \mathcal{L}, \quad \ldots, \quad L_m \in \mathcal{L},$$

the vectors $e(L_1)$, $e(L_2)$,..., $e(L_m)$ are linearly independent;

(ii) for each set $Y \subset E$ with $\mathrm{card}(Y) < \mathrm{card}(E)$ and for each point $x \in E \setminus Y$, there exists a hyperplane $L \in \mathcal{L}$ passing through x and not intersecting Y.

Applying Theorem 5 to E and $\mathcal{L}$, we conclude that, for any natural number $k \geq m$, the given family $\mathcal{L}$ contains a k-homogeneous covering of E.

Example 9. Let $E = \mathbf{R}^2$ and let $\mathcal{L}$ be a family of straight lines in E satisfying the condition

$$\mathrm{card}(\{l \in \mathcal{L} : x \in l\}) = \mathrm{card}(E) = \mathbf{c}$$

for every point $x \in E$. Again, Theorem 5 is applicable in this situation. We thus obtain that, for any natural number $k \geq 2$, the family $\mathcal{L}$ contains a k-homogeneous covering of E.

In connection with Example 9, the following problem of combinatorial (or, if one prefers, of set-theoretical) geometry seems to be of some interest.

Problem. Let $k \geq 2$ be a natural number and let $\mathcal{L}$ be a family of straight lines in the plane $\mathbf{R}^2$. Find necessary and sufficient conditions under which $\mathcal{L}$ contains a k-homogeneous covering of $\mathbf{R}^2$.

As far as we know, this problem remains open.

Example 10. Let $k \geq 2$ be a natural number, let $E = \mathbf{R}^2$, and let $\mathcal{L}$ be a family of non-degenerate irreducible algebraic curves in E. It can be checked that, for any pairwise distinct curves

$$L_1 \in \mathcal{L}, \quad L_2 \in \mathcal{L}, \quad \ldots, \quad L_k \in \mathcal{L},$$

the relation

$$\mathrm{card}(L_1 \cap L_2 \cap \ldots \cap L_k) < \omega < \mathrm{card}(E)$$

is fulfilled. Suppose also that the following condition is satisfied:

for each set $Y \subset E$ with $\mathrm{card}(Y) < \mathrm{card}(E)$ and for each point $x \in E \setminus Y$, there exists a curve $L \in \mathcal{L}$ passing through x and not intersecting Y.

Then, applying Theorem 5 to E and $\mathcal{L}$, we obtain that $\mathcal{L}$ contains a k-homogeneous covering of E.

Example 11. Let $E = S$ be a two-dimensional sphere in the space $\mathbf{R}^3$ and let L be a circle on S whose radius is smaller than the radius of S. Consider the family $\mathcal{L}$ of all those circles on S which are congruent to L. Clearly, Theorem 5 can be applied to S and $\mathcal{L}$. We thus get that, for any natural number $k \geq 2$, there exists a k-homogeneous covering of S, all members of which belong to $\mathcal{L}$.

The same result remains true for the family of all those circles on S whose radii are equal to the radius of S (in this case we cannot apply Theorem 5, so some additional technical details occur, but they are not difficult).

Several exercises are presented below, which yield further information around the topic discussed in this chapter.

EXERCISES

1. Give a detailed proof of Theorem 2 by using the method of transfinite induction.

2. Show that there exists no 1-homogeneous covering of $\mathbf{R}^2$ (respectively, of $\mathbf{S}_2$) with circles.

   Infer this fact from the more general result stating that there exists no 1-homogeneous covering of $\mathbf{R}^2$ (respectively, of $\mathbf{S}_2$) with Jordan curves (recall that a Jordan curve is any homeomorphic image of the unit circle $\mathbf{S}_1$).

   Conclude that if $k \geq 2$ is a natural number, then no k-homogeneous covering of $\mathbf{R}^2$ with Jordan curves is decomposable.

   On the other hand, show that there exists an effective partition of the plane $\mathbf{A}^2$ into countably many circles (cf. Example 3 of this chapter).

3. Give an effective example of a 1-homogeneous covering of the space $\mathbf{R}^3$ with circles.

   Remark 1. Let us stress that the circles from Exercise 3 are not required to be of one and the same radius. A beautiful effective construction of such a 1-homogeneous covering is presented in [337] (see also [198]; further interesting results may be found in [58], [153], and [358]).

4. Give an effective example of a 1-homogeneous covering of $\mathbf{R}^3$ with straight lines, no two of which are parallel.

5*. Let I be a set of cardinality continuum $\mathbf{c}$ and let $\{r_i : i \in I\}$ be a family of strictly positive real numbers.

By using the method of transfinite induction, prove that there exists a partition $\{C_i : i \in I\}$ of $\mathbf{R}^3$ such that every C_i is a circle of radius r_i.

In particular, putting $r_i = r > 0$ for all $i \in I$, conclude that there exists a partition of $\mathbf{R}^3$ into pairwise congruent circles (cf. [61], [192], [198]).

6. Let r be a strictly positive real number not exceeding $1/2$.

Show that there exists a partition of the three-dimensional closed unit ball

$$\mathbf{B}_3 = \{(x_1, x_2, x_3) : x_1^2 + x_2^2 + x_3^2 \leq 1\} \subset \mathbf{R}^3$$

into circles with radius r.

7. Give an effective example of a family $\mathcal{F}$ of pairwise congruent circles in $\mathbf{R}^2$ such that the set $\{C \in \mathcal{F} : x \in C\}$ is countably infinite for every point x of $\mathbf{R}^2$.

In other words, demonstrate that the required family $\mathcal{F}$ is an ω-homogeneous covering of $\mathbf{R}^2$ with pairwise congruent circles.

Also, consider the analogous problem and give an effective example of a closed subset of $\mathbf{R}^2$ which meets every straight line of $\mathbf{R}^2$ in precisely ω many points.

8. Let $k \geq 3$ be a natural number.

Prove that there exists a subset X of $\mathbf{R}^2$ satisfying the following two conditions:

(a) X is a $(k, 1)$-Sierpiński set;

(b) the unit circle $\mathbf{S}_1$ has no common points with X.

9. Let k be a nonzero natural number.

Show effectively that there exist two sets $X \subset \mathbf{R}^2$ and $Y \subset \mathbf{R}^2$ such that both of them are of cardinality continuum $\mathbf{c}$ and

$$\mathrm{card}(f(X) \cap g(Y)) = 2k$$

for any two isometric transformations (i.e., motions) f and g of $\mathbf{R}^2$.

10*. By using the method of transfinite induction, demonstrate that there exist two sets $A \subset \mathbf{R}^2$ and $B \subset \mathbf{R}^2$ such that both of them are of cardinality continuum $\mathbf{c}$ and

$$\mathrm{card}(f(A) \cap g(B)) = 1$$

for any two isometric transformations f and g of $\mathbf{R}^2$.

Remark 2. The result formulated in Exercise 10 is due to Sierpiński. As far as we know, the question whether there exists an effective example of two sets A and B having the above-mentioned property still remains open.

11*. Prove that there exists a set $X \subset \mathbf{R}^2$ satisfying the following two conditions:

(a) every line in $\mathbf{R}^2$ meets X in exactly two points (i.e., X is a Mazurkiewicz set);

(b) every circle lying in $\mathbf{R}^2$ meets X in exactly three points.

Let $\mathcal{C}$ denote the 2-homogeneous family of circles described in Example 2 of this chapter.

Applying conditions (a) and (b), deduce that there exists a 3-homogeneous family $\mathcal{L}$ of circles in $\mathbf{R}^2$ such that:

(c) all circles from $\mathcal{L}$ are of radius $1/2$;

(d) for any isometric transformation h of $\mathbf{R}^2$, the two families $h(\mathcal{C})$ and $\mathcal{L}$ have infinitely many common circles.

Remark 3. The result stated in Exercise 11 shows that a 5-homogeneous covering of $\mathbf{R}^2$ with circles of radius $1/2$ cannot be trivially obtained by using a 3-homogeneous covering of $\mathbf{R}^2$ with circles of radius $1/2$ and a 2-homogeneous covering of $\mathbf{R}^2$ with circles of the same radius.

12*. Give detailed transfinite constructions of Mazurkiewicz sets with the additional properties indicated in Example 5 of this chapter.

Namely, demonstrate that:

(a) there exists a Mazurkiewicz set X in $\mathbf{R}^2$ which is nowhere dense and also is of λ_2-measure zero;

(b) there exists a Mazurkiewicz set Y in $\mathbf{R}^2$ which is λ_2-thick and hence is not measurable with respect to λ_2.

For establishing (a), start with a family

$$\{l_i : i \in I\} \cup \{l'_i : i \in I\}$$

of straight lines in $\mathbf{R}^2$ such that $\mathrm{card}(I) = \mathbf{c}$, all lines l_i are distinct and horizontal, all lines l'_i are distinct and vertical, the union $\cup\{l_i \cup l'_i : i \in I\}$ is nowhere dense in $\mathbf{R}^2$, and

$$\lambda_2(\cup\{l_i \cup l'_i : i \in I\}) = 0.$$

Try to construct the required Mazurkiewicz set X as a subset of the union $\cup\{l_i \cup l'_i : i \in I\}$.

For establishing (b), take into account the fact that the family of all Borel subsets of $\mathbf{R}^2$ with strictly positive λ_2-measure is of cardinality $\mathbf{c}$, define the required Mazurkiewicz set Y by using the method of transfinite recursion, and check that Fubini's theorem is not applicable to Y.

Remark 4. Several works should be mentioned, which are devoted to constructions of Mazurkiewicz sets in $\mathbf{R}^2$ with additional properties of geometric nature. In this connection, see, for instance, [55] and [307].

13. Give a proof of Theorem 6.

14. Demonstrate the validity of the facts indicated in Example 7.

15. Return to Example 4 of the present chapter and consider again the Hopf mapping $\phi = (\phi_1, \phi_2, \phi_3) : \mathbf{S}_3 \to \mathbf{S}_2$ defined by the formulas

$$\phi_1(x_1, x_2, x_3, x_4) = x_1^2 + x_2^2 - x_3^2 - x_4^2,$$

$$\phi_2(x_1, x_2, x_3, x_4) = 2(x_1x_4 + x_2x_3),$$

$$\phi_3(x_1, x_2, x_3, x_4) = 2(x_2x_4 - x_1x_3),$$

where (x_1, x_2, x_3, x_4) is any point of the unit sphere $\mathbf{S}_3 \subset \mathbf{R}^4$, so

$$x_1^2 + x_2^2 + x_3^2 + x_4^2 = 1.$$

Express $(x_1, x_2, x_3, x_4) \in \mathbf{S}_3$ in the following form:

$$x_1 = r\cos(\alpha), \quad x_2 = r\sin(\alpha), \quad x_3 = t\cos(\beta), \quad x_4 = t\sin(\beta),$$

where $r \geq 0, \ t \geq 0, \ \alpha \in [0, 2\pi], \ \beta \in [0, 2\pi], \ r^2 + t^2 = 1.$

Now, let $(a, b, c) \in \mathbf{R}^3$ be an arbitrary point of the sphere $\mathbf{S}_2$.

Supposing that $\phi(x_1, x_2, x_3, x_4) = (a, b, c)$, verify that

$$r^2 - t^2 = a, \quad 2rt\sin(\alpha + \beta) = b, \quad -2rt\cos(\alpha + \beta) = c,$$

and infer from the above formulas that $\phi^{-1}((a, b, c))$ is a great circle of the sphere $\mathbf{S}_3$ (for this purpose, it suffices to indicate two distinct hyperplanes in $\mathbf{R}^4$ each of which passes through the origin of $\mathbf{R}^4$ and contains $\phi^{-1}((a, b, c))$).

16*. Let Z be a subset of the plane $\mathbf{R}^2$ such that, for every straight line l in $\mathbf{R}^2$, one has $\mathrm{card}(Z \cap l) = \mathbf{c}$.

Prove (by using the method of transfinite induction) that there exists a Mazurkiewicz set entirely contained in Z.

Also, give an effective example of a set $Y \subset \mathbf{R}^2$ satisfying the following two conditions:

(a) for every straight line l in $\mathbf{R}^2$, the equality $\mathrm{card}(Y \cap l) = \omega$ holds true;

(b) Y is contained in the union of a countable family of straight lines in $\mathbf{R}^2$.

Deduce from (b) that there exists no Mazurkiewicz set contained in Y.

17*. Let μ be a σ-finite measure on $\mathbf{R}^2$ invariant (or, more generally, quasi-invariant) with respect to the group of all translations of $\mathbf{R}^2$.

Prove that, for any μ-measurable Mazurkiewicz set Z, the equality $\mu(Z) = 0$ is valid.

CHAPTER 8

Three-colorings of the Euclidean plane and associated triangles of a prescribed type

In this chapter we intend to show that, for any three-coloring of the Euclidean plane $\mathbf{R}^2$, there are continuum many triangles of a prescribed type (e.g., acute-angled, right-angled, obtuse-angled, isosceles) such that the vertices of each of them carry all three colors. Before proving these results, we shall introduce several preliminary notions which will be useful for our further purposes.

It should be noticed that there are many interesting geometric problems connected with various colorings of all points of the plane $\mathbf{R}^2$ (or, more generally, connected with colorings of all points of Euclidean spaces of higher dimension). For instance, one may mention an old and still open problem of Hadwiger and Nelson which requires to find the precise value of the chromatic number of $\mathbf{R}^2$, and also a lot of other questions of analogous types (cf. [63], [117], [131], [132], [271], [288]). As has already been said above, in this chapter we would like to consider several questions of somewhat similar nature.

First, let us present the definition that will be basic throughout the chapter (it makes sense to compare Chapter 4 in which we were dealing with colorings of all edges of the full graphs produced by the points of it-subsets of the Euclidean space $\mathbf{R}^m$).

A three-coloring of the plane $\mathbf{R}^2$ is defined as any surjection g of $\mathbf{R}^2$ onto the three-element set $\{1, 2, 3\}$ (of course, here the natural numbers 1, 2, and 3 are treated as three distinct colors).

Example 1. Let a mapping

$$g : \mathbf{R}^2 \to \{1, 2, 3\}$$

DOI: 10.1201/9781003458708-8

be defined as follows:

(a) the range of the restriction of g to the line $\mathbf{R} \times \{0\}$ coincides with the set $\{1, 2\}$;

(b) the range of the restriction of g to the set $\mathbf{R}^2 \setminus (\mathbf{R} \times \{0\})$ coincides with the singleton $\{3\}$.

It is easy to see that, for this coloring g, there are no three collinear points in $\mathbf{R}^2$ which have pairwise different colors.

Example 2. Let a mapping

$$f : \mathbf{R}^2 \to \{1, 2, 3\}$$

be defined as follows:

(a) $f((0,0)) = 3$, i.e., the color 3 is assigned to the origin $(0,0)$ of $\mathbf{R}^2$;

(b) if l is any straight line in $\mathbf{R}^2$ passing through the origin $(0,0)$, then the range of the restriction of f to $l \setminus \{(0,0)\}$ coincides either with $\{1\}$ or with $\{2\}$.

Again, it can readily be checked that, for this coloring f, there exist no three collinear points in $\mathbf{R}^2$ which have pairwise different colors.

As usual, the triangle with vertices $x \in \mathbf{R}^2$, $y \in \mathbf{R}^2$, and $z \in \mathbf{R}^2$ will be denoted by $[x, y, z]$. We always assume below that

$$x \neq y \neq z \neq x,$$

but we do not exclude that $[x, y, z]$ may be a degenerate triangle; in other words, the points x, y, z may be collinear, i.e., lying on some straight line in $\mathbf{R}^2$.

Let g be a three-coloring of $\mathbf{R}^2$. A triangle $[x, y, z]$ is said to be associated with g (in short, g-associated) if the equality

$$g(\{x, y, z\}) = \{1, 2, 3\}$$

holds true (notice that in the book [2] and in the paper [263] such $[x, y, z]$ is called a rainbow triangle; undoubtedly, their terminology is more visual).

In connection with various three-colorings of $\mathbf{R}^2$, the following question naturally arises:

How many triangles of a prescribed type (for instance, acute-angled, right-angled, obtuse-angled, isosceles) are associated with a given three-coloring g of the plane $\mathbf{R}^2$?

In what follows we will be trying to obtain the answer to this question.

The proofs of the first four theorems formulated below will be done by using one and the same scheme. Namely, if a three-coloring g of $\mathbf{R}^2$ is fixed and we already have a family of g-associated triangles of a prescribed type, such that the total number of triangles of this family is strictly less than the cardinality continuum (as usual, denoted by the symbol $\mathbf{c}$), then we indicate one more triangle of the prescribed type, which is also associated with g. Such an approach leads us to the required result. Namely, we can conclude that there are continuum many g-associated triangles of the prescribed type.

Let us first consider the case of acute-angled triangles. In this case, we need the following simple auxiliary proposition.

Lemma 1. *Let g be a three-coloring of the plane $\mathbf{R}^2$ such that, for some three distinct and collinear points x, y and z in $\mathbf{R}^2$, one has*

$$g(x) = 1, \quad g(y) = 2, \quad g(z) = 3,$$

and let ε be an arbitrary strictly positive real number.

Then there exists an injective family $\{\triangle_i : i \in I\}$ of g-associated triangles such that:

(1) $\mathrm{card}(I) = \mathbf{c}$*;*

(2) all triangles $\triangle_i$ $(i \in I)$ are acute-angled;

(3) in any triangle $\triangle_i$ $(i \in I)$ the measure of at least one internal angle does not exceed ε;

(4) for each index $i \in I$, at least one vertex of $\triangle_i$ belongs to $\{x, y, z\}$.

We omit the easy proof of this lemma and leave it to the reader (see Exercise 1).

Theorem 1. *For every three-coloring g of $\mathbf{R}^2$, there are continuum many g-associated acute-angled triangles.*

Proof. Fix a three-coloring g of $\mathbf{R}^2$. According to Lemma 1, we may assume without loss of generality that no straight line in $\mathbf{R}^2$ carries all three colorings 1, 2 and 3. Further, we can choose a non-degenerate triangle $[A, B, C]$ such that

$$g(A) = 1, \quad g(B) = 2, \quad g(C) = 3.$$

Now, we would like to recall the reader the following interesting fact of elementary plane geometry:

Every non-degenerate triangle in $\mathbf{R}^2$ admits a dissection (actually, triangulation) into 7 smaller acute-angled triangles.

In connection with this fact, see Exercise 2 of the present chapter.

Let us consider one of such dissections of our triangle $[A, B, C]$. All vertices of the indicated dissection of $[A, B, C]$ carry some colors from $\{1, 2, 3\}$. Since no straight line carries all colors, the dissection satisfies the conditions of Sperner's well-known combinatorial lemma (see, for instance, [4], [5], [82], [240] or Exercise 19 from Appendix 1).

So, in view of Sperner's lemma, we easily get that one of the triangles in this dissection is necessarily g-associated. In particular, it may happen that there are some other g-associated small acute-angled triangles in the above-mentioned dissection of $[A, B, C]$. Recall, by the way, that the total number of g-associated small acute-angled triangles in $[A, B, C]$ must be always odd, so is not equal to zero. However, this more delicate fact is not required for our further purposes.

Now, slightly moving the vertices of this dissection (except for the vertices A, B and C which remain fixed), we come to another dissection of $[A, B, C]$ into 7 acute-angled triangles.

Again, according to Sperner's lemma, there exists a triangle of the new dissection of $[A, B, C]$, which is also g-associated. Let us underline that there are continuum many possibilities for slightly changed dissections of $[A, B, C]$ of the described type. Moreover, if we consider any two such dissections of $[A, B, C]$, then all triangles of the first dissection differ from all triangles of the second dissection. The latter circumstance directly leads to the desired result, i.e., there are continuum many acute-angled triangles which are associated with the given three-coloring g.

Theorem 1 has thus been proved. □

Remark 1. It should be noticed that 7 is a precise estimate for the number of acute-angled triangles in the above-mentioned dissections of $[A, B, C]$. Indeed, it can be proved that if $[A, B, C]$ is a right-angled triangle or an obtuse-angled triangle, then $[A, B, C]$ does not admit any dissection into less than 7 acute-angled triangles.

Let us formulate several analogs of Theorem 1 for other types of g-associated triangles in the plane $\mathbf{R}^2$.

Theorem 2. *For every three-coloring g of the plane $\mathbf{R}^2$, there are continuum many g-associated right-angled triangles.*

Theorem 3. *For every three-coloring g of the plane $\mathbf{R}^2$, there are continuum many g-associated obtuse-angled triangles.*

Theorem 4. *For every three-coloring g of the plane $\mathbf{R}^2$, there are continuum many g-associated isosceles triangles.*

As we have already remarked earlier, the proofs of Theorems 2, 3, and 4 follow the scheme of adding a new g-associated triangle of a prescribed type (right-angled, obtuse-angled, isosceles) to a family of g-associated triangles of the same type, assuming that this family is of cardinality strictly less than $\mathbf{c}$.

The analogous question can be posed for triangles similar to a given one. However, this situation substantially differs from the cases considered above. To explain it, we need one more notion.

Let $\triangle$ be an arbitrary non-degenerate triangle in the plane $\mathbf{R}^2$ and let Z be a subset of $\mathbf{R}^2$.

We shall say that Z is $\triangle$-closed if, for any two distinct points $x \in Z$ and $y \in Z$, the third vertex z of any triangle $[x, y, z]$ which is similar to $\triangle$, also belongs to Z.

The next auxiliary proposition is valid.

Lemma 2. *Let $\triangle$ be an arbitrary non-degenerate triangle in the plane $\mathbf{R}^2$ and let x_0 and y_0 be any two distinct points in $\mathbf{R}^2$.*

Then there exists a set $Z \subset \mathbf{R}^2$ such that:

(1) Z is at most countable;

(2) Z is $\triangle$-closed;

(3) $x_0 \in Z$ and $y_0 \in Z$.

Proof. As usual, we denote by $\mathbf{N}$ the set of all natural numbers (recall that $\mathbf{N}$ is often identified with the least ordinal number ω).

Now, we are going to recursively construct an increasing (by the standard inclusion relation) sequence $\{Z_n : n \in \mathbf{N}\}$ of subsets of $\mathbf{R}^2$.

In order to carry out our plan, we first put

$$Z_0 = \{x_0, y_0\}.$$

Suppose now that, for a natural number n, the set $Z_n \subset \mathbf{R}^2$ has already been constructed, such that $\mathrm{card}(Z_n) \leq \mathrm{card}(\mathbf{N})$. We then define

$$Z'_n = \{z' \in \mathbf{R}^2 : (\exists x \in Z_n)(\exists y \in Z_n)(\text{the triangle } [x, y, z'] \text{ is similar to } \triangle)\}$$

and we put

$$Z_{n+1} = Z_n \cup Z'_n.$$

Proceeding in this manner, the required increasing sequence of sets $\{Z_n : n \in \mathbf{N}\}$ will be determined by ordinary recursion.

Further, it is easy to show by induction on n that all sets Z_n are at most countable. So, denoting

$$Z = \cup\{Z_n : n \in \mathbf{N}\},$$

we finally come to the subset Z of $\mathbf{R}^2$ which also is at most countable.

It can readily be checked that Z is $\triangle$-closed. Indeed, let x and y be any two distinct points from Z. Obviously, there exists a natural index n such that $x \in Z_n$ and $y \in Z_n$. According to the definition of Z_{n+1}, all vertices of any triangle $[x, y, z']$ similar to $\triangle$ belong to Z_{n+1}. In view of the relation

$$Z_{n+1} = Z_n \cup Z'_n \subset Z,$$

we conclude that Z is a $\triangle$-closed subset of $\mathbf{R}^2$.

This completes the proof of Lemma 2. □

By using the above lemma, it is not difficult to obtain the following statement.

Theorem 5. *For every given non-degenerate triangle* $[x, y, z]$ *in the plane* $\mathbf{R}^2$*, there exists a three-coloring* g *of* $\mathbf{R}^2$ *such that no triangle in* $\mathbf{R}^2$ *similar to* $[x, y, z]$ *is associated with* g.

Proof. Fix a non-degenerate triangle $\triangle = [x, y, z]$ and take the set Z which is described in Lemma 2. Then equip each point of Z either with color 1 or with color 2. Since Z contains x_0 and y_0, we may assume that x_0 is colored with 1 and y_0 is colored with 2. Further, equip all the points of $\mathbf{R}^2 \setminus Z$ with color 3. Notice that, by virtue of the countability of Z, the set $\mathbf{R}^2 \setminus Z$ is nonempty (moreover, $\mathbf{R}^2 \setminus Z$ has cardinality $\mathbf{c}$). Therefore, for any color from $\{1, 2, 3\}$, there are points in $\mathbf{R}^2$ which carry this color. In other words, we obtain a certain three-coloring g of $\mathbf{R}^2$. Now, taking into account the properties (1), (2) and (3) of the set Z (see Lemma 2), it is easy to check that no triangle in $\mathbf{R}^2$ similar to $\triangle$ can be associated with g.

Theorem 5 has thus been proved. □

Remark 2. Theorem 5 can be generalized to the case of a multi-dimensional Euclidean space (cf. Exercise 12 of the present chapter).

Remark 3. In connection with the proof of Theorem 1, one may pose the question of whether any simplex in the Euclidean space $\mathbf{R}^m$ $(m \geq 3)$ admits a dissection into finitely many acute-angled or non-obtuse-angled simplexes. Here a simplex is called acute-angled (respectively, non-obtuse-angled) if the angles of all its two-dimensional faces are acute (respectively, non-obtuse). This question is closely connected with Hadwiger's old problem which asks whether any simplex admits a dissection into finitely many orthogonal simplexes (see [130]; cf. also [46], [155], [156], [178], [236], [346], [347], [351]). Notice that every orthogonal simplex is a non-obtuse simplex. As far as we know, the above-mentioned problem of Hadwiger still remains unsolved for $m > 5$.

Remark 4. There are various weak axiomatic systems of geometry which provide us with corresponding models of the Euclidean plane (in particular, in such models the plane can be incomplete, and so on). As a rule, those axiomatic systems belong to the first-order logic. The problem that has been formulated at the beginning of this chapter makes sense for the planes in weaker logical systems. Namely, if the plane is colored with three colors $\{1, 2, 3\}$, then the question about the existence of an associated triangle of a prescribed type remains meaningful. Of course, in such a situation one cannot speak of the cardinality of the family of all associated triangles, because the concept of a cardinal number is outside of first-order logic. However, in some cases the existence of associated acute-angled (respectively, obtuse-angled, right-angled) triangles can be demonstrated. For more details, see the recent extensive paper [280], where the existence of rainbow triangles of a prescribed type is established within the framework of certain weak geometries.

EXERCISES

1. Give a detailed proof of Lemma 1.

2*. Let $[A, B, C]$ be a triangle in the plane $\mathbf{R}^2$ whose greatest angle is at the vertex B, and let $D \in [A, C]$ be such that $[B, D]$ is the altitude of $[A, B, C]$. Take a point $E \in]B, D[$ sufficiently near to B and take two points $A' \in [A, B]$ and $C' \in [C, B]$ so that

$$E \in [A', C'] \ \& \ [A', C'] \text{ is parallel to } [A, C].$$

Further, take two points $D' \in [A, D]$ and $D'' \in [C, D]$ so that

$$||D' - A'|| = ||D' - E||,$$

$$||D'' - C'|| = ||D'' - E||.$$

Consider the following seven triangles:

$$[B, A', E], \quad [B, C', E], \quad [D', A', E], \quad [D'', C', E],$$

$$[D', D'', E], \quad [A, A', D'], \quad [C, C', D''].$$

Keeping in mind that the point E is closely near to B and slightly moving the points A' and C' to B, obtain a triangulation of $[A, B, C]$ into seven acute-angled triangles.

Also, demonstrate that 7 is the minimum number of those acute-angled triangles in which any right-angled (obtuse-angled) triangle can be dissected.

3. Give a proof of Theorem 2.

For this purpose, consider separately the case when there are three distinct collinear points of pairwise different colors, and the case when such three points do not exist.

4. Give a proof of Theorem 3.

For this purpose, consider again separately the case when there are three distinct collinear points of pairwise different colors, and the case when such three points do not exist.

5*. Try to show the validity of the following slightly strengthened version of Theorem 1.

If g is an arbitrary 3-coloring of the plane $\mathbf{R}^2$, then there exists a family $\{\Delta_i : i \in I\}$ of triangles in $\mathbf{R}^2$ such that:

(a) $\text{card}(I) = \mathbf{c}$;

(b) each triangle Δ_i $(i \in I)$ is associated with g;

(c) all triangles Δ_i $(i \in I)$ are acute-angled;

(d) for any two distinct indices $i \in I$ and $j \in I$, the triangles Δ_i and Δ_j have at most one common vertex.

For proving this result, argue analogously to the proof of Theorem 1.

6*. Give a proof of Theorem 4.

Argue as follows. Let g be a 3-coloring of the plane $\mathbf{R}^2$ and let $\{\Delta_i : i \in I\}$ be a family of triangles in $\mathbf{R}^2$ which satisfies the conditions (a), (b), (c) and (d) of Exercise 5. Take any Δ_i from this family and put

$$\Delta_i = [A_i, B_i, C_i].$$

Further, denote by O_i the circumcenter of Δ_i and consider the following three triangles:

$$[O_i, A_i, B_i], \quad [O_i, B_i, C_i], \quad [O_i, C_i, A_i].$$

Verify that exactly one of these three triangles is associated with g. Denote such a triangle by Δ'_i and consider the family $\{\Delta'_i : i \in I\}$.

Finally, demonstrate that:

(a′) all triangles from $\{\Delta'_i : i \in I\}$ are pairwise different;

(b′) all of them are isosceles;

(c′) all of them are associated with g.

In particular, conclude that the assertion of Theorem 4 holds true.

7. In the formulation of Theorem 5 take as $\triangle$ any equilateral triangle and present a more concrete 3-coloring g of the plane $\mathbf{R}^2$ such that no triangle similar to $\triangle$ is g-associated.

In this case, use the colors 1 and 2 for coloring some discrete regular system of points in $\mathbf{R}^2$ and use the color 3 for coloring all other points of $\mathbf{R}^2$.

8. Work in **ZF** set theory and prove that there are continuum many 3-colorings g of the plane $\mathbf{R}^2$ such that no equilateral triangle is associated with g.

For this purpose, apply the result of the previous exercise.

9. Give a detailed explanation of the fact that Theorems 1, 2, 3, and 4 cannot be generalized to the case of the Euclidean space $\mathbf{R}^3$ and all 4-colorings of $\mathbf{R}^3$ (for the precise definition of a 4-coloring of $\mathbf{R}^3$, see Exercise 12 below).

10. Consider the plane $\mathbf{A}^2$ over the field $\mathbf{A}$ of all algebraic real numbers (cf. Example 3 from Chapter 7). Let g be an arbitrary 3-coloring of $\mathbf{A}^2$.

Show that:

(a) there are infinitely many g-associated acute-angled triangles;

(b) there are infinitely many g-associated obtuse-angled triangles;

(c) there are infinitely many g-associated right-angled triangles;

(d) there are infinitely many g-associated isosceles triangles.

11. Verify that in the hyperbolic plane there is a non-degenerate triangle which does not have a circumcenter, i.e., there does not exist a circle passing through all vertices of this triangle.

Infer from this fact that, in certain weak systems of axioms for plane geometry, the existence of a non-degenerate triangle which is g-associated with a 3-coloring g of the plane, does not logically imply the existence of a g-associated isosceles triangle.

12*. By definition, an $(m+1)$-coloring of the Euclidean space $\mathbf{R}^m$ is any surjective mapping

$$g : \mathbf{R}^m \to \{1, 2, ..., m, m+1\}.$$

Let $m \geq 2$ be a natural number and let S be a regular m-dimensional simplex in $\mathbf{R}^m$.

Demonstrate that there are $2^{\mathbf{c}}$ many $(m+1)$-colorings g of $\mathbf{R}^m$ such that no simplex in $\mathbf{R}^m$ similar to S is g-associated.

For this purpose, use an argument analogous to the proof of Theorem 5 of the present chapter.

13*. An m-simplex S in the Euclidean space $\mathbf{R}^m$ $(m \geq 2)$ is called isosceles if there exists a vertex of S, all incident edges of which are pairwise congruent.

An $(m+1)$-coloring of the same space $\mathbf{R}^m$ is called admissible if there are no three distinct collinear points in $\mathbf{R}^m$ colored by pairwise different colors.

Consider an arbitrary admissible coloring of $\mathbf{R}^m$:

$$g : \mathbf{R}^m \to \{1, 2, ..., m+1\}.$$

Demonstrate for this coloring that:

(a) if $k < m$ is a natural number, then there are no $(k+2)$ points in $\mathbf{R}^m$ which lie in a k-dimensional affine manifold of $\mathbf{R}^m$ and carry pairwise different colors;

(b) there exist continuum many isosceles m-simplices in $\mathbf{R}^m$ which are associated with g.

For (a), use the method of induction on k.

For (b), utilize an argument similar to the proof of Theorem 1 and also based on Sperner's lemma (see Exercise 19 from Appendix 1).

Remark 5. Assuming uncountable forms of the Axiom of Choice, it can be shown that there are many admissible $(m + 1)$-colorings of the space $\mathbf{R}^m$, where $m \geq 2$. In this connection, see Chapter 15 where non-constructive non-Archimedean valuations are applied to dissections of the unit square $[0, 1]^2$ into finitely many triangles of equal areas.

Note that, up to now, no description of the admissible $(m+1)$-colorings of $\mathbf{R}^m$ is known.

Remark 6. In view of our definition, an m-simplex S in the space $\mathbf{R}^m$ is isosceles if there exists at least one vertex x of S such that all edges of S containing x have the same length.

Other notions of an isosceles m-simplex can also be introduced. Let k be a nonzero natural number not exceeding $m - 1$.

We may say that an m-simplex S is k-isosceles if there exists at least one vertex x of S such that all k-faces of S containing x are pairwise congruent.

In particular, for $k = 1$, we get our definition of an isosceles m-simplex.

It should be underlined that even in the space $\mathbf{R}^3$, the family of 1-isosceles 3-simplexes differs from the family of 2-isosceles 3-simplexes. More precisely, there exist many 1-isosceles 3-simplexes which are not 2-isosceles and, conversely, there exist many 2-isosceles 3-simplexes which are not 1-isosceles.

In addition to the said above, if $\triangle$ is an arbitrary acute-angled triangle in the plane $\mathbf{R}^2$, then always there exists a 3-simplex S in $\mathbf{R}^3$, all 2-faces of which are congruent to this $\triangle$. Consequently, if all sides of $\triangle$ have pairwise distinct lengths, then S is a 2-isosceles 3-simplex and, simultaneously, is not a 1-isosceles 3-simplex.

14. Give an example of a 4-coloring g of the space $\mathbf{R}^3$ such that there exists no g-associated non-degenerate tetrahedron in $\mathbf{R}^3$ (cf. Exercise 9 of this chapter).

15*. Consider the hypersphere $\mathbf{S}_{m-1}$ of the Euclidean space $\mathbf{R}^m$, where $m \geq 3$, and try to formulate and prove the results, analogous to the presented ones in this chapter, for spherical geometry on $\mathbf{S}_{m-1}$.

16. Verify that if an m-simplex S in the space $\mathbf{R}^m$ can be dissected into finitely many orthogonal m-simplexes, than the same S can be dissected into finitely many isosceles m-simplexes.

CHAPTER 9

Chromatic numbers of graphs associated with point sets in Euclidean space

In Exercise 17 from Appendix 5 the notion of a bichromatic graph is mentioned and a certain characterization of such graphs is presented.

Recall that, according to this well-known characterization, a graph (V, E) with $\text{card}(V) \geq 2$ is bichromatic if and only if E does not contain a simple (reduced) cycle of an odd length.

It is also indicated in the same Appendix 5 that the above characterization heavily relies on the Axiom of Choice (**AC**). Here we wish to develop this topic in connection with certain graphs associated with point sets in the Euclidean space $\mathbf{R}^m$, where $m \geq 1$.

For any natural number $k \geq 2$, the more general notion of a k-chromatic graph is also known (see, e.g., [87], [131], [132], [135], [275], and Exercise 18 from Appendix 5). Let us remind this notion.

A graph $\Gamma = (V, E)$ is called k-chromatic if k is the least natural number having the following property:

The set V of all vertices of Γ can be represented in the form

$$V = V_1 \cup V_2 \cup ... \cup V_k,$$

where the sets V_i $(i = 1, 2, ..., k)$ are nonempty and pairwise disjoint, and the endpoints of every edge $e \in E$ belong to different sets V_i and V_j (of course, here the indices i and j depend on e).

Obviously, in the special case $k = 2$, we come to the definition of a bichromatic graph.

DOI: 10.1201/9781003458708-9

If (V, E) is a finite graph, then its chromatic number is always finite (more precisely, the chromatic number does not exceed the cardinality of V). At the same time, very simple examples show that many infinite graphs do not have finite chromatic numbers. For instance, it is easy to see that no infinite full graph possesses a finite chromatic number.

Remark 1. Taking into account the above circumstance, the chromatic number of an arbitrary graph (V, E) is defined in **ZFC** set theory as the least cardinal number κ having the property that there exists a partition $\{V_i : i \in \kappa\}$ of V such that the endpoints of every edge $e \in E$ belong to different sets V_i and V_j (again, the indices i and j depend on e).

The question naturally arises to find a criterion which guarantees that a given infinite graph (V, E) has a finite chromatic number.

In this direction, the fundamental old result due to de Bruijn and Erdös [87] should be mentioned.

Theorem 1. *Let $k \geq 2$ be a natural number.*

For any graph (V, E), the following two assertions are equivalent:

(1) each finite subgraph of (V, E) has chromatic number less than or equal to k;

(2) the chromatic number of (V, E) exists and does not exceed k.

Proof. In fact, this statement is a particular case of the so-called compactness theorem of Malcev and Gödel from model theory (see, e.g., [59], [81], [237], [259] or [296]). But, for the sake of completeness and in order to avoid a metamathematical argument, we will give here a direct mathematical proof of this theorem.

The implication (2) $\Rightarrow$ (1) is trivial. So it only remains to establish the implication (1) $\Rightarrow$ (2). For this purpose, we need Tychonov's celebrated theorem which states that any topological product of quasi-compact spaces is quasi-compact, too (see [82], [240]). Actually, in what follows we will use a very special case of the above-mentioned theorem.

Let us denote by Φ the family of all those functions which act from the set V into the set $\{1, 2, ..., k\}$. Each function belonging to Φ may be treated as a coloring of V by colors from $\{1, 2, ..., k\}$. Equip $\{1, 2, ..., k\}$ with the discrete topology and, accordingly, equip Φ with the standard product topology. In view of Tychonov's theorem, Φ becomes a compact space.

Let f be an arbitrary function belonging to Φ and let X be a subset of V. Recall that the symbol $[X]^2$ stands for the family of all two-element subsets of X.

We shall say that f provides a nice coloring for X if the restriction $f|X$ is such that there exists no edge of the subgraph $(X, E \cap [X]^2)$ of (V, E), both endpoints of which are colored by one color from $\{1, 2, ..., k\}$.

Further, let $[V]^{<\omega}$ be the family of all finite subsets of V and let, for every set $X \in [V]^{<\omega}$, the symbol Φ_X denote the family of all those functions $f \in \Phi$ for which the restriction $f|X$ gives a nice coloring of X.

It is not difficult to verify that the family $\{\Phi_X : X \in [V]^{<\omega}\}$ consists of nonempty closed subsets of Φ and is centered. Consequently, by virtue of the compactness of Φ, there exists at least one function $g \in \Phi$ belonging to all members of this family. Now, it is clear that g yields a k-coloring of V such that the endpoints of every edge of (V, E) are colored by different colors.

This completes the proof of Theorem 1. □

As a straightforward consequence of the above theorem, we have that if the chromatic number of a graph (V, E) strictly exceeds a natural number k, then there exists a finite subgraph of (V, E) whose chromatic number also strictly exceeds k.

The general problem of finding the chromatic number (if it exists) of a given graph (finite or infinite) naturally arises in many interesting and important geometric topics.

The following two examples illustrate this circumstance.

Example 1. Let X be some finite subset of the Euclidean space $\mathbf{R}^m$ with $\mathrm{card}(X) \geq 2$. We may canonically associate with this X a certain graph (V, E). Namely, we first put $V = X$. Then we take any two distinct points $x \in V$ and $y \in V$ and put $\{x, y\} \in E$ if and only if the line segment $[x, y]$ is a diameter of X, i.e.,

$$||x - y|| = \max\{||z - z'|| : z \in X,\ z' \in X\}.$$

As we already know, (V, E) is k-chromatic for some natural number k (because our set X is finite). This fact directly implies that X can be represented in the form of the union of pairwise disjoint sets,

$$X = X_1 \cup X_2 \cup ... \cup X_k,$$

where the diameters of all X_i $(i = 1, 2, ..., k)$ are strictly less than the diameter of X.

In connection with Example 1, it should be recalled that the problem of decomposing a bounded set $X \subset \mathbf{R}^m$ with $\mathrm{card}(X) \geq 2$ into subsets of strictly smaller diameter goes back to Borsuk and has a long and rather dramatic history (see, for instance, [32], [34], [36], [38], [39], [40], [154], [157], [287], [289]). During many years, the conjecture remained open according to which every set in $\mathbf{R}^m$ of finite diameter $d > 0$ can be decomposed into $m + 1$ many parts such that each of them has diameter strictly smaller than d. This conjecture was confirmed for lower dimensions m, e.g., for $m \leq 3$. But finally it turned out that, for sufficiently large dimensions m, the conjecture fails to be true even for finite sets $X \subset \mathbf{R}^m$. Moreover, it was established that the growth of the number of needed parts of strictly smaller diameters is non-polynomial with respect to m, as m tends to infinity. For more details, we again refer the reader to the above-mentioned works (or to Chapter 13 of this book).

Example 2. Consider again the space $\mathbf{R}^m$ and associate with it the graph (V, E) as follows. First, put $V = \mathbf{R}^m$. Further, for any two points $x \in V$ and $y \in V$, stipulate $\{x, y\} \in E$ if and only if $||x - y|| = 1$. A simple geometric argument shows that this graph has a finite chromatic number $k = k(m)$ (see Exercise 1). For low dimensions m, some estimates of $k(m)$ can be indicated. Namely, if $m = 2$, then

$$4 \leq k(m) \leq 7.$$

The above inequalities are not difficult to prove (in Exercise 2 for this chapter the reader is asked to check them). The famous Hadwiger–Nelson problem requires to find the precise value of $k(2)$. Quite recently, it was established with the aid of a certain computer program that

$$k(2) \geq 5.$$

The latter inequality is an essential progress in solving the Hadwiger–Nelson problem. Actually, it was described a subset Z of $\mathbf{R}^2$ with $\mathrm{card}(Z) = 1581$ such that the chromatic number of the graph $(Z, E \cap [Z]^2)$ is greater than or equal to 5. For more details about the latter fact, see [117].

In the case of $\mathbf{R}^3$, a delicate argument is needed for establishing the inequalities

$$6 \leq k(3) \leq 15.$$

The proofs of the latter inequalities can be found in [63], [288].

As regards to the values $k(m)$, where m is sufficiently large, no more or less precise estimates (in terms of m) are known.

Now, we want to touch upon some set-theoretical ideas in connection with Example 2. Obviously, the problem of finding the chromatic number $k(m)$ is equivalent to the problem of finding the minimal natural number k such that the space $\mathbf{R}^m$ admits a decomposition

$$\mathbf{R}^m = X_1 \cup X_2 \cup ... \cup X_k,$$

where no couple of points from X_i $(i = 1, 2, ..., k)$ realizes distance 1. In the above-mentioned decomposition, the sets X_i are not limited in the sense of their descriptive structure, so they can be arbitrary subsets of $\mathbf{R}^m$. But mathematicians already know that such subsets may possess extraordinary and even pathological properties which are implied by certain set-theoretic assumptions connected with various forms of the Axiom of Choice (**AC**), the Continuum Hypothesis (**CH**), or Martin's Axiom (**MA**). So, a priori it may happen that the chromatic number $k(m)$ somehow depends on the class of admissible decompositions of the space $\mathbf{R}^m$. Although this seems to be unbelievable in the concrete situation of $k(m)$, in other cases of graphs, which are associated with the Euclidean structure, certain paradoxical phenomena can occur.

Indeed, as was demonstrated by S. Shelah and A. Soifer in the series of their works [313], [319], [320], the behavior of the chromatic numbers of some graphs over $\mathbf{R}^m$ substantially depends on a set-theoretical base chosen for $\mathbf{R}^m$ and its subsets. Briefly speaking, the values of the chromatic numbers of such graphs can vary and are closely connected with descriptive properties of those subsets of $\mathbf{R}^m$ which are allowed to participate in the corresponding decompositions.

Actually, we will see below that a radical difference occurs when we restrict ourselves to λ_m-measurable subsets of $\mathbf{R}^m$, where λ_m denotes the m-dimensional Lebesgue measure on $\mathbf{R}^m$, and when no restrictions about the descriptive structure of subsets of $\mathbf{R}^m$ are made.

Thus, we recognize the phenomenal role of measurable and nonmeasurable sets in the combinatorial structure of a finite-dimensional Euclidean space (cf. Chapter 11 of the present book).

In this context, it seems reasonable to recall the classical construction of G. Vitali [353] that leads, with the aid of uncountable forms of **AC**, to the existence of Lebesgue nonmeasurable subsets of the real line $\mathbf{R}$ (see also Chapter 11 where a much more general result is discussed).

Let $\mathbf{Q}$ denote, as usual, the set of all rational numbers. Clearly, $(\mathbf{Q}, +)$ is a countable subgroup of the additive group $(\mathbf{R}, +)$. Let us introduce a binary relation $W \subset \mathbf{R} \times \mathbf{R}$ defined by the formula

$$(x, y) \in W \Leftrightarrow x - y \in \mathbf{Q}.$$

It is easy to see that this W is an equivalence relation on $\mathbf{R}$. The graph of the relation W is a rather simple subset of the Euclidean plane $\mathbf{R}^2$. Namely, this graph can be represented as the union of a certain countable family of straight lines lying in $\mathbf{R}^2$ and parallel to the standard line

$$l = \{(x, y) \in \mathbf{R}^2 \ : \ y - x = 0\}.$$

Let us denote by $\{V_i \ : \ i \in I\}$ the partition of $\mathbf{R}$ canonically associated with W. For this partition, we have

$$\mathrm{card}(I) = \mathbf{c}, \qquad (\forall i \in I)(\mathrm{card}(V_i) = \omega),$$

where, as earlier, ω denotes the least infinite cardinal (ordinal) number and $\mathbf{c}$ stands for the cardinality of the continuum.

The family $\{V_i \ : \ i \in I\}$ is usually called the Vitali partition of $\mathbf{R}$ (see [353]; cf. also [137], [150], [151], [355]).

Let X be an arbitrary selector of the Vitali partition. In other words, let X be a subset of $\mathbf{R}$ satisfying the relation

$$\mathrm{card}(X \cap V_i) = 1$$

for each index $i \in I$. Clearly, the existence of such a selector follows immediately from the Axiom of Choice.

In the sequel, X will be called a Vitali subset of the real line (this terminology is commonly used in modern real analysis).

Vitali was the first mathematician to prove, in 1905, that X is not measurable with respect to the classical Lebesgue measure λ $(= \lambda_1)$ on $\mathbf{R}$ (see again [353]).

In order to establish this important fact, let us observe that the following two relations are obviously true:

(1) $\cup\{X + q \ : \ q \in \mathbf{Q}\} = \mathbf{R}$;

(2) if $q \in \mathbf{Q}$, $r \in \mathbf{Q}$ and $q \neq r$, then $(X + q) \cap (X + r) = \emptyset$.

Suppose, for a moment, that X is a λ-measurable set. Then, since λ is invariant under the group $\mathbf{Q}$ and relation (1) holds, we must have $\lambda(X) > 0$. Evidently, there exists a natural number n such that

$$\lambda(X \cap [-n, n]) > 0.$$

Let us put $Y = X \cap [-n, n]$ and consider the set

$$Z = \cup\{Y + q \ : \ q \in \mathbf{Q}, \ |q| < 1\}.$$

Taking into account relation (2) and the invariance of λ with respect to $\mathbf{Q}$, we see that $\lambda(Z) = +\infty$. On the other hand, it is clear that Z is a bounded subset of the real line, so we must have the inequality $\lambda(Z) < +\infty$. Thus, we come to a contradiction which gives us the desired result.

Remark 2. A similar argument (however, slightly more complicated) shows that no Vitali subset of $\mathbf{R}$ has the Baire property with respect to the standard Euclidean topology of $\mathbf{R}$. Necessary information about the Baire property of subsets of a general topological space may be found, for instance, in [82], [151], and [240].

The analogous construction can be carried out for the m-dimensional Euclidean space $\mathbf{R}^m$ and for its everywhere dense countable subgroup $\mathbf{Q}^m$. In particular, the Vitali construction successfully works for the Euclidean plane $\mathbf{R}^2$ and for its "rational part" $\mathbf{Q}^2$. In this manner, we obtain (of course, with the aid of the Axiom of Choice) a subset of $\mathbf{R}^2$ nonmeasurable with respect to the standard two-dimensional Lebesgue measure λ_2 on $\mathbf{R}^2$. However, it should be noticed that, by virtue of an isomorphism between the measure structures of λ_1 and λ_2, the two-dimensional Vitali construction turns out to be superfluous for proving the existence of a λ_2-nonmeasurable set in $\mathbf{R}^2$ (in this connection, see Exercise 8).

Furthermore, the argument presented above enables one to prove a more general statement. In order to formulate it, we need the notion of a set of Vitali's type.

Let Γ be a subgroup of the additive group $\mathbf{R}^m$, where $m \geq 1$. Consider the partition of $\mathbf{R}^m$ canonically associated with the equivalence relation

$$x \in \mathbf{R}^m \ \& \ y \in \mathbf{R}^m \ \& \ x - y \in \Gamma.$$

Let X be any selector of this partition. We shall say in our further considerations that X is a Γ-selector (or that X is a set of Vitali's type with respect to the group $\Gamma \subset \mathbf{R}^m$).

An appropriate modification of the preceding argument leads to the following result.

Theorem 2. *Let Γ be a countable everywhere dense subgroup of the additive group $\mathbf{R}^m$, where $m \geq 1$, and let μ be a measure defined on some σ-algebra of subsets of $\mathbf{R}^m$. Suppose that these three conditions are satisfied:*

(1) μ is an invariant measure with respect to Γ; in other words, for all $g \in \Gamma$ and for all $Z \in \mathrm{dom}(\mu)$, we have $g + Z \in \mathrm{dom}(\mu)$ and $\mu(g + Z) = \mu(Z)$;

(2) $[0,1]^m \in \mathrm{dom}(\mu)$;

(3) $0 < \mu([0,1]^m) < +\infty$.

Then every Γ-selector is nonmeasurable with respect to μ.

The proof of this result is very similar to the argument presented above and is left to the reader.

We also want to remark that conditions (1), (2), and (3) of Theorem 2 imply the following fact:

The completion of the given measure μ is an extension of some measure on $\mathbf{R}^m$ which is proportional to the m-dimensional Lebesgue measure λ_m. More precisely, the completion of μ is an extension of the measure $t\lambda_m$, where

$$t = \mu([0,1]^m).$$

Now, following the main idea of Vitali's construction (with some necessary modifications), let us define a concrete graph (V, E) within **ZF** set theory.

First of all, let us put $V = \mathbf{R}$. Then take any two points $x \in V$ and $y \in V$ and put $\{x, y\} \in E$ if and only if

$$x - y \in \mathbf{Q} + 2^{1/2} \quad \vee \quad y - x \in \mathbf{Q} + 2^{1/2}.$$

Our goal is to show, within **ZFC** set theory, that the graph (V, E) is bichromatic.

For this purpose, introduce the countable group

$$G = \mathbf{Q} + (2^{1/2})\mathbf{Z} \subset \mathbf{R},$$

where $\mathbf{Z}$ denotes, as usual, the ring of all integers, and consider the equivalence relation in $\mathbf{R}$ canonically associated with G. As known, the equivalence classes coincide with the G-orbits in $\mathbf{R}$. Let S be a selector of the quotient set $\mathbf{R}/G$. According to Theorem 2, the set S is not measurable with respect to the Lebesgue measure λ on $\mathbf{R}$ and, moreover, S turns out to be absolutely nonmeasurable with respect to the class of all those measures on $\mathbf{R}$ which are translation invariant extensions of λ.

Clearly, the selector S produces a certain function

$$f : \mathbf{R} \to \mathbf{R}$$

defined by the formula

$$f(v) \in S \cap (G + v) \qquad (v \in \mathbf{R}).$$

It directly follows from the definition of this f that if $v \in \mathbf{R}$, then

$$v - f(v) = q + 2^{1/2}n$$

for some uniquely determined rational number q and for some uniquely determined integer n. Evidently, we may write

$$n = 2k + l,$$

where $k \in \mathbf{Z}$ and $l \in \{0, 1\}$. Thus, we are able to associate to each $v \in V$ the index (color) $l = l(v) \in \{0, 1\}$. Since the latter index ranges over the two-element set

$$\tau = \{0, 1\},$$

we get the partition of V $(= \mathbf{R})$ into its two subsets V_0 and V_1 which correspond, respectively, to the elements of τ. Actually, here τ plays the role of the set of two colors for an appropriate coloring of V.

Now, we can formulate and prove the following statement due to Shelah and Soifer (see [313], [320]).

Theorem 3. *For each edge $e \in E$ of the above-mentioned graph (V, E), one has $e \cap V_0 \neq \emptyset$ and $e \cap V_1 \neq \emptyset$.*

Consequently, (V, E) is a bichromatic graph.

Proof. Take any two vertices v and u from V for which $\{v, u\} \in E$. Obviously, we may write

$$v - f(v) = q + 2^{1/2}n \in G,$$
$$u - f(u) = q' + 2^{1/2}n' \in G,$$

where

$$q \in \mathbf{Q}, \quad q' \in \mathbf{Q}, \quad n \in \mathbf{Z}, \quad n' \in \mathbf{Z}.$$

Suppose to the contrary that both v and u have the same color from τ. In view of $\{v, u\} \in E$, the difference $v - u$ satisfies the relation

$$v - u \in \mathbf{Q} + 2^{1/2} \quad \vee \quad u - v \in \mathbf{Q} + 2^{1/2}.$$

and, in particular, $v - u$ belongs to the group G. It is easy to see that

$$(v - u) - (f(v) - f(u)) \in G,$$

so we obtain

$$f(v) - f(u) \in G.$$

Further, $f(v) \in S$ and $f(u) \in S$. Since S is a selector of the quotient set $\mathbf{R}/G$, we must have $f(v) = f(u)$ and, therefore,

$$v - u = (q - q') + 2^{1/2}(n - n').$$

The last equality directly leads to a contradiction. For example, assuming that

$$v - u \in \mathbf{Q} + 2^{1/2},$$

we come to the relation

$$q'' + 2^{1/2} = (q - q') + 2^{1/2}(n - n')$$

for some rational number q''. Since $2^{1/2}$ is irrational, we then infer that

$$q'' = q - q', \quad n - n' = 1,$$

which shows, in particular, that v and u cannot be colored by the same color from τ (indeed, if n is odd, then n' is even; similarly, if n is even, then n' is odd).

The impossibility of the case $u - v \in \mathbf{Q} + 2^{1/2}$ can be established quite analogously, because it yields $n' - n = 1$.

Theorem 3 has thus been proved. □

Now, let us try to examine the value of the chromatic number of the same graph (V, E) in a certain weak fragment of **ZFC** set theory.

Notice that (V, E) is defined effectively, i.e., without the aid of **AC**. So, it makes sense to pose the question whether the chromatic number of (V, E) preserves its value in various extensions of **ZF** set theory or in extensions of **ZF** & **DC** set theory. The latter theory is more relevant for classical mathematical analysis and elementary point set topology (see [23], [137], [150], [316] for further information about the principle of dependent choices **DC** which is a slightly strengthened form of the countable version of **AC**).

Some important remarks should be made in this context. First of all, notice that even the existence of a chromatic number (defined as the least cardinal with the desired property) for all infinite graphs cannot be established without the help of **AC**. Of course, assuming **AC**, the question becomes trivial, because in any nonempty family of cardinal numbers always there is the least one. But if we work in a weaker fragment of **ZFC** theory, then it may happen that a given infinite graph does not possess the chromatic number or its chromatic number has another value in a different weaker fragment of **ZFC**.

To be more concrete, let us consider the theory

$$\mathbf{ZF}\ \&\ \mathbf{DC}\ \&\ (\text{every subset of } \mathbf{R} \text{ is Lebesgue measurable}).$$

According to Solovay's classical result (see [321]), this theory is equiconsistent with the theory

$$\mathbf{ZF}\ \&\ \mathbf{DC}\ \&\ (\text{there exists a strongly inaccessible cardinal number}).$$

For further details, we refer the reader to [151], [321].

We now are going to show that if all subsets of $\mathbf{R}$ are Lebesgue measurable, then the same graph (V, E) either does not possess the chromatic number or its chromatic number is uncountable (cf. Remark 1 made at the beginning of the present chapter).

To establish this circumstance, we need one simple auxiliary proposition from classical real analysis.

Lemma 1. *If a set $X \subset \mathbf{R}$ is λ-measurable and has strictly positive λ-measure, then there are two points $x \in X$ and $y \in X$ which are adjacent in the above-mentioned graph (V, E).*

Proof. Since $\lambda(X) > 0$, there are density points in X (see, e.g., [49]). Let z be one of such points. There exists an open interval $D \subset \mathbf{R}$ whose midpoint coincides with z and for which

$$\lambda(X \cap D) > (3/5)\lambda(D).$$

Notice also that in this inequality the length $d = \lambda(D)$ of D can be taken arbitrarily small. Further, choose a rational number q so that

$$2^{1/2} < q < 2^{1/2} + d/10$$

and consider the point $t = q - 2^{1/2}$ in $\mathbf{R}$. It is easy to verify that

$$\lambda(((D \cap X) + t) \cap (D \cap X)) > 0.$$

Consequently, we get

$$((D \cap X) + t) \cap (D \cap X) \neq \emptyset,$$

$$(X + t) \cap X \neq \emptyset.$$

This implies that there are points $x \in X$ and $y \in X$ for which $x - y = t$ and, therefore, $\{x, y\} \in E$.

Lemma 1 has thus been proved. □

Now, we are ready to formulate and prove the following statement.

Theorem 4. *Working in $\mathbf{ZF}$ & $\mathbf{DC}$ theory, assume that all subsets of the real line $\mathbf{R}$ are measurable with respect to the Lebesgue measure λ $(= \lambda_1)$.*

Then the disjunction of these two assertions holds true:

(1) the graph (V, E) of Theorem 3 does not possess the chromatic number;

(2) the chromatic number of (V, E) is uncountable.

Proof. If (V, E) does not possess the chromatic number, then we are done. Assume now that the chromatic number exists for (V, E) and let us show that it cannot be countable. In this case we argue on the contrary, i.e., we suppose that there exists a countable partition $\{X_n : n < \omega\}$ of V such that the endpoints of any edge $e \in E$ lie in different sets X_n and X_k, where n and k depend, of course, on e. At the same time, we have

$$V = \mathbf{R} = \cup\{X_n : n < \omega\},$$

from which it follows that at least one X_k is of strictly positive λ-measure. By virtue of Lemma 1, we get that there are two adjacent vertices from X_k, contradicting the definition of $\{X_n : n < \omega\}$.

The contradiction obtained ends the proof of Theorem 4. □

It is natural to ask whether the above construction can be generalized to Euclidean spaces of higher dimension. It turns out that the answer is positive, because a very similar argument is applicable to the space $\mathbf{R}^m$, where $m \geq 2$. To illustrate this circumstance, we restrict ourselves to the case of the plane $\mathbf{R}^2$ (leaving the case of a multi-dimensional Euclidean space to the reader).

Analogously to the situation with $\mathbf{R}$, we are going to define within **ZF** set theory a concrete graph (V, E) over $\mathbf{R}^2$. First of all, let us put

$$V = \mathbf{R}^2 = \mathbf{R} \times \mathbf{R},$$

so the set of vertices of our graph coincides with the entire plane $\mathbf{R}^2$. Then let us fix the following two points (vectors) in $\mathbf{R}^2$:

$$e_1 = (2^{1/2}, 0), \qquad e_2 = (0, 2^{1/2}).$$

Further, by using these two vectors, we define the set E of edges. Namely, for any two distinct points $u \in \mathbf{R}^2$ and $v \in \mathbf{R}^2$, we stipulate $\{u, v\} \in E$ if and only if at least one of these four relations is valid:

(a) $u - v \in \mathbf{Q}^2 + e_1$;

(b) $v - u \in \mathbf{Q}^2 + e_1$;

(c) $u - v \in \mathbf{Q}^2 + e_2$;

(d) $v - u \in \mathbf{Q}^2 + e_2$.

Proceeding in such a manner, we come to the required graph (V, E).

Our goal is to show, within **ZFC** set theory, that the chromatic number of the introduced graph does not exceed 4.

For this purpose, define the countable group

$$G = \mathbf{Q}^2 + (2^{1/2})\mathbf{Z}^2 \subset \mathbf{R}^2,$$

where $\mathbf{Z}$ denotes again the set of all integers, and consider the equivalence relation in $\mathbf{R}^2$ canonically associated with G. As we know, the corresponding equivalence classes coincide with the G-orbits in $\mathbf{R}^2$. Let S be a selector of $\mathbf{R}^2/G$. According to Theorem 2, the set S is not measurable with respect to the Lebesgue measure λ_2 on $\mathbf{R}^2$ and, moreover, S turns out to be absolutely nonmeasurable with respect to the class of all those measures on $\mathbf{R}^2$ which are translation invariant extensions of λ_2.

As in the case of $\mathbf{R}$, the selector S produces a certain function

$$f : \mathbf{R}^2 \to \mathbf{R}^2$$

defined by the formula

$$f(v) \in S \cap (G + v) \qquad (v \in \mathbf{R}^2).$$

It directly follows from the definition of this f that if $v \in \mathbf{R}^2$, then

$$v - f(v) = (q_1 + 2^{1/2}n_1, q_2 + 2^{1/2}n_2)$$

for some uniquely determined rational numbers q_1 and q_2, and for some uniquely determined integers n_1 and n_2.

Consider the pair (n_1, n_2). Evidently, we may write

$$n_1 = 2k_1 + l_1, \qquad n_2 = 2k_2 + l_2,$$

where

$$k_1 \in \mathbf{Z}, \quad k_2 \in \mathbf{Z}, \quad l_1 \in \{0, 1\}, \quad l_2 \in \{0, 1\}.$$

Thus, we are able to associate to each $v \in V$ the pair (l_1, l_2). Since all such pairs range over the fixed four-element set

$$\tau = \{(0, 0), \ (0, 1), \ (1, 0), \ (1, 1)\},$$

we have the partition of V into its four subsets

$$V_{00}, \quad V_{01}, \quad V_{10}, \quad V_{11},$$

which correspond, respectively, to the elements of τ (as before, τ plays the role of the set of colors for an appropriate coloring of V).

Now, we can prove the following statement due to Shelah and Soifer (see [313], [320]).

Theorem 5. *There exists no edge of the above-mentioned graph* (V, E) *whose vertices belong to exactly one member of the family*

$$\{V_{00}, \quad V_{01}, \quad V_{10}, \quad V_{11}\}.$$

Consequently, the chromatic number of (V, E) *does not exceed* 4.

Proof. Take any two vertices v and u from V for which $\{v, u\} \in E$. Obviously, we may write

$$v - f(v) = (q_1 + 2^{1/2}n_1, q_2 + 2^{1/2}n_2) \in G,$$

$$u - f(u) = (q_1' + 2^{1/2}n_1', q_2' + 2^{1/2}n_2') \in G,$$

where

$$(q_1, q_2) \in \mathbf{Q}^2, \quad (q_1', q_2') \in \mathbf{Q}^2, \quad (n_1, n_2) \in \mathbf{Z}^2, \quad (n_1', n_2') \in \mathbf{Z}^2.$$

Suppose to the contrary that both v and u have the same color. In view of $\{u, v\} \in E$, the difference $v - u$ satisfies one of the relations (a), (b), (c), (d) and, in particular, $v - u$ belongs to the group G. It is easy to see that

$$(v - u) - (f(v) - f(u)) \in G,$$

so we get

$$f(v) - f(u) \in G.$$

Further, $f(v) \in S$ and $f(u) \in S$. Since S is a selector of the quotient set $\mathbf{R}^2/G$, we must have $f(v) = f(u)$ and, therefore,

$$v - u = ((q_1 - q_1') + 2^{1/2}(n_1 - n_1'), (q_2 - q_2') + 2^{1/2}(n_2 - n_2')).$$

The last equality immediately leads to a contradiction. For example, assuming that (a) holds true for $v - u$, we come to the relation

$$(q_1'' - 2^{1/2}, q_2'') = ((q_1 - q_1') + 2^{1/2}(n_1 - n_1'), (q_2 - q_2') + 2^{1/2}(n_2 - n_2'))$$

for some rational numbers q_1'' and q_2''. Since $2^{1/2}$ is irrational, we infer that

$$q_1'' = q_1 - q_1', \quad n_1 - n_1' = -1, \quad n_2 = n_2',$$

which shows, in particular, that v and u cannot be colored by the same color from τ (cf. the final part of the proof of Theorem 3).

Analogously, the impossibility of relations (b), (c), and (d) can be established.

Theorem 5 has thus been proved. □

Now, we need the next simple auxiliary proposition which is a straightforward two-dimensional analogue of Lemma 1.

Lemma 2. *If a set $X \subset \mathbf{R}^2$ is λ_2-measurable and has strictly positive λ_2-measure, then there are two points $x \in X$ and $y \in X$ which are adjacent in the graph (V, E).*

Proof. Since $\lambda_2(X) > 0$, there are density points in X (see, e.g., [49]). Let z be one of such points. There exists a square $D \subset \mathbf{R}^2$ whose center coincides with z, whose sides are respectively parallel to $\mathbf{R} \times \{0\}$ and $\{0\} \times \mathbf{R}$, and for which

$$\lambda_2(X \cap D) > (3/5)\lambda_2(D).$$

Notice that in this inequality the length d of a side of D can be taken arbitrarily small. Now, choose again a rational number q so that

$$2^{1/2} < q < 2^{1/2} + d/10.$$

and consider the vector $w = (0, q - 2^{1/2})$ in $\mathbf{R}^2$. It is easy to verify that

$$\lambda_2(((D \cap X) + w) \cap (D \cap X)) > 0.$$

Consequently, we have

$$((D \cap X) + w) \cap (D \cap X) \neq \emptyset,$$

$$(X + w) \cap X \neq \emptyset.$$

This implies that there are points $x \in X$ and $y \in X$ for which $x - y = w$ and, therefore, $\{x, y\} \in E$.

The proof of Lemma 2 is thus complete. □

Finally, we are ready to prove the following statement (see again the articles [313] and [320]).

Theorem 6. *Working in* **ZF** & **DC** *theory, assume that all subsets of the plane* $\mathbf{R}^2$ *are measurable with respect to the Lebesgue measure* λ_2.

Then the disjunction of these two assertions holds true:

(1) the graph (V, E) *of Theorem 5 does not possess the chromatic number;*

(2) the chromatic number of (V, E) *is uncountable.*

Proof. The argument is almost identical with the proof of Theorem 4. If (V, E) does not possess the chromatic number, then we are done. In the case of the existence of the chromatic number for (V, E), let us show that it cannot be countable. Arguing to the contrary, suppose that the chromatic number of (V, E) is countable and consider a countable partition $\{X_n : n < \omega\}$ of V such that no edge from E connects two vertices lying in one member of this partition. At the same time, we have

$$\mathbf{R}^2 = V = \cup\{X_n : n < \omega\},$$

from which it follows that at least one set X_k is of strictly positive λ_2-measure. By virtue of Lemma 2, we get that there are two adjacent vertices from this X_k, contradicting the definition of $\{X_n : n < \omega\}$.

The contradiction obtained ends the proof of Theorem 6. □

EXERCISES

1. Under the notation of Example 2, demonstrate within **ZF** theory that the chromatic number $k(m)$ exists for any dimension $m \geq 1$.

For this purpose, consider an appropriate decomposition of the space $\mathbf{R}^m$ into pairwise congruent m-dimensional cubes.

2*. Work in **ZF** theory and check the validity of the following relations:

(a) $k(1) = 2$;

(b) $k(2) \geq 4$;

(c) $k(2) \leq 7$.

For establishing (b), suppose otherwise, i.e., suppose that there exists a covering $\{A, B, C\}$ of the plane $\mathbf{R}^2$ such that no set from $\{A, B, C\}$ realizes distance 1, and show that there are two points x and y in $\mathbf{R}^2$ for which

$$||x - y|| = 3^{1/2},$$
$$\{x, y\} \notin [A]^2 \cup [B]^2 \cup [C]^2.$$

Then obtain a contradiction.

For establishing (c), consider a tiling of $\mathbf{R}^2$ by pairwise congruent hexagons with an appropriate diameter.

3. Work in **ZF** theory and, for any natural number $m \geq 2$, verify the validity of the inequality $k(m) \geq m + 2$ (cf. (b) of Exercise 2).

4. Give a detailed proof of Theorem 2.

5*. Preserving the notation of Theorem 2, show that the completion of the measure μ is an extension of some measure on $\mathbf{R}^m$ which is proportional to λ_m. In other words, the completion of μ is an extension of the measure $t\lambda_m$ where $t = \mu([0, 1]^m)$.

For this purpose, assume without loss of generality that $\mu([0, 1]^m) = 1$ and step by step verify that

$$\mu([p_1, q_1] \times [p_2, q_2] \times ... \times [p_m, q_m]) = (q_1 - p_1)(q_2 - p_2)...(q_m - p_m),$$

for any two sequences $(p_1, p_2, ..., p_m)$ and $(q_1, q_2, ..., q_m)$ of rational numbers, where

$$p_1 < q_1, \quad p_2 < q_2, \quad . . . , \quad p_m < q_m.$$

Finally, conclude that μ extends the restriction of λ_m to the Borel σ-algebra of $\mathbf{R}^m$.

6*. Formulate and prove an appropriate analogue of Theorem 2 in terms of the Baire property.

For this purpose, take into account the following topological fact:

If a set $X \subset \mathbf{R}^m$ has the Baire property and is not of first category in $\mathbf{R}^m$, then there exists a real $\varepsilon > 0$ such that

$$X \cap (X + h) \neq \emptyset$$

for every $h \in \mathbf{R}^m$ with $||h|| < \varepsilon$.

7*. Show that the cardinality of the quotient set $\mathbf{R}/\mathbf{Q}$ is equal to $\mathbf{c}$, but this equality cannot be established without appealing to uncountable forms of **AC**.

More precisely, work in **ZF** & **DC** set theory and show that if $\mathbf{R}/\mathbf{Q}$ can be linearly ordered, then there exists a Lebesgue nonmeasurable subset of $\mathbf{R}$.

Remark 3. The result of Exercise 7 is due to Sierpiński (see [316] and the references therein).

8. Work in **ZF** & **DC** theory and demonstrate that if there exists a subset of $\mathbf{R}$ nonmeasurable with respect to λ, then there exists a λ_m-nonmeasurable subset of the space $\mathbf{R}^m$, where $m \geq 2$.

Do this in two different manners:

(a) by using Fubini's classical theorem;

(b) by using the well-known fact that λ_1 and λ_m are isomorphic measures (in the same theory).

9*. Work in **ZF** theory and give an example of a graph (V, E) such that:

(a) $V = \mathbf{R}^m$, where $m \geq 3$;

(b) the chromatic number of (V, E) does not exceed 2^m in **ZFC** set theory;

(c) in **ZF** & **DC** theory, if all subsets of $\mathbf{R}^m$ are measurable with respect to λ_m, then either (V, E) does not possess the chromatic number or its chromatic number is uncountable.

For this purpose, argue similarly to the proofs of Theorem 5 and Theorem 6 of the present chapter.

Remark 4. In connection with Exercise 9, see again [313], [319], [320].

CHAPTER 10

The Szemerédi–Trotter theorem and its applications

In the Euclidean plane $\mathbf{R}^2$ consider two families of geometric objects:

P - a finite family (system) of points;

$\mathcal{L}$ - a finite family (system) of straight lines.

For these two families we introduce the following notation:

$$p = \text{card}(P), \quad l = \text{card}(\mathcal{L}).$$

The above notation will be used throughout this chapter.

According to the standard terminology, we shall say that a point $x \in P$ and a line $y \in \mathcal{L}$ are incident if $x \in y$.

In this case, (x, y) will be called an incidence pair (with respect to the two given families P and $\mathcal{L}$).

The symbol $J(P, \mathcal{L})$ will denote the total number of incidence pairs with respect to P and $\mathcal{L}$, i.e., we have the equality

$$J(P, \mathcal{L}) = \text{card}(\{(x, y) : x \in P, y \in \mathcal{L}, x \in y\}).$$

Observe that the relation

$$J(P, \mathcal{L}) = J(P, \mathcal{L}_1) + J(P, \mathcal{L}_2)$$

is valid whenever $\{\mathcal{L}_1, \mathcal{L}_2\}$ is a partition of $\mathcal{L}$.

Analogously, the relation

$$J(P, \mathcal{L}) = J(P_1, \mathcal{L}) + J(P_2, \mathcal{L})$$

is valid whenever $\{P_1, P_2\}$ is a partition of P.

DOI: 10.1201/9781003458708-10

Obviously, we may write $J(P, \mathcal{L}) \leq p \cdot l$, but this trivial estimate is very far from being more or less precise.

The main goal of the present chapter is to give a delicate proof of the more precise estimate of $J(P, \mathcal{L})$ which was obtained by Szemerédi and Trotter in their remarkable work [336].

For our further purposes, we need one auxiliary proposition concerning the crossing number of a finite graph (see Exercise 12 from Appendix 5). This proposition is due to Székely [335].

Lemma 1. *Let $G = (V, E)$ be a finite graph with $v = \mathrm{card}(V) \geq 3$ and suppose that $4v \leq e = \mathrm{card}(E)$.*

Then the following inequality holds true:

$$c(V, E) \geq \frac{e^3}{64v^2},$$

where the symbol $c(V, E) = c_G$ stands for the crossing number of $G = (V, E)$.

Proof. Denote $q = 4v/e$. According to our assumption, $q \leq 1$. We will be dealing with random subgraphs of G. More precisely, H is a random subgraph of G if all vertices of H are taken mutually independently from V, each with probability q, and all edges of H are induced by corresponding edges of G.

For such a random graph H, we have its basic parameters:

$v_H =$ the number of vertices of H;

$e_H =$ the number of edges of H;

$c_H =$ the crossing number of H.

Moreover, for these three random variables v_H, e_H and c_H the inequality

$$c_H \geq e_H - 3v_H$$

is valid (see Exercise 16 of Appendix 5).

Denoting the (mathematical) expectation of random variables by the standard symbol $\mathbf{E}$, we readily infer from the above inequality that

$$\mathbf{E}(c_H) \geq \mathbf{E}(e_H) - 3\mathbf{E}(v_H).$$

Now, keeping in mind the relations

$$\mathbf{E}(v_H) = qv, \quad \mathbf{E}(e_H) = q^2 e, \quad \mathbf{E}(c_H) \leq q^4 c_G$$

(see Exercise 1 of this chapter), we obtain

$$q^4 c_G \geq q^2 e - 3qv.$$

Remembering that $q = 4v/e$ and making some straightforward calculations, we finally come to the required inequality

$$c(V, E) \geq \frac{e^3}{64v^2}.$$

This completes the proof of Lemma 1. □

By using this lemma, we are able to deduce the following important statement due to Szemerédi and Trotter [336].

Theorem 1. *Under the notation introduced at the beginning of this chapter, the formula*

$$J(P, \mathcal{L}) = O((pl)^{2/3} + p + l)$$

holds true (and, moreover, the above formula is asymptotically precise).

Proof. First, consider the case when each line from $\mathcal{L}$ contains at least two points from P.

Take any line $y \in \mathcal{L}$. There are some points on y which belong to P. Obviously, those points divide y into two rays and finitely many line segments (without common interior points).

We will be dealing with the graph G whose vertices are all points of P and whose edges are all the above-mentioned line segments (when y ranges over $\mathcal{L}$).

Let e denote the number of all edges of G. It is easy to see that

$$J(P, \mathcal{L}) = l + e.$$

Here only two cases are possible.

Case 1. $e \leq 4p$.

In this case, the required formula is trivially valid, because

$$J(P, \mathcal{L}) = l + e \leq l + 4p \leq 4(p + l).$$

Case 2. $e \geq 4p$.

In this case, applying Lemma 1 to the graph G, we get

$$c_G \geq \frac{e^3}{64p^2},$$

where c_G denotes the crossing number of G. At the same time, it is clear that $c_G \leq l^2$, whence it follows

$$e \leq 4l^{2/3}p^{2/3},$$

$$J(P, \mathcal{L}) = l + e \leq l + 4l^{2/3}p^{2/3}.$$

We see that the required formula is valid again.

Now, consider the general case of P and $\mathcal{L}$.

Let $\mathcal{L}_1$ denote the subfamily of $\mathcal{L}$ consisting of all those lines which contain at most one point from P. Denote also by $\mathcal{L}_2$ the subfamily of $\mathcal{L}$ consisting of all those lines which contain at least two points from P.

Obviously, we may write

$$J(P, \mathcal{L}) = J(P, \mathcal{L}_1) + J(P, \mathcal{L}_2).$$

Further, introduce the notation

$$k = \mathrm{card}(\mathcal{L}_1).$$

Then $k \leq l$ and the relation

$$J(P, \mathcal{L}) \leq k + J(P, \mathcal{L}_2)$$

is readily verified.

For P and $\mathcal{L}_2$, we have already proved that

$$J(P, \mathcal{L}_2) \leq d((p(l-k))^{2/3} + p + (l-k)),$$

where $d > 0$ is some absolute real constant. Therefore,

$$J(P, \mathcal{L}) \leq k + d((p(l-k))^{2/3} + p + (l-k)).$$

Keeping in mind the trivial inequality

$$k + d((p(l-k))^{2/3} + p + (l-k)) \leq (d+1)((pl)^{2/3} + p + l),$$

we finally obtain

$$J(P, \mathcal{L}) \leq (d+1)((pl)^{2/3} + p + l),$$

where $d + 1 > 0$ is also an absolute real constant.

The Szemerédi–Trotter theorem has thus been proved. □

The next two examples show that the upper estimate in the Szemerédi–Trotter formula is precise in a certain sense (at least, asymptotically).

Example 1. Let $p \geq 2$ be a natural number and let P be a set of points in general position in the plane $\mathbf{R}^2$, such that $\text{card}(P) = p$. Let $\mathcal{L}$ denote the family of all lines determined by P. Then we obviously have

$$\text{card}(\mathcal{L}) = p(p-1)/2, \quad J(P, \mathcal{L}) = p(p-1).$$

Therefore, the order of growth of $J(P, \mathcal{L})$ is the same as the order of growth of

$$(pl)^{2/3} + p + l,$$

where $p = \text{card}(P)$ and $l = \text{card}(\mathcal{L})$.

Example 2. Let $k \geq 1$ be a natural number and let $p = 4k^3$. Consider in the plane $\mathbf{R}^2$ the set of points

$$P = \{(i, j) : i = 0, 1, ..., k-1;\ j = 0, 1, ..., 4k^2 - 1\}.$$

Obviously, we have the equality $\text{card}(P) = p$.

Also, consider the set $\mathcal{L}$ of lines in $\mathbf{R}^2$ which are given, respectively, by the formulae

$$y = ax + b \quad (a = 0, 1, 2, ..., 2k-1;\ b = 0, 1, 2, ..., 2k^2 - 1).$$

Clearly, $\text{card}(\mathcal{L}) = p$. Pick an index

$$i \in \{0, 1, ..., k-1\}$$

and take an arbitrary line $y = ax + b$ from $\mathcal{L}$. In view of the inequalities

$$ai + b < ak + b < 2k^2 + 2k^2 = 4k^2,$$

we infer that the line $y = ax + b$ contains some point $(i, j) \in P$. This fact implies the relation

$$J(P, \mathcal{L}) \geq k \cdot \mathrm{card}(\mathcal{L}) = (p/4)^{1/3} p = (1/4)^{1/3} p^{4/3},$$

which again shows the asymptotical exactness of the Szemerédi–Trotter formula.

Naturally, the Szemerédi–Trotter theorem has many applications and useful geometric consequences. One of them will be discussed below (some other applications are presented in exercises for this chapter).

Let P be a finite family of points in $\mathbf{R}^2$ with $\mathrm{card}(P) = p$.

If all points from P are in general position, then the number of straight lines determined by P is obviously equal to $p(p-1)/2$.

On the other hand, if all points from P are collinear, then there is only one line determined by P. More generally, if the number of lines determined by P is small (say, it does not exceed some constant $c > 0$), then there exists a line determined by P and containing at least $\lfloor p/c \rfloor$ points.

It turns out that the above two extreme possibilities are typical. More precisely, it was demonstrated by Beck [17] that there exist two absolute real constants $a > 0$ and $b > 0$ such that, for any natural number $p \geq 2$ and for any point set $P \subset \mathbf{R}^2$ with $\mathrm{card}(P) = p$, the disjunction of these two assertions holds true:

(1) there is a straight line determined by P and containing at least $\lfloor p/a \rfloor$ points from P;

(2) there are at least $\lfloor p^2/b \rfloor$ straight lines determined by P.

This disjunction can be deduced starting from the Szemerédi–Trotter theorem. For this purpose, we need several auxiliary statements.

Lemma 2. *Let P be a nonempty finite point set in $\mathbf{R}^2$ with $\mathrm{card}(P) = p$, let $k \in [2, p]$ be a natural number, and let $l = l(k, P)$ denote the number of all those lines determined by P, each of which contains at least k points from P.*

Then the formula

$$l = l(k, P) = O(p^2/k^3 + p/k)$$

is valid.

Proof. Let $\mathcal{L} = \mathcal{L}(k, P)$ denote the family of all those lines determined by P, each of which contains at least k points from P. Thus, we have by definition

$$l = l(k, P) = \mathrm{card}(\mathcal{L}).$$

According to the Szemerédi–Trotter theorem, we may write

$$J(P, \mathcal{L}) = O((pl)^{2/3} + p + l).$$

In view of the trivial inequality

$$l \leq p(p-1)/2 \leq p^2,$$

we get $(pl)^{2/3} \geq l$, so

$$J(P, \mathcal{L}) = O((pl)^{2/3} + p).$$

Also, it is clear that

$$k \cdot l \leq J(P, \mathcal{L}) = O((pl)^{2/3} + p).$$

Consider now the two possible cases.

Case 1. $l \leq p^{1/2}$.

In this case, we have

$$(pl)^{2/3} \leq p, \qquad kl \leq O(2p) = O(p),$$

whence it follows that $l = O(p/k)$.

Case 2. $p^{1/2} \leq l$.

In this case, we have

$$p \leq (pl)^{2/3}, \qquad kl \leq O(2(pl)^{2/3}) = O((pl)^{2/3}),$$

whence it follows that $l = O(p^2/k^3)$.

Summarizing all the facts above, we obtain the desired estimate

$$l = l(k, P) = O(p^2/k^3 + p/k).$$

This completes the proof of Lemma 2. □

Remark 1. If $k \leq p^{1/2}$, then $p^2/k^3 \geq p/k$, and in this situation Lemma 2 immediately gives the estimate

$$l(k, P) = O(p^2/k^3).$$

Similarly, if $k \geq p^{1/2}$, then $p^2/k^3 \leq p/k$, and in this situation Lemma 2 immediately provides the estimate

$$l(k, P) = O(p/k).$$

We will use the above almost trivial observation in our further considerations.

Lemma 3. *Let P be a finite point set in $\mathbf{R}^2$ with $\text{card}(P) = p \geq 2$, let k be a natural number from the closed interval $[2, p^{1/2}]$, and let $\mathcal{L}$ denote the family of all those lines determined by P, each of which contains at least k points from P and at most $p^{1/2}$ points from P. Further, denote by $\mathcal{P}$ the family of all those two-element subsets of P which lie on some line from $\mathcal{L}$.*

Then the formula

$$\text{card}(\mathcal{P}) = O(p^2/k)$$

holds true.

Proof. Consider the following closed intervals:

$$[k, 2k],\ [2k, 4k],\ \ldots,\ [2^{i-1}k, 2^i k],\ \ldots,\ [2^{r-1}k, 2^r k],$$

where r is a natural number such that $2^{r-1}k \le p^{1/2} \le 2^r k$.

For any natural number $i \in [1, r]$, let $\mathcal{L}_i$ stand for the family of all those lines determined by P, each of which contains at least $2^{i-1}k$ points from P and at most $2^i k$ points from P. Obviously, we have

$$\mathcal{L} \subset \cup\{\mathcal{L}_i : 1 \le i \le r\}.$$

Further, for any natural number $i \in [1, r]$, denote by $\mathcal{P}_i$ the family of all those two-element subsets of P which lie on some line from $\mathcal{L}_i$. Clearly, we may write

$$\mathrm{card}(\mathcal{P}) \le \sum\{\mathrm{card}(\mathcal{P}_i) : 1 \le i \le r\} \le$$

$$\sum\{2^{i-1}k(2^i k - 1)\mathrm{card}(\mathcal{L}_i) : 1 \le i \le r\}.$$

By virtue of Lemma 2 and Remark 1, we have

$$\mathrm{card}(\mathcal{L}_i) \le O(p^2/(2^{3(i-1)}k^3)) = (1/2^{3i})O(p^2/k^3).$$

Consequently,

$$2^{i-1}k(2^i k - 1)\mathrm{card}(\mathcal{L}_i) = (1/2^i)O(p^2/k),$$

whence it follows that

$$\mathrm{card}(\mathcal{P}) \le (1/2 + 1/4 + 1/8 + ... + 1/2^r)O(p^2/k) = O(p^2/k),$$

which ends the proof of Lemma 3. □

The next auxiliary proposition is analogous to Lemma 3.

Lemma 4. *Let P be a finite point set in $\mathbf{R}^2$ with $\mathrm{card}(P) = p \ge 2$, let k be a natural number from the closed interval $[p^{1/2}, p]$, and let $\mathcal{L}$ denote the family of all those lines determined by P, each of which contains at least $p^{1/2}$ points from P and at most k points from P. Further, denote by $\mathcal{P}$ the family of all those two-element subsets of P which lie on some line from $\mathcal{L}$.*

Then the formula

$$\mathrm{card}(\mathcal{P}) = O(pk)$$

holds true.

Proof. Consider the following closed intervals:

$$[p^{1/2}, 2p^{1/2}],\ [2p^{1/2}, 4p^{1/2}],\ \ldots,\ [2^{i-1}p^{1/2}, 2^i p^{1/2}],\ \ldots,\ [2^{r-1}p^{1/2}, 2^r p^{1/2}],$$

where r is a natural number such that

$$2^{r-1}p^{1/2} \le k \le 2^r p^{1/2}.$$

For every natural number $i \in [1, r]$, let $\mathcal{L}_i$ stand for the family of all those lines determined by P, each of which contains at least $2^{i-1}p^{1/2}$ points from P and at most $2^i p^{1/2}$ points from P. Obviously, we have

$$\mathcal{L} \subset \cup\{\mathcal{L}_i : 1 \leq i \leq r\}.$$

Further, for any natural number $i \in [1, r]$, denote by $\mathcal{P}_i$ the family of all those two-element subsets of P which lie on some line from $\mathcal{L}_i$. Clearly, we may write

$$\mathrm{card}(\mathcal{P}) \leq \sum\{\mathrm{card}(\mathcal{P}_i) : 1 \leq i \leq r\} \leq$$

$$\sum\{2^{i-1}p^{1/2}(2^i p^{1/2} - 1)\mathrm{card}(\mathcal{L}_i) : 1 \leq i \leq r\}.$$

By virtue of Lemma 2 and Remark 1, we get

$$\mathrm{card}(\mathcal{L}_i) \leq O(p/(2^{i-1}p^{1/2})).$$

Consequently,

$$2^{i-1}p^{1/2}(2^i p^{1/2} - 1)\mathrm{card}(\mathcal{L}_i) \leq 2^{i-1}p^{1/2}O(p),$$

whence it follows (in view of $2^r p^{1/2} \leq 2k$) that

$$\mathrm{card}(\mathcal{P}) \leq (1 + 2 + ... + 2^{r-1})p^{1/2}O(p) \leq (2k - p^{1/2})O(p) \leq O(kp).$$

Lemma 4 has thus been proved. □

The next auxiliary statement plays a key role for establishing Beck's remarkable result.

Lemma 5. *Let $c \geq 2$ be a real constant and let P be a finite point set in $\mathbf{R}^2$ with $\mathrm{card}(P) = p \geq 2$. Denote by $\mathcal{L}$ the family of all those lines determined by P, each of which contains at least c and at most p/c points from P.*

Then the formula

$$\mathrm{card}(\mathcal{L}) = O(p^2/c)$$

is valid.

Proof. The required result easily follows from Lemmas 3 and 4. Indeed, let us denote:

$\mathcal{L}'$ = the family of all those lines determined by P, each of which contains at least c and at most $p^{1/2}$ points from P;

$\mathcal{L}''$ = the family of all those lines determined by P, each of which contains at least $p^{1/2}$ and at most p/c points from P.

Clearly, we have the equality

$$\mathcal{L} = \mathcal{L}' \cup \mathcal{L}''.$$

Now, according to Lemma 3, we may write

$$\mathrm{card}(\mathcal{L}') = O(p^2/c).$$

Similarly, according to Lemma 4, we may write

$$\mathrm{card}(\mathcal{L}'') = O((p/c) \cdot p) = O(p^2/c).$$

Taking into account both facts above, we obtain the desired equality

$$\mathrm{card}(\mathcal{L}) = O(p^2/c),$$

and the proof of Lemma 5 is complete. □

Now, we are able to establish Beck's theorem [17].

Theorem 2. *There exist two absolute real constants $a > 0$ and $b > 0$ such that, for any natural number $p \geq 2$ and for any point set $P \subset \mathbf{R}^2$ with* $\mathrm{card}(P) = p$, *the disjunction of these two assertions holds:*

(1) there is a line determined by P and containing at least p/a distinct points from P;

(2) there are at least p^2/b distinct lines determined by P.

Proof. Let $c \geq 2$ be a real constant and let P be a point set in $\mathbf{R}^2$ with $\mathrm{card}(P) = p$. As before, denote by $\mathcal{L}$ the family of all those lines determined by P, each of which contains at least c and at most p/c points from P. Then, by virtue of Lemma 5, we may write

$$\mathrm{card}(\mathcal{L}) = O(p^2/c).$$

By definition, the latter formula means that

$$\mathrm{card}(\mathcal{L}) \leq d \cdot (p^2/c)$$

for some absolute constant $d > 0$. Now, take $c = 8d$. Then

$$\mathrm{card}(\mathcal{L}) \leq p^2/8.$$

Further, denote by $\mathcal{L}^*$ the family of all those lines determined by P which do not belong to $\mathcal{L}$ ($\mathcal{L}^*$ is supposed to be injective: every line from $\mathcal{L}^*$ is determined by some two-element subset of P, and here we assume for a while that distinct two-element subsets of P determine distinct lines). Since $p \geq 2$, we must have

$$\mathrm{card}(\mathcal{L}^*) \geq p(p-1)/2 - p^2/8 \geq p^2/8.$$

If there exists a line in $\mathcal{L}^*$ which contains more than p/c points from P, then we are done (putting $a = c$). Otherwise, any line from $\mathcal{L}^*$ contains fewer than c points of P. In this case, it becomes clear that there are at least p^2/b pairwise distinct lines in $\mathcal{L}^*$ determined by P, where $b = 4c(c-1)$.

Theorem 2 has thus been proved. □

Remark 2. It is easy to see that the previous theorem can be formulated in a slightly different manner which reads as follows.

There exists an absolute real constant $c \geq 2$ such that, for any natural number $p \geq 2$ and for any point set $P \subset \mathbf{R}^2$ with $\mathrm{card}(P) = p$, the disjunction of these two assertions holds:

(*) there is a line determined by P and containing at least p/c distinct points from P;

(**) there are at least p^2/c distinct lines determined by P.

Indeed, it suffices to put $c = \max(a, b)$, where a and b are absolute constants of Theorem 2.

Theorem 3. *There exists an absolute real constant $c \geq 2$ such that, for any natural number $p \geq 2$ and for any non-collinear point set $P \subset \mathbf{R}^2$ with* $\mathrm{card}(P) = p$*, a point $x \in P$ can be found belonging to $\lfloor p/c \rfloor$ distinct lines determined by P.*

Proof. According to Remark 2, there exists an absolute real constant $c \geq 2$ such that, for any natural number $p \geq 2$ and for any point set $P \subset \mathbf{R}^2$ with $\mathrm{card}(P) = p$, the disjunction of these two assertions is valid:

(1) there is a line determined by P and containing at least $\lfloor p/c \rfloor$ distinct points from P;

(2) there are at least p^2/c distinct lines determined by P.

Denote by $\mathcal{L} = \mathcal{L}(P)$ the family of all lines determined by P and put

$$l = \mathrm{card}(\mathcal{L}).$$

Suppose first that (1) is true and let y be a line from $\mathcal{L}$ containing at least $\lfloor p/c \rfloor$ points from P, which we denote by $z_1, z_2, ..., z_k$, where $k \geq \lfloor p/c \rfloor$.

Since P is not collinear, there is a point $z \in P$ not belonging to y. Then it is clear that z belongs to all lines

$$\{tz+(1-t)z_1 : t \in \mathbf{R}\}, \quad \{tz+(1-t)z_2 : t \in \mathbf{R}\}, \quad \dots, \quad \{tz+(1-t)z_k : t \in \mathbf{R}\},$$

which are determined by P, and the number of these lines is greater than or equal to $\lfloor p/c \rfloor$.

Suppose now that (2) is true. We may write

$$t_1 + t_2 + ... + t_p = r_1 + r_2 + ... + r_l,$$

where t_i denotes the number of those lines from $\mathcal{L}$ which pass through the i-th point of P, and r_j denotes the number of those points from P which lie on the j-th line from $\mathcal{L}$.

Observe that, for each natural index $j \in [1, l]$, the inequality $r_j \geq 2$ is valid and, according to our assumption, $l \geq p^2/c$. Consequently,

$$t_1 + t_2 + ... + t_p \geq 2(p^2/c).$$

The last relation immediately implies that there exists t_i such that

$$t_i \geq 2\lfloor p/c \rfloor \geq \lfloor p/c \rfloor,$$

which ends the proof of Theorem 3. □

Remark 3. Recall that, for a given set P of points in $\mathbf{R}^2$, an ordinary line determined by P is any line containing exactly two distinct points from P. In connection with the above theorem, it makes sense to note that, according to one conjecture of Dirac and Motzkin, if P is a non-collinear point set in $\mathbf{R}^2$ and $p = \text{card}(P) > 13$, then the number of all ordinary lines determined by P is greater than or equal to $p/2$. To this moment the conjecture is not proved and is not disproved either. Nevertheless, one may suppose that if a finite non-collinear set $P \subset \mathbf{R}^2$ contains many points, then there are also many ordinary lines determined by P. Indeed, as was established by Green and Tao [116], the number of such lines is greater than or equal to $p/2$ whenever p is sufficiently large. For some purely combinatorial aspects of this theme inspired by the well-known Sylvester problem [334], see Appendix 5.

Remark 4. An interesting multi-dimensional version of the Szemerédi–Trotter theorem was obtained in [323].

EXERCISES

1. Suppose that a nonempty finite graph $G = (V, E)$ is given.

The phrase that H is a random subgraph of G means that all vertices of H are taken mutually independently from V, each with probability q, where q is a fixed positive real number not exceeding 1, and all edges of H are induced by corresponding edges of G. More precisely, using the notation

$$v = \text{card}(V), \qquad V = \{1, 2, ..., v\},$$

one may introduce the space of elementary events $\Omega = \mathcal{P}(V)$ equipped with a probability measure μ defined as follows. The μ-probability that a set

$$W = \{i_1, i_2, ..., i_w\} \subset V$$

is the set of all vertices of a random graph H is equal to $q^w(1-q)^{v-w}$ (clearly, here $w = \text{card}(W)$).

Thus, for any random subgraph H of G, one has its basic parameters:

v_H = the number of vertices of H;

e_H = the number of edges of H;

c_H = the crossing number of H.

In fact, these parameters are random variables on the described probability space.

Denoting the (mathematical) expectation of random variables by the standard symbol $\mathbf{E}$, verify in detail the validity of the relations

$$\mathbf{E}(v_H) = qv,$$
$$\mathbf{E}(e_H) = q^2 e,$$
$$\mathbf{E}(c_H) \leq q^4 c_G,$$

and complete the proof of Lemma 1.

2*. For any natural number $k \in [2, p]$, denote again by the symbol $\mathcal{L}(k, P)$ the family of all those lines determined by P, each of which contains at least k points from P.

Show that

$$J(P, \mathcal{L}(k, P)) = O(p^2/k^2 + p).$$

For this purpose, apply the Szemerédi-Trotter theorem and Lemma 2.

3. Let A be a nonempty finite subset of $\mathbf{R}$ such that $\mathrm{card}(A) = n$, and let

$$A + A = \{x + y : x \in A, y \in A\}.$$

Check that the inequality

$$\mathrm{card}(A + A) \geq n$$

takes place.

Further, give an example of a set $B \subset \mathbf{R}$ such that $\mathrm{card}(B) = n$ and

$$\mathrm{card}(B + B) \leq 2n - 1.$$

Therefore, in this case one has $\mathrm{card}(B + B) = O(n)$.

For this purpose, consider an arithmetic sequence in $\mathbf{R}$ of length n.

4. Let A be a nonempty finite subset of $\mathbf{R}$ such that $\mathrm{card}(A) = n$, and let

$$A \cdot A = \{x \cdot y : x \in A, y \in A\}.$$

Verify that the inequality

$$\mathrm{card}(A \cdot A) \geq n$$

takes place.

Give an example of a set $B \subset \mathbf{R}$ such that $\mathrm{card}(B) = n$ and

$$\mathrm{card}(B \cdot B) \leq 2n - 1.$$

Therefore, in this case one has $\mathrm{card}(B \cdot B) = O(n)$.

For this purpose, consider a geometric sequence in $\mathbf{R}$ of length n.

5*. Let A be as in Exercise 4 and suppose that $0 \notin A$. In the Euclidean plane $\mathbf{R}^2$ consider the point set

$$P = (A + A) \times (A \cdot A).$$

By using the results of the two previous exercises, verify that

$$\operatorname{card}(P) \geq n^2.$$

Further, for any two numbers $a_i \in A$ and $a_j \in A$, introduce the straight line $l_{i,j}$ in $\mathbf{R}^2$ determined by the equation

$$y = a_i(x - a_j),$$

and denote by $\mathcal{L}$ the family of all such lines. Clearly, the equality $\operatorname{card}(\mathcal{L}) = n^2$ holds true.

Check that each line from $\mathcal{L}$ contains at least n distinct points from P.

Taking into account Lemma 2, deduce the relation

$$n^2 = O(p^2/n^3 + p/n),$$

where $p = \operatorname{card}(P)$.

Finally, remembering that $p \geq n^2$, obtain that $n^2 = O(p^2/n^3)$ and conclude that $n^{5/2} = O(p)$.

Remark 5. The equality $n^{5/2} = O(p)$ is known as the Elekes theorem. This result shows, in particular, that the two relations

$$\operatorname{card}(A + A) = O(n), \qquad \operatorname{card}(A \cdot A) = O(n)$$

cannot be valid simultaneously.

6*. Let A and B be any two nonempty subsets of $\mathbf{R}$.

Demonstrate that the inequality

$$\operatorname{card}(A + B) \geq \operatorname{card}(A) + \operatorname{card}(B) - 1$$

holds true.

For this purpose, consider separately the following two cases:

(i) one of the given sets A and B is infinite;

(ii) both given sets A and B are finite.

In the second case use induction on $\operatorname{card}(A) + \operatorname{card}(B)$.

Show also that, in general, the above inequality does not hold for nonempty subsets A and B of a commutative group $(\Gamma, +)$ and try to describe those groups $(\Gamma, +)$ for which this inequality still remains valid.

CHAPTER 11

Minkowski's theorem, number theory, and nonmeasurable sets

In previous sections of this book we were concerned with various combinatorial aspects of Euclidean geometry. Our considerations therein were mostly of finite or discrete character, so we did not systematically use the Axiom of Choice (**AC**) or any of its well-known logical equivalents. In this respect, it should be underlined that there is a big gap between classical mathematics (with its traditional disciplines such as number theory, elementary algebra, Euclidean geometry, mathematical analysis on the real line **R**, etc.) and contemporary mathematics which studies much more abstract structures with diverse sophisticated interrelations between them. One of the features substantially separating classical and modern mathematics can be expressed by those forms of **AC** which are exploited in mathematical reasonings. Classical mathematical objects and structures either do not need this axiom or appeal to rather weak forms of it, e.g., the countable version of **AC** (cf. [137]). Contemporary mathematical theories such as set theory itself, universal algebra, general topology, functional analysis, measure theory, etc., are heavily based on uncountable forms of **AC** and, in fact, cannot exist without using them. On the other hand, strong forms of **AC** have very unpleasant corollaries and produce a number of unexpected, extraordinary or paradoxical situations.

In this context, measure theory provides us with many examples of subsets of the space $\mathbf{R}^m$ $(m \geq 1)$, which cannot be measurable in any reasonable sense and, consequently, they must be treated as somewhat pathological subsets of $\mathbf{R}^m$. Without question, the most striking result in this direction is the famous Banach–Tarski paradox (see [13]). But it should be mentioned once more that the first example of subsets of this type is attributed to Vitali. In 1905 he proved, by applying an ingenious and elegant argument based

 DOI: 10.1201/9781003458708-11

on an uncountable form of **AC**, that there exists a subset of **R** which is nonmeasurable in the Lebesgue sense and does not have the Baire property (see his work [353]; cf. also [108], [130], [316], [355]). However, we shall see in this chapter that a few years earlier Minkowski also was very close to the idea of a nonmeasurable point set and his motivation was closely connected with the standard notion of convexity in the Euclidean space $\mathbf{R}^m$.

First of all, we would like to recall the reader that Minkowski (one of the founders of the general theory of convex sets) successfully applied his methods to various problems concerning profound properties of natural numbers. In this respect, it must be noticed that Minkowski's new ideology became basic for further development of the so-called geometric number theory in which purely arithmetic (or algebraic) properties of natural numbers are studied from the geometric point of view.

Namely, in 1896 Minkowski published his celebrated work [260] in which he suggested geometric solutions to some classical questions of number theory and extensively developed beautiful geometric methods in this theory. Of course, the central place in his methodology was occupied by convex subsets of the space $\mathbf{R}^m$ ($m \geq 2$) with their deep and nice properties.

Among many other important results, Minkowski was able to prove in [260] the following fundamental statement.

Minkowski's Theorem. *If C is a compact convex body in the space $\mathbf{R}^m$, symmetric with respect to the origin of $\mathbf{R}^m$ and having volume greater than or equal to 2^m, then C contains at least two nonzero points of the lattice $\mathbf{Z}^m$, where $\mathbf{Z}$ is the set of all integers.*

This statement was then generalized in various directions (see, e.g., [121], [123] and the references therein; see also Theorem 2 of this chapter). The argument used by Minkowski in the proof of his theorem is purely group-theoretical and measure-theoretical. Actually, by applying a similar argument, it can be established that there exists a subset of $\mathbf{R}^m$ ($m \geq 1$) nonmeasurable in the Lebesgue sense (see, for instance, Theorem 5 containing a much stronger result). In this context, it is interesting to note that more or less precise measure-theoretical concepts were introduced some years later after Minkowski's theorem was stated. Namely, at the very beginning of the twentieth century Lebesgue gave an ingenious construction of his measure $\lambda = \lambda_1$ on $\mathbf{R} = \mathbf{R}^1$, and the existence of subsets of $\mathbf{R}$ nonmeasurable with respect to λ was first rigorously proved by Vitali in 1905.

Remark 1. An analogous astonishing situation can be seen in connection with the famous Poincaré theorem on recurrent points for a ground (base) space E equipped with a finite measure μ and with a single transformation $g : E \to E$ which preserves μ. The above-mentioned theorem was demonstrated by Poincaré before the basic concepts of measure theory (e.g., the countable additivity of measures) were introduced in real analysis. The precise formulation of Poincaré's theorem is given in Exercise 12 for this chapter and its proof is also outlined therein.

One of the goals of the present chapter is to consider some simple applications of Minkowski's theorem to classical number theory. Another goal is to illustrate the role of Minkowski's method in proving the existence of nonmeasurable subsets of Euclidean spaces and, more generally, in establishing the existence of nonmeasurable sets in an abstract space equipped with a transformation group and with some measure which is assumed to be invariant under the group. In the sequel, we will introduce the concept of a weakly metrically transitive invariant measure and we will show how such a measure necessarily produces absolutely nonmeasurable sets with respect to the class of all its invariant extensions.

We begin with several well-known notions concerning spaces endowed with transformation groups. These notions and related concepts are useful in many fields of mathematics.

We recall that a space with a transformation group is any pair (E, G), where E is a base (ground) set and G is some group of transformations of E; in other words, G consists of bijections acting from E onto itself and, in addition to this, G forms a group with respect to the operation $\circ$ of composition of mappings.

Such spaces play a crucial role in contemporary geometry and topology. Moreover, from the modern point of view, any geometry over E can be described in terms of invariants of an appropriate transformation group G acting in E. It worth remembering that this approach was first formulated in the famous Erlangen Program by F. Klein (see [233]).

Recall that a transformation group G acts transitively on E (in E) if, for any two points $x \in E$ and $y \in E$, there exists $g \in G$ such that $g(x) = y$.

Equivalently, G acts transitively on E (in E) if a unique G-orbit in E is E itself.

We also recall that a transformation group G acts freely on E (in E) if, for any $x \in E$, $g \in G$ and $h \in G$, the relation $g(x) = h(x)$ implies $g = h$.

Equivalently, G acts freely on E (in E) if, for any $x \in E$ and $g \in G$, the equality $g(x) = x$ implies $g = \mathrm{Id}_E$ (here Id_E denotes, as usual, the identity mapping of E onto itself).

Let E be a nonempty set, G be a group of transformations of E, and let μ be a measure defined on a G-invariant σ-algebra $\mathrm{dom}(\mu)$ of subsets of E and invariant (or, more generally, quasi-invariant) under all transformations from G. In this case, we briefly say that μ is a G-invariant (G-quasi-invariant) measure (cf. [141], [355]). We would like to recall the reader that the G-quasi-invariance of μ is much weaker than its G-invariance, because the G-quasi-invariance of μ just means that G transforms the class of all μ-measure zero sets in E onto itself.

Let X be some subset of a ground (base) space E equipped with a G-invariant (G-quasi-invariant) measure μ.

We shall say that X is G-thick (with respect to μ) if there exists a countable subset (equivalently, countable subgroup) H of G such that

$$\mu(E \setminus \cup\{h(X) : h \in H\}) = 0.$$

We shall say that X is G-thin (with respect to μ) if

$$(\forall g \in G)(\forall h \in G)(g \neq h \Rightarrow \mu(g(X) \cap h(X)) = 0).$$

In particular, if a group G acts freely in a space E, then every singleton in E is G-thin in E.

Example 1. If $E = \mathbf{R}^2$ and $G = \{0\} \times \mathbf{R}$, then the graph $\mathrm{Gr}(f)$ of any partial function $f : \mathbf{R} \to \mathbf{R}$ is a G-thin subset of E with respect to every G-quasi-invariant measure on E. On the other hand, it may happen that, for some such partial function f, the graph $\mathrm{Gr}(f)$ is $\mathbf{R}^2$-thick with respect to a certain $\mathbf{R}^2$-invariant measure on E extending the standard two-dimensional Lebesgue measure λ_2 (in this connection, see [182]). Moreover, it may happen that there exists a countable family $\{g_n : n < \omega\}$ of isometric transformations (i.e., motions) of $\mathbf{R}^2$ such that

$$\mathbf{R}^2 = \cup\{g_n(\mathrm{Gr}(f)) : n < \omega\}.$$

For details concerning the latter intriguing fact, see e.g. [182] and references therein.

Example 2. Suppose that μ is a σ-finite G-invariant (or, more generally, G-quasi-invariant) measure on E metrically transitive (i.e., ergodic) with respect to G, which means that, for any μ-measurable set Z, the relation

$$(\forall g \in G)(\mu(g(Z) \triangle Z) = 0)$$

implies the disjunction $\mu(Z) = 0 \ \vee \ \mu(E \setminus Z) = 0$.

Then it can easily be checked that every set $X \in \mathrm{dom}(\mu)$ with $\mu(X) > 0$ is G-thick in E. Conversely, if every μ-measurable set X with $\mu(X) > 0$ is G-thick in E, then μ is metrically transitive with respect to G.

In this context, let us recall that the left (right) Haar measure ν on a σ-compact locally compact topological group $(H, \cdot)$ is metrically transitive with respect to the group of all left (right) translations of H (see, e.g., [141]). Moreover, it turns out that the same ν is metrically transitive with respect to each everywhere dense subgroup of H.

Example 3. Suppose that a group G of transformations of a given space E is countable and suppose, in addition, that G acts almost freely in E with respect to some G-invariant (G-quasi-invariant) measure μ on E, i.e., for any two distinct transformations $g \in G$ and $h \in G$, we have

$$\mu(\{x \in E : g(x) = h(x)\}) = 0.$$

Consider the partition of E into all G-orbits of the points of E. Let X be an arbitrary selector of this partition (in short, G-selector). Then it is not difficult to verify that X is simultaneously G-thick and G-thin in E. The details of such a verification are left to the reader.

The following statement is essentially due to Minkowski (cf. [260]).

Theorem 1. *Let (E, G) be a space equipped with a transformation group and let μ be a σ-finite G-invariant measure on E. Suppose that X is a G-thick subset of E and Y is a G-thin subset of E.*

If both these sets are measurable with respect to μ, then $\mu(Y) \leq \mu(X)$.

Proof. Only two cases are possible.

Case 1. The group G is uncountable. In this case there is nothing to prove, because the μ-measurable G-thin subset Y of E is necessarily of μ-measure zero (in view of the σ-finiteness and G-invariance of μ and the almost disjointness of the family $\{g(Y) : g \in G\}$). Here we may refer to the general fact, according to which if μ is an arbitrary σ-finite measure on E, then there exists no uncountable almost disjoint family of μ-measurable sets, all of which are of strictly positive μ-measure.

Case 2. The group G is countable. In this case, we obviously have

$$\mu(E \setminus \cup\{g(X) : g \in G\}) = 0$$

and, consequently, we can write

$$\mu(Y) = \mu(Y \cap (\cup\{g(X) : g \in G\})) \leq \sum_{g \in G} \mu(g(X) \cap Y) =$$

$$\sum_{g \in G} \mu(X \cap g^{-1}(Y)) \leq \mu(X),$$

which yields the required result.

Theorem 1 has thus been proved. □

Lemma 1. *Let (E, G) be a space endowed with a complete G-invariant (G-quasi-invariant) measure μ and let H be a subgroup of G.*

Denote by $\{Hg_i : i \in I\}$ the partition of G into the (right) equivalence classes modulo H.

Then the following two assertions are valid:

(1) *if a set $X \subset E$ is G-thick with respect to μ, then the set*

$$X' = \cup\{g_i(X) \ : \ i \in I\}$$

is H-thick with respect to μ;

(2) *if* $\mathrm{card}(I) \leq \omega$ *and a set $Y \subset E$ is G-thin with respect to μ, then the set*

$$Y' = \cup\{g_i(Y) \ : \ i \in I\}$$

is H-thin with respect to μ.

The proof is not hard and is left to the reader.

Now, we are able to demonstrate the classical Minkowski theorem (in a slightly more general formulation). As usual, the symbol λ_m stands for the standard m-dimensional Lebesgue measure on the Euclidean space $\mathbf{R}^m$.

Theorem 2. *Let H be a discrete group of translations of the space $\mathbf{R}^m$, where $m \geq 1$, and let T be a compact convex body in $\mathbf{R}^m$ symmetric with respect to the origin of $\mathbf{R}^m$. Furthermore, let X be an arbitrary λ_m-measurable H-thick subset of $\mathbf{R}^m$.*

If the inequality $\lambda_m(T) \geq 2^m \lambda_m(X)$ is satisfied, then there exists a nonzero vector $h \in H \cap T$.

Proof. First, let us consider the case when $\lambda_m(T) > 2^m \lambda_m(X)$. In this case, we may write

$$\lambda_m(T/2) > \lambda_m(X).$$

Applying Theorem 1 to the sets $T/2$ and X, we infer that $T/2$ cannot be H-thin in the space $\mathbf{R}^m$. Consequently, there exist two distinct vectors h' and h'' in H such that

$$(h' + T/2) \cap (h'' + T/2) \neq \emptyset.$$

This implies that

$$h' - h'' \in (T/2) + (-T/2) = T/2 + T/2 = T,$$

because T is convex and symmetric with respect to the origin of $\mathbf{R}^m$. Thus, the nonzero vector $h' - h''$ lies in $H \cap T$ and is the required one (clearly, the same can be said on the vector $h'' - h'$).

Suppose now that $\lambda_m(T) = 2^m \lambda_m(X)$ or, equivalently,

$$\lambda_m(T/2) = \lambda_m(X).$$

For a natural number $k > 0$, consider the closed $(1/k)$-neighborhood of $T/2$ which is also convex and symmetric with respect to the origin. Denoting this neighborhood by T_k and applying the obtained result to T_k, we get a nonzero vector $h_k \in H \cap T_k$. Now, since our group H is discrete (and, consequently, closed in $\mathbf{R}^n$), it is easy to see that there exists a vector $h \in H$ such that $h = h_k$ for infinitely many numbers k. Keeping in mind the relation

$$T = \cap\{T_k : k \in \mathbf{N} \setminus \{0\}\},$$

we obviously have $h \neq 0$ and $h \in H \cap T$, which ends the proof of Theorem 2. □

For a while, let us turn our attention to several consequences of Theorem 2 in classical number theory.

Let a, b, c be some integers and let $c \neq 0$. As usual, we will write

$$a \equiv b \ (\mathrm{mod}(c))$$

whenever c is a divisor of the difference $a - b$. Recall that $\equiv$ yields an equivalence relation in the set $\mathbf{Z}$ of all integers (when c is fixed).

Take any two integers a_1 and a_2 and consider a linear form $u : \mathbf{R}^2 \to \mathbf{R}$ on the Euclidean plane $\mathbf{R}^2$, defined by

$$x \to a_1x_1 + a_2x_2 \qquad (x = (x_1, x_2) \in \mathbf{R}^2).$$

The following result concerning this form is due to A. Thue.

Theorem 3. *For any natural number $p \geq 1$, there exist integers z_1 and z_2 such that:*

1) $|z_1| + |z_2| > 0$, i.e the vector $z = (z_1, z_2) \in \mathbf{R}^2$ is nonzero;
2) $|z_1| \leq p^{1/2}$ and $|z_2| \leq p^{1/2}$;
3) $u(z_1, z_2) = a_1z_1 + a_2z_2 \equiv 0 \pmod{p}$.

Proof. Let us denote

$$T = \{(x_1, x_2) \in \mathbf{R}^2 \ : \ |x_1| \leq p^{1/2}, \ |x_2| \leq p^{1/2}\}.$$

Clearly, T is a square whose center coincides with the origin $(0, 0)$ of the plane $\mathbf{R}^2 = \mathbf{R} \times \mathbf{R}$. In particular, T is convex and symmetric with respect to the origin of $\mathbf{R}^2$. Moreover, we have the equality $\lambda_2(T) = 4p$, where λ_2 is the standard two-dimensional Lebesgue measure on $\mathbf{R}^2$. Further, we put:

G = the group of all those translations of $\mathbf{R}^2$, both coordinates of which are integers;

H = the subgroup of G consisting of all those elements $(x_1, x_2) \in G$ for which $u(x_1, x_2) = a_1x_1 + a_2x_2 \equiv 0 \pmod{p}$.

It can directly be verified that $(G : H) \leq p$, where $(G : H)$ denotes, as usual, the index of H in G, i.e., $(G : H)$ is the number of equivalence classes in G modulo H.

Now, let $\{Hg_i : 1 \leq i \leq k\}$ be the partition of G into equivalence classes modulo H. As mentioned above, we have the inequality $k \leq p$. According to Lemma 1, the set

$$X = \cup\{g_i([0, 1]^2) : 1 \leq i \leq k\}$$

is H-thick in the plane $\mathbf{R}^2$. In addition to this, we may write

$$\lambda_2(T) = 4p \geq 4k \geq 2^2\lambda_2(X).$$

So, by virtue of the Minkowski theorem, there exists a nonzero vector

$$z = (z_1, z_2) \in H \cap T.$$

Obviously, the coordinates z_1 and z_2 of this vector are the required integers.

This ends the proof of Theorem 3. □

We recommend the reader formulate and prove an appropriate analogue of Theorem 3 for the m-dimensional Euclidean space $\mathbf{R}^m$.

For our further purposes, we need a purely technical lemma. Actually, its formulation and proof goes back to L. Euler.

Lemma 2. *Let x_1, x_2, x_3, x_4, y_1, y_2, y_3, y_4 be any eight elements of a commutative ring (equipped with the standard algebraic operations + and ·).*

Then the following equality holds true:

$$(x_1^2 + x_2^2 + x_3^2 + x_4^2)(y_1^2 + y_2^2 + y_3^2 + y_4^2) = z_1^2 + z_2^2 + z_3^2 + z_4^2,$$

where

$$z_1 = x_1y_1 + x_2y_2 + x_3y_3 + x_4y_4,$$

$$z_2 = x_1y_2 - x_2y_1 + x_3y_4 - x_4y_3,$$

$$z_3 = x_1y_3 - x_3y_1 + x_4y_2 - x_2y_4,$$

$$z_4 = x_1y_4 - x_4y_1 + x_2y_3 - x_3y_2.$$

The proof of this equality is reduced to its straightforward checking. From the geometric point of view (more precisely, from the point of view of classical theory of quaternions), the equality simply indicates that the norm of the product of any two quaternions coincides with the product of their norms.

Lemma 3. *Let p be an arbitrary prime natural number.*

Then there exist two natural numbers a and b such that the equality

$$a^2 + b^2 + 1 \equiv 0 \ (\mathrm{mod}(p))$$

is valid.

Proof. If $p = 2$, then the assertion is trivial. Suppose now that $p > 2$ and consider the following two sequences of integers:

$$0^2,\ 1^2,\ \ldots,\ m^2,\ \ldots,\ ((p-1)/2)^2,$$

$$-1 - 0^2,\ -1 - 1^2,\ \ldots,\ -1 - m^2,\ \ldots,\ -1 - ((p-1)/2)^2.$$

Taking any two distinct terms m^2 and l^2 from the first sequence, we see that their difference $m^2 - l^2 = (m+l)(m-l)$ is not divisible by the prime p, because of the inequalities

$$0 < m + l < p, \quad 0 < |m - l| < p.$$

Thus, if we successively divide by p all terms of the first sequence, we will obtain pairwise distinct remainders. The analogous circumstance may be observed for the second sequence. But, the total amount of integers in both sequences equals $p + 1$, i.e., is strictly greater than p. Hence there exist a member a^2 from the first sequence and a member $-1 - b^2$ from the second sequence, such that the difference

$$a^2 - (-1 - b^2) = a^2 + b^2 + 1$$

is divisible by p, i.e., $a^2 + b^2 + 1 \equiv 0 \ (\mathrm{mod}(p))$.

This completes the proof of Lemma 3. □

Lemma 4. *The volume of the four-dimensional unit ball* $\mathbf{B}_4$ *in the Euclidean space* $\mathbf{R}^4$ *is equal to* $\pi^2/2$.

Proof. Here the following general formula can be applied:

$$\lambda_m(\mathbf{B}_m) = \frac{\pi^{m/2}}{\Gamma((m/2)+1)},$$

where λ_m denotes the usual m-dimensional Lebesgue measure on the Euclidean space $\mathbf{R}^m$ and $\mathbf{B}_m$ is the unit ball in this space. The above-mentioned formula is well known from the standard course of mathematical analysis. But, for our further purpose, we do not need the whole power of this formula. Indeed, let us remind its elementary analog for the three-dimensional ball $B(0,r) \subset \mathbf{R}^3$ with center 0 and radius r, namely, $\lambda_3(B(0,r)) = (4/3)\pi r^3$.

Applying Fubini's theorem to the vertical sections of $\mathbf{B}_4$ by hyperplanes, we obtain

$$\lambda_4(\mathbf{B}_4) = \int_{-1}^{1} (4/3)\pi(1-x^2)^{3/2}dx.$$

An easy calculation of this integral leads to the required equality

$$\lambda_4(\mathbf{B}_4) = \pi^2/2.$$

Lemma 4 has thus been proved. □

Remark 2. It worth noticing that $\lambda_5(\mathbf{B}_5)$ attains the maximum value among all the volumes $\lambda_m(\mathbf{B}_m)$, where m ranges over the set $\mathbf{N}$.

Now, we are able to demonstrate the following classical result in number theory, due to Lagrange.

Theorem 4. *Every natural number can be represented as the sum of four squares of integers.*

Proof. According to the fundamental theorem of arithmetic, any natural number is a finite product of some primes. So, by virtue of Lemma 2, it suffices to show that every prime number can be represented as the sum of four squares of integers. In order to do this, take a prime p and, by using Lemma 3, find two integers a and b for which

$$a^2 + b^2 + 1 \equiv 0 \ (\mathrm{mod}(p)).$$

Let ε be some strictly positive real number, the precise value of which will not play an essential role in our further argument. In the Euclidean space $\mathbf{R}^4$, consider the closed ball T with center 0 and radius

$$r = (4-\varepsilon)^{1/4}p^{1/2}.$$

Of course, we assume here that $\varepsilon < 4$. Denote by G the group of all those translations of $\mathbf{R}^4$ whose coordinates are integers (in fact, $G = \mathbf{Z}^4$). Obviously, this group is discrete in $\mathbf{R}^4$. Define two linear forms u_1 and u_2 on $\mathbf{R}^4$ by putting

$$u_1(x) = u_1(x_1, x_2, x_3, x_4) = ax_1 + bx_2 - x_3,$$

$$u_2(x) = u_2(x_1, x_2, x_3, x_4) = bx_1 - ax_2 - x_4$$

for all vectors $x = (x_1, x_2, x_3, x_4)$ from $\mathbf{R}^4$. In addition to this, introduce the subgroup H of G consisting of all those elements $x \in G$ for which

$$u_1(x) \equiv 0 \ (\mathrm{mod}(p)), \qquad u_2(x) \equiv 0 \ (\mathrm{mod}(p)).$$

It is not hard to verify that $(G : H) \leq p^2$.

Now, consider the partition $\{Hg_i : 1 \leq i \leq k\}$ of G into equivalence classes modulo H. So we have the inequality $k \leq p^2$. Putting

$$X = \cup\{g_i([0,1]^4) : 1 \leq i \leq k\}$$

and applying Lemma 1, we infer that X is an H-thick subset of $\mathbf{R}^4$. Also, according to Lemma 4, we may write

$$\lambda_4(T) = (\pi^2/2)(4 - \varepsilon)p^2.$$

Hence, taking $\varepsilon > 0$ sufficiently small, we obtain

$$\lambda_4(T) \geq (\pi^2/2)(4 - \varepsilon)k > 2^4 \lambda_4(X).$$

Consequently, by virtue of the Minkowski theorem, there exists a nonzero vector

$$z = (z_1, z_2, z_3, z_4) \in H \cap T.$$

For such a z, we obviously have

$$0 < z_1^2 + z_2^2 + z_3^2 + z_4^2 \leq (4 - \varepsilon)^{1/2} p < 2p.$$

At the same time, the relations

$$(az_1 + bz_2)^2 \equiv z_3^2 \ (\mathrm{mod}(p)), \quad (bz_1 - az_2)^2 \equiv z_4^2 \ (\mathrm{mod}(p))$$

are valid in view of the definition of H. From these two relations, remembering that $a^2 + b^2 \equiv -1 \ (\mathrm{mod}(p))$, we come to the relation

$$z_1^2 + z_2^2 + z_3^2 + z_4^2 \equiv 0 \ (\mathrm{mod}(p)),$$

which means that the prime p divides the sum $z_1^2 + z_2^2 + z_3^2 + z_4^2$. But, as was mentioned earlier, $z_1^2 + z_2^2 + z_3^2 + z_4^2 < 2p$, so we conclude that

$$z_1^2 + z_2^2 + z_3^2 + z_4^2 = p.$$

The Lagrange theorem has thus been proved. □

Now, we turn our attention to those λ_m-nonmeasurable sets in $\mathbf{R}^m$ which appear as a byproduct of Minkowski's method.

Let (E, G) be a space equipped with a transformation group and let μ be a σ-finite G-invariant (or, more generally, G-quasi-invariant) measure on E.

We shall say that μ is weakly metrically transitive (or weakly ergodic) with respect to G if, for an arbitrary real $\varepsilon > 0$, there exist a μ-measurable set X with $\mu(X) < \varepsilon$ and a countable subset (equivalently, countable subgroup) H of G such that

$$\mu(E \setminus \cup\{h(X) : h \in H\}) = 0.$$

It is not hard to verify that any σ-finite non-atomic G-invariant metrically transitive measure μ on E is weakly metrically transitive. The converse assertion is not true in general. Indeed, it directly follows from the above definition that every G-invariant extension of a G-invariant weakly metrically transitive measure is also weakly metrically transitive. But the analogous statement fails to be true for G-invariant metrically transitive measures (see Exercise 15 of this chapter).

Example 4. Suppose that a σ-finite G-invariant metrically transitive measure μ on E has at least one atom A, i.e., the set $A \in \mathrm{dom}(\mu)$ is such that

$$0 < \mu(A) < +\infty,$$

$$(\forall X \in \mathrm{dom}(\mu))(X \subset A \Rightarrow (\mu(X) = 0 \vee \mu(A \setminus X) = 0)).$$

Then it is not difficult to describe the structure of μ. Namely, in this case the space E admits a representation $E = \cup\{A_i : i \in I\}$ for which the following four relations are satisfied:

(1) the set I is at most countable and the family $\{A_i : i \in I\}$ is disjoint;

(2) each set A_i $(i \in I)$ is an atom of μ;

(3) for any two indices $i \in I$ and $j \in I$, there exists a transformation $g \in G$ such that $\mu(g(A_i) \triangle A_j) = 0$;

(4) for every transformation $h \in G$, the family $\{h(A_i) : i \in I\}$ almost coincides with the family $\{A_i : i \in I\}$, i.e., each member of the first family almost coincides with some member of the second family, and conversely.

In other words, we obtain a certain G-invariant lattice of atoms of our measure μ, which is similar to the standard lattice $\{z+[0,1]^m : z \in \mathbf{Z}^m\}$ of $\mathbf{R}^m$ endowed with the discrete group $\mathbf{Z}^m$ and with the restriction of λ_m to the σ-algebra generated by this lattice.

Let (E, G) be a space equipped with a transformation group and let $\mathcal{M}$ be a class of G-invariant (G-quasi-invariant) measures on E (we would like to stress that, in general, the measures ν from $\mathcal{M}$ are defined on different σ-algebras of subsets of E).

We shall say that a set $X \subset E$ is absolutely nonmeasurable with respect to $\mathcal{M}$ if there exists no measure from $\mathcal{M}$ for which X is measurable, i.e.,

$$(\forall \nu \in \mathcal{M})(X \notin \mathrm{dom}(\nu)).$$

Now, let μ be some G-invariant (G-quasi-invariant) measure on E.

We shall say that a set $X \subset E$ is G-absolutely nonmeasurable with respect to μ if X is absolutely nonmeasurable with respect to the class $\mathcal{M}(\mu)$ of all those G-invariant (G-quasi-invariant) measures on E which extend μ.

The next example highlights close relationships between convex sets and absolutely nonmeasurable sets.

Example 5. Let E be a real infinite-dimensional separable Banach space, G be the group of all translations of E, and let B be a bounded convex body in E (i.e., B is a closed bounded convex set in E with nonempty interior). It can be proved that B is absolutely nonmeasurable with respect to the class of all nonzero σ-finite G-invariant (more generally, G-quasi-invariant) measures on E. In particular, the above fact readily implies that E does not admit any nonzero σ-finite G-quasi-invariant Borel measure, so in E there is no suitable analogue of the standard Lebesgue measure λ on $\mathbf{R}$. By using an algebraic isomorphism between the additive groups $(E, +)$ and $(\mathbf{R}, +)$, it can also be shown that there exists a subset of $\mathbf{R}$ which is absolutely nonmeasurable with respect to the class of all nonzero σ-finite $\mathbf{R}$-invariant (more generally, $\mathbf{R}$-quasi-invariant) measures on $\mathbf{R}$.

Example 6. Let us put:

$E =$ the real line $\mathbf{R}$,

$G =$ the group of all translations of $\mathbf{R}$,

$\mu =$ the Lebesgue measure on $\mathbf{R}$ (i.e., $\mu = \lambda_1 = \lambda$).

Recall that, by definition, a Vitali subset of $\mathbf{R}$ is any selector of $\mathbf{R}/\mathbf{Q}$ (here $\mathbf{Q}$ denotes, as usual, the group of all rational numbers). It is well known that every Vitali set is absolutely nonmeasurable with respect to the class of all those G-invariant measures on E which extend μ (see, for instance, [130], [182], and [355]).

The next statement may be treated as a substantial generalization of the previous example.

Theorem 5. *Let G be a countable group of transformations of E and let μ be a nonzero σ-finite G-invariant weakly metrically transitive measure on E. Suppose also that G acts almost freely in E with respect to μ.*

Then every G-selector in E is G-absolutely nonmeasurable with respect to μ.

Proof. Take an arbitrary G-selector X in E. We have to verify that for any G-invariant measure μ' on E extending the original measure μ, the set X is not μ'-measurable. Suppose otherwise, i.e., $X \in \mathrm{dom}(\mu')$ for some G-invariant extension μ' of μ. Then, keeping in mind the countability of G and the equality

$$E = \cup\{g(X) : g \in G\},$$

we infer that $\mu'(X) > 0$. Let us denote $\mu'(X) = \varepsilon$. According to our assumption, there exist a μ-measurable set Y with $\mu(Y) < \varepsilon$ and a countable family $H \subset G$ such that

$$\mu(E \setminus \cup\{h(Y) : h \in H\}) = 0.$$

In other words, the set Y turns out to be a G-thick subset of E with respect to μ. Since μ' extends μ, the same Y is G-thick with respect to μ' as well. On the other hand, the selector X is a G-thin subset of E with respect to μ and hence with respect to μ' (see Example 3). In view of Theorem 1, the relation

$$\varepsilon = \mu'(X) \leq \mu'(Y) = \mu(Y) < \varepsilon$$

must be valid, which yields a contradiction.

This contradiction completes the proof of Theorem 5. □

Remark 3. Clearly, Theorem 5 may be regarded as a generalization of Vitali's theorem [353] stating the existence of subsets of **R** nonmeasurable in the Lebesgue sense. Indeed, put in the formulation of Theorem 5:

E = the real line **R**,

G = the additive group $(\mathbf{Q}, +)$ of all rational numbers,

μ = the Lebesgue measure on **R** (i.e., $\mu = \lambda_1 = \lambda$).

Then we directly come to Vitali's classical result on the existence of λ-nonmeasurable sets in **R**. Therefore, Theorem 5 may be regarded as an abstract version of the above-mentioned classical result. Some other generalized versions of Vitali's theorem are presented in [141], [182], [355].

It turns out that the notion of weak metrical transitivity for measures on $\mathbf{R}^m$ is closely connected with the existence of absolutely nonmeasurable subsets of $\mathbf{R}^m$. More precisely, we have the following statement.

Theorem 6. *Let H be a countable group of isometric transformations of the Euclidean space $\mathbf{R}^m$, where $m \geq 1$.*

The following three assertions are equivalent:

(1) the Lebesgue measure λ_m is weakly metrically transitive with respect to H;

(2) every H-selector in $\mathbf{R}^m$ is absolutely nonmeasurable with respect to λ_m;

(3) H is non-discrete in the group of all isometries of $\mathbf{R}^m$.

To establish the validity of Theorem 6, one needs specific techniques of the theory of measurable selectors. Here we omit the proof of this theorem and refer the reader to [229], where some additional information on properties of absolutely nonmeasurable sets can also be found.

Remark 4. The direct analogue of Theorem 6 holds true for a Haar measure. Let us give the precise formulation of this analogue. Let $(G, \cdot)$ be an uncountable locally compact Polish group, μ be the left Haar measure on G, and let H be a countable subgroup of G. We may consider H as a group of transformations of G. Namely, each $h \in H$ is identified with the mapping $h : G \to G$ (we preserve the same notation h) defined by the standard formula

$$h(x) = h \cdot x \quad (x \in G).$$

In this way, we come to the partition of G into all left H-orbits.

The following three assertions are equivalent:

(a) μ is weakly metrically transitive with respect to H;

(b) every H-selector in G is H-absolutely nonmeasurable with respect to μ;

(c) H is non-discrete in G.

For the proof of the equivalence of (a), (b) and (c) we again refer the reader to [229].

EXERCISES

1. In Example 2 of this chapter the two definitions of metrical transitivity (i.e., ergodicity) are presented.

   Check the equivalence of these definitions.

2. Verify those properties of a G-selector X, which are pointed out in Example 3.

3. Demonstrate that the group of all motions of the space $\mathbf{R}^m$ $(m \geq 1)$ acts almost freely in $\mathbf{R}^m$ with respect to the Lebesgue measure λ_m.

   By using this fact, show that if G is an uncountable group of motions of $\mathbf{R}^m$ and X is a G-selector, then either X is nonmeasurable with respect to λ_m or $\lambda_m(X) = 0$.

   More generally, prove that if an uncountable group G of transformations of a ground space E acts almost freely in E with respect to a σ-finite G-quasi-invariant measure μ on E, then any G-selector is either μ-nonmeasurable or has μ-measure zero.

4. Give a proof of Lemma 1.

5. Try to generalize Theorem 3 to the case of $\mathbf{R}^m$.

6. Check that no natural number of the form $8k + 7$, where $k \in \mathbf{N}$, can be represented as the sum of three squares of integers.

7. Let p be an arbitrary natural number greater than or equal to 2.

   Show that the following two assertions are equivalent:

   (a) p is a prime number;

   (b) $(p - 1)! + 1$ is divisible by p.

   This classical result is widely known as Wilson's theorem.

   The implication (b) $\Rightarrow$ (a) is almost trivial.

   To establish the validity of (a) $\Rightarrow$ (b), assume (a) and consider the set

$$G = \{1, 2, ..., p - 1\}$$

   as a group with respect to the multiplication operation modulo p.

   Another (geometric) way for establishing the implication (a) $\Rightarrow$ (b) is as follows. Let $p > 2$ be a prime number and let V_p denote the set of all vertices of a regular (convex) p-gon P in the plane $\mathbf{R}^2$. Observe that there are precisely $(p - 1)!/2$ polygonal cycles with the same set V_p of vertices, and among such cycles there are precisely $(p-1)/2$ regular ones (i.e., having pairwise congruent sides). Let G_p denote the group of all those rotations g of $\mathbf{R}^2$ which transform P onto itself. Obviously, G_p a cyclic group of order p. Observe that:

   (i) if L is a regular cycle (e.g., if L is the boundary of P), then $g(L) = L$ for any $g \in G_p$;

   (ii) if L is a non-regular cycle, then there exist exactly p many cycles which are G_p-congruent to L.

   Consequently, the total number of all non-regular cycles is divisible by p, which means that $((p - 1)! - (p - 1))/2$ is divisible by p. Since p is odd, one can conclude that p divides $(p - 1)! + 1$.

   Check that the unique prime number q which divides $(q-1)!-1$ is $q = 2$.

   Deduce from Wilson's theorem that if p is a prime number of the form $p = 4n + 1$ for some $n \in \mathbf{N}$, then

$$((2n)!)^2 + 1 \equiv 0 \ (\mathrm{mod}(p)).$$

8*. Demonstrate that, for an odd prime number p, the following two assertions are equivalent:

   (a) p has the form $p = 4n + 1$, where $n \in \mathbf{N}$;

   (b) $p = x^2 + y^2$ for some integers x and y.

The implication (b) $\Rightarrow$ (a) is almost trivial.

To prove the implication (a) $\Rightarrow$ (b), argue as follows.

Assume (a), denote $c = (2n)!$ and keep in mind that

$$c^2 + 1 \equiv 0(\mathrm{mod}(p))$$

(see the previous exercise). Then introduce the set

$$T = \{(x_1, x_2) \in \mathbf{R}^2 \ : \ |x_1| \leq p^{1/2}, \ |x_2| \leq p^{1/2}\}$$

and define a linear mapping $u : \mathbf{R}^2 \to \mathbf{R}$ by the formula

$$u(x_1, x_2) = x_1 + cx_2 \qquad ((x_1, x_2) \in \mathbf{R}^2).$$

Further, applying Theorem 3 show that there exists a point

$$(x, y) \in (T \cap \mathbf{Z}^2) \setminus \{(0, 0)\}$$

satisfying the relation $u(x, y) \equiv 0 \ (\mathrm{mod}(p))$.

Finally, check that $x^2 + y^2 \equiv 0 \ (\mathrm{mod}(p))$ and taking into account the fact that $x^2 + y^2 < 2p$, deduce the desired equality $x^2 + y^2 = p$.

9*. Let $p \neq 2$ be a prime natural number such that $mp = x^2 + y^2$, where x, y, m are integers and $1 < m < p$.

Demonstrate that p admits a representation $p = x_0^2 + y_0^2$, where x_0 and y_0 are some integers.

For this purpose, consider the following two cases.

(a) m is an even number. In this case, either both x and y are even or both of them are odd. So one can write

$$(m/2)p = ((x + y)/2)^2 + ((x - y)/2)^2,$$

where $m/2 < m$.

(b) m is an odd number. In this case, express x and y as

$$x = mk_1 + r_1, \quad y = mk_2 + r_2,$$

where k_1, r_1, k_2, r_2 are integers and

$$0 \leq |r_1| < m/2, \quad 0 \leq |r_2| < m/2.$$

Observe that $r_1^2 + r_2^2 = mn$, where n is a natural number and $0 < n < m/2$. Consequently,

$$mp \cdot mn = (x^2 + y^2)(r_1^2 + r_2^2) = z^2 + t^2,$$

where both z and t are integers divisible by m.

Conclude from the latter representation that $pn = x_1^2 + y_1^2$ for some integers x_1 and y_1.

Repeating the above argument finitely many times, obtain the desired result.

10*. Let n be an arbitrary natural number.

Prove that the following two assertions are equivalent:

(a) n is representable as a sum of two squares of integers;

(b) any prime number p taking part in the canonical expansion of n and having the form $p = 4k + 3$ is in an even power in this expansion.

For establishing the above-mentioned equivalence, first note that the implication (b) $\Rightarrow$ (a) is a direct consequence of the two preceding exercises.

To show the validity of (a) $\Rightarrow$ (b), assume (a) and argue as follows.

Let $p = 4k+3$ be a prime number and let $n = p^{2r+1}l$, where r and l are natural numbers and l is not divisible by p.

Supposing that $n = x^2+y^2$ for some integers x and y, infer that $x = p^t a$ and $y = p^t b$, where $t \leq r$ is a natural number, a and b are integers not divisible by p. Then deduce from this fact that $x_1^2 + y_1^2 = mp$, where

$$x_1 \in \mathbf{Z}, \quad y_1 \in \mathbf{Z}, \quad m \in \mathbf{N}, \quad 0 < |x_1| < p/2, \quad 0 < |y_1| < p/2.$$

Finally, keeping in mind Exercise 9, conclude that p is representable as the sum of two squares of integers, which is impossible in view of the assumption $p = 4k + 3$ (see Exercise 8).

The obtained contradiction yields (a) $\Rightarrow$ (b).

Remark 5. Recall that the classical results given in Exercises 8, 9, and 10 are due to P. Fermat and L. Euler (see [270] for further extensive information about representations of natural numbers as sums of powers of integers).

11. Let $h(t)$ be a polynomial over $\mathbf{R}$, all values of which are non-negative.

Prove that there exist two polynomials $f(t)$ and $g(t)$ over $\mathbf{R}$ such that

$$h(t) = (f(t))^2 + (g(t))^2.$$

Also, show that for real polynomials of two real variables the analogous fact is not true in general. Moreover, consider the real polynomial

$$P(x,y) = 1 + x^2y^4 + x^4y^2 - 3x^2y^2$$

and demonstrate that all of its values are non-negative, but it cannot be represented as a sum of squares of real polynomials in x and y.

Remark 6. In connection with the previous exercise, let us recall that Hilbert's 17th problem requires to establish that any real rational function of several real variables, which takes only non-negative values, can always be represented as a sum of squares of real rational functions (see [142]). A positive solution of this problem was first obtained by E. Artin (for more details, see [10], [70]; in [296] a solution to this problem is given by using some metamathematical methods).

12*. Let E be a set equipped with a finite measure μ and let $g : E \to E$ be a transformation of E which preserves μ, i.e., for any set $X \in \mathrm{dom}(\mu)$, one has

$$g^{-1}(X) \in \mathrm{dom}(\mu), \quad \mu(g^{-1}(X)) = \mu(X).$$

Let k be a natural number and let A be an arbitrary μ-measurable set in E. Denote

$$R_k(A) = \{x \in A : (\exists n \geq k)(g^n(x) \in A)\}$$

and consider a sequence of μ-measurable sets

$$g^{-m_1}(A \setminus R_k(A)), \ g^{-m_2}(A \setminus R_k(A)), \ \ldots, \ g^{-m_i}(A \setminus R_k(A)), \ \ldots,$$

where $|m_i - m_j| \geq k$ for any two distinct natural indices i and j.

Check that the members of this sequence are pairwise disjoint, whence it follows that

$$\mu(A \setminus R_k(A)) = 0.$$

Deduce from this fact that $\mu(A \setminus \cap\{R_k(A) : k < \omega\}) = 0$.

Finally, conclude that μ-almost all points of A infinitely many times return to A via appropriate iterations of g.

Remark 7. This result is Poincaré's celebrated theorem on recurrence, which found a lot of applications in various fields of mathematics and physics. For a substantial generalization of the above-mentioned theorem, see [100]. The generalized version of the recurrence theorem was successfully applied to difficult problems in combinatorial number theory (cf. [101]).

13. Let (E_1, G_1) and (E_2, G_2) be two spaces equipped with groups of transformations G_1 and G_2, respectively. Let μ_1 be a σ-finite G_1-quasi-invariant measure on E_1 and let μ_2 be a σ-finite G_2-quasi-invariant measure on E_2. Suppose that μ_1 is weakly metrically transitive with respect to G_1 and μ_2 is weakly metrically transitive with respect to G_2.

Show that the product measure $\mu_1 \otimes \mu_2$ is weakly metrically transitive with respect to the product group $G_1 \times G_2$.

14. Give a proof of the fact indicated in Example 4.

15*. As in Example 6, introduce the notation:

$E =$ the real line $\mathbf{R}$,

$G =$ the group of all translations of $\mathbf{R}$,

$\mu =$ the Lebesgue measure on $\mathbf{R}$ (i.e., $\mu = \lambda_1 = \lambda$).

Obviously, one may identify G with the additive group $(\mathbf{R}, +)$. It is well known that μ is metrically transitive with respect to the additive group $(\mathbf{Q}, +)$ of all rational numbers (as well as with respect to any everywhere dense subgroup of $\mathbf{R}$).

By applying the method of transfinite induction, prove that there exist two subsets A and B of $\mathbf{R}$ satisfying the following four conditions:

(1) $A \cap B = \emptyset$ and $A \cup B = \mathbf{R}$;

(2) $\text{card}(A) = \text{card}(B) = \mathbf{c}$, where $\mathbf{c}$ denotes, as usual, the cardinality of the continuum;

(3) $\mu_*(A) = \mu_*(B) = 0$, where μ_* denotes the inner measure associated with μ;

(4) both sets A and B are almost $\mathbf{R}$-invariant, which means that, for all $h \in \mathbf{R}$, the inequalities

$$\text{card}((h + A)\triangle A) < \mathbf{c}, \quad \text{card}((h + B)\triangle B) < \mathbf{c}$$

hold true.

Further, by using these two sets, show that the original measure μ can be extended to an $\mathbf{R}$-invariant measure μ' on $\mathbf{R}$ which fails to be metrically transitive with respect to the group $\mathbf{R}$ (hence, with respect to the group $\mathbf{Q} \subset \mathbf{R}$).

For this purpose, introduce the σ-algebra

$$\mathcal{S} = \sigma(\{A, B\} \cup \text{dom}(\mu))$$

generated by A, B and $\text{dom}(\mu)$, and consider the functional μ' defined on $\mathcal{S}$ by the formula

$$\mu'((A \cap X) \cup (B \cap Y)) = (1/2)(\mu(X) + \mu(Y)),$$

where X and Y are any sets from $\text{dom}(\mu)$.

Check that this definition of μ' is correct and μ' is a measure extending μ.

Then show that μ' can be extended to an $\mathbf{R}$-invariant measure μ'' on $\mathbf{R}$ such that all subsets of $\mathbf{R}$, whose cardinalities are strictly less than $\mathbf{c}$, belong to $\text{dom}(\mu'')$.

Finally, observe that μ'' is not metrically transitive in view of the almost $\mathbf{R}$-invariance of the μ''-measurable sets A and B which both are of strictly positive μ''-measure; moreover, these sets satisfy the equalities

$$\mu''(A) = \mu''(B) = +\infty.$$

At the same time, μ'' is an extension of μ, so μ'' is weakly metrically transitive.

16*. Let $p > 1$ be an odd natural number. Define the structure of a graph over the set $\mathbf{R}$ of all real numbers. First of all, put $V = \mathbf{R}$. Secondly, if r and t are two distinct real numbers, then put $\{r, t\} \in E$ if and only if there exists an integer k such that $|r - t| = p^k$.

Verify that the graph (V, E) contains no simple cycles with odd lengths.

By starting with this fact, infer that (V, E) is a bichromatic graph (see Exercise 17 from Appendix 5), so $\mathbf{R}$ admits a partition $\{X, Y\}$ such that any edge from E has one of its endpoints in X and the other endpoint in Y.

Demonstrate that, for each real $\varepsilon > 0$, there exists $h \in \mathbf{R}$ satisfying the following two relations:

(i) $|h| < \varepsilon$;

(ii) $h + X = Y$ and, consequently, $(h + X) \cap X = \emptyset$.

Deduce from (i) and (ii) that the sets X and Y are not measurable in the Lebesgue sense, i.e., $X \notin \mathrm{dom}(\lambda)$ and $Y \notin \mathrm{dom}(\lambda)$.

For this purpose, utilize the translation invariance of λ and the well-known theorem from real analysis, stating that if $Z \subset \mathbf{R}$ is a Lebesgue measurable set of strictly positive λ-measure, then $(r+Z) \cap Z \neq \emptyset$ for all sufficiently small $r \in \mathbf{R}$ (see, e.g., [141] where a much stronger statement is formulated for Haar measurable sets of strictly positive measure).

Remark 8. The result of Exercise 16 was obtained in [343]. In particular, it shows that the equivalence between the assertions (a) and (b) of Exercise 17 from Appendix 5 cannot be proved in **ZF** theory enriched by the countable form of the Axiom of Choice. Indeed, there exist certain models of this enriched theory, in which all subsets of the real line $\mathbf{R}$ turn out to be Lebesgue measurable. Recall that the latter remarkable result is due to Solovay [321] (see also [151] where a detailed proof is given).

17*. Let X be a λ_m-measurable subset of the space $\mathbf{R}^m$, and suppose that $\lambda_m(X) > k$, where k is a natural number.

Demonstrate that there exists a vector $h \in \mathbf{R}^m$ such that

$$\mathrm{card}((X + h) \cap \mathbf{Z}^m) \geq k + 1.$$

This result is known as Blichfeldt's theorem.

Deduce from Blichfeldt's theorem that if a bounded λ_m-measurable subset Y of $\mathbf{R}^m$ satisfies the inequality $\lambda_m(Y) < 1$, then there exists a vector $g \in \mathbf{R}^m$ such that $(Y + g) \cap \mathbf{Z}^m = \emptyset$.

To obtain the second result, take some m-dimensional cube $K \subset \mathbf{R}^m$ which satisfies the relation $\mathrm{int}(K) \supset Y$ and all vertices of which belong to $\mathbf{Z}^m$. Then apply Blichfeldt's theorem to the set $\mathrm{int}(K) \setminus Y$.

CHAPTER 12

Tarski's plank problem

Let C be a compact convex body in the space $\mathbf{R}^m$, where $m \geq 2$, and let z_1 and z_2 be two distinct boundary points of C, i.e., $z_1 \in \mathrm{bd}(C)$, $z_2 \in \mathrm{bd}(C)$, and $z_1 \neq z_2$.

The chord $[z_1, z_2]$ of C is called an affine diameter of C (cf. Exercise 1 from Chapter 3) if there exist two distinct supporting hyperplanes Γ_1 and Γ_2 of C such that

$$z_1 \in \Gamma_1, \quad z_2 \in \Gamma_2, \quad \Gamma_1 \text{ and } \Gamma_2 \text{ are parallel.}$$

It is not hard to check that any chord $[z'_1, z'_2]$ of C such that $||z'_1 - z'_2||$ equals the diameter of C, in the standard Euclidean metric of $\mathbf{R}^m$, is simultaneously an affine diameter of C (see Exercise 1 of this chapter).

In the sequel, we will use the following simple auxiliary statement.

Lemma 1. *Let C be a compact convex body in the space $\mathbf{R}^m$, where $m \geq 2$, and let e be any nonzero vector in $\mathbf{R}^m$.*

There exist two points $z_1 \in \mathrm{bd}(C)$ and $z_2 \in \mathrm{bd}(C)$ such that the vector $z_1 - z_2$ is parallel to e and the chord $[z_1, z_2]$ is an affine diameter of C.

The proof of this statement is left to the reader (it makes sense to keep in mind the hint given in Exercise 2 of the present chapter).

Recall that the width $w(C)$ of C is defined as the infimum of all distances between two distinct parallel supporting hyperplanes of C.

In view of the compactness of C, this infimum is attained by at least one couple of two distinct parallel supporting hyperplanes of C. In other words, always there are two parallel supporting hyperplanes Γ_1 and Γ_2 of C such that

$$\mathrm{dist}(\Gamma_1, \Gamma_2) = w(C).$$

Clearly, if $z_1 \in \Gamma_1 \cap C$ and $z_2 \in \Gamma_2 \cap C$, then the chord $[z_1, z_2]$ is an affine diameter of C (corresponding to $w(C)$).

Recall also that an open strip H in the space $\mathbf{R}^m$ is the set of all those points of $\mathbf{R}^m$ which lie strictly between two distinct parallel affine hyperplanes

DOI: 10.1201/9781003458708-12

G_1 and G_2 in $\mathbf{R}^m$. In the sequel, we shall say that H is determined by G_1 and G_2. So we may write

$$H = H(G_1, G_2).$$

Analogously, we may define the closed strip $\mathrm{cl}(H)$ determined by G_1 and G_2. We will be dealing below only with closed strips determined by couples of distinct parallel affine hyperplanes in $\mathbf{R}^m$ (cf. Chapter 2 where some strips in $\mathbf{R}^2$ were considered in connection with ot-subsets of $\mathbf{R}^2$).

According to the standard definition, the width of a strip H is the distance between the boundary hyperplanes of H.

The following beautiful problem was posed by A. Tarski (see [338]).

Let C be a compact convex body in the space $\mathbf{R}^m$ ($m \geq 2$) covered by finitely many strips

$$H_1, \ H_2, \ \ldots, \ H_n$$

whose widths are, respectively, $w_1, w_2, \ldots, w_n$. Does the inequality

$$w(C) \leq w_1 + w_2 + \ldots + w_n$$

take place?

Tarski himself proved that the answer is positive in the case of a closed disc C in the Euclidean plane $\mathbf{R}^2$ (see Exercise 3). Later, a complete solution of this problem was obtained by Th. Bang [14] and W. Fenchel [95].

Since the formulation of Tarski's question undoubtedly belongs to the area of convex and combinatorial geometry, we intend to devote the present chapter to this problem and its solution.

Lemma 2. *Let C be a compact convex body in the space $\mathbf{R}^m$, whose width is $w(C)$, and let e be a nonzero vector in $\mathbf{R}^m$ whose length $||e||$ is strictly less than $w(C)/2$.*

Then the width of the convex body

$$C_1 = (C + e) \cap (C - e)$$

is greater than or equal to $w(C) - 2||e||$.

Proof. By virtue of Lemma 1, there exists a chord $[z_1, z_2]$ of C which is an affine diameter of C parallel to e. We may assume, without loss of generality, that $z_1 - z_2 = te$, where $t > 0$. Denote

$$d = ||z_1 - z_2||.$$

Evidently, we have the inequality $d \geq w(C)$ and the equality $d = t||e||$.

Further, let us put:

f_1 = the homothety of $\mathbf{R}^m$ with coefficient $(d - 2||e||)/d$, which satisfies $f_1(z_1) = z_1 - e$;

f_2 = the homothety of $\mathbf{R}^m$ with coefficient $(d - 2||e||)/d$, which satisfies $f_2(z_2) = z_2 + e$.

It can readily be checked that the homotheties f_1 and f_2 have a common fixed point (i.e., common center) lying on the segment $[z_1, z_2]$, so these homotheties are identical to each other and, in particular,

$$f_1(C) = f_2(C).$$

Consequently, we may introduce the notation

$$T = f_1(C) = f_2(C).$$

Clearly, for the width $w(T)$ of T, we get

$$w(T) = \frac{d - 2||e||}{d} w(C).$$

Therefore,

$$w(T) = \frac{d - 2||e||}{d} w(C) \geq w(C) - 2||e||.$$

At the same time, we obviously have the inclusions

$$f_1(C) \subset C - e, \qquad f_2(C) \subset C + e,$$

whence it follows that

$$T = f_1(C) \cap f_2(C) \subset (C + e) \cap C - e),$$

$$w((C + e) \cap (C - e)) \geq w(T) \geq w(C) - 2||e||,$$

which completes the proof of Lemma 2. □

In our further considerations, we will be dealing with the elements of the set $\{-1, 1\}^n$, where $n > 0$ is a natural number. Each element ε of $\{-1, 1\}^n$ has the form

$$\varepsilon = (\varepsilon_1, \varepsilon_2, ..., \varepsilon_n),$$

where $\varepsilon_i \in \{-1, 1\}$ for any natural index $i \in [1, n]$.

Using this notation, we can formulate the next auxiliary statement.

Lemma 3. *Let C be a compact convex body in the Euclidean space $\mathbf{R}^m$ and let $e_1, e_2, \ldots, e_n$ be nonzero vectors in $\mathbf{R}^m$ such that*

$$||e_1|| + ||e_2|| + ... + ||e_n|| < w(C)/2.$$

Then the following inequality is valid:

$$w(\cap\{C + \varepsilon_1 e_1 + \varepsilon_2 e_2 + ... + \varepsilon_n e_n : \varepsilon \in \{-1, 1\}^n\})$$

$$\geq w(C) - 2(||e_1|| + ||e_2|| + ... + ||e_n||).$$

Proof. We argue by induction on n. If $n = 1$, then it suffices to refer to Lemma 2. Suppose now that the assertion holds true for a natural number n and let us verify its validity for $n + 1$. Take a compact convex body $C \subset \mathbf{R}^m$ and a family of nonzero vectors

$$\{e_1,\ e_2, ...,\ e_n,\ e_{n+1}\} \subset \mathbf{R}^m$$

such that

$$||e_1|| + ||e_2|| + ... + ||e_n|| + ||e_{n+1}|| < w(C)/2.$$

Denote

$$C_n = \cap\{C + \varepsilon_1 e_1 + \varepsilon_2 e_2 + ... + \varepsilon_n e_n : \varepsilon \in \{-1, 1\}^n\}.$$

By virtue of the inductive assumption, we have

$$w(C_n) \geq w(C) - 2(||e_1|| + ||e_2|| + ... + ||e_n||) > 2||e_{n+1}||.$$

Further, according to Lemma 2,

$$w((C_n + e_{n+1}) \cap (C_n - e_{n+1})) \geq w(C_n) - 2||e_{n+1}||$$

$$\geq w(C) - 2(||e_1|| + ||e_2|| + ... + ||e_n|| + ||e_{n+1}||).$$

At the same time, it is clear that

$$(C_n + e_{n+1}) \cap (C_n - e_{n+1}) = C_{n+1},$$

where

$$C_{n+1} = \cap\{C + \varepsilon_1 e_1 + \varepsilon_2 e_2 + ... + \varepsilon_n e_n + \varepsilon_{n+1} e_{n+1} : \varepsilon \in \{-1, 1\}^{n+1}\}.$$

This ends the proof of Lemma 3. □

Let Γ_1 and Γ_2 be two distinct parallel affine hyperplanes in the space $\mathbf{R}^m$ and let H be the strip determined by these hyperplanes. Also, let us assume that Γ_1 is a positive hyperplane and Γ_2 is a negative hyperplane. In this case, we shall say that the strip H is oriented by its pair (Γ_1, Γ_2) (it makes sense to say that $+1$ is assigned to Γ_1 and -1 is assigned to Γ_2).

Define a vector $e \in \mathbf{R}^m$ by the following two conditions:

(*) e is perpendicular to both these hyperplanes and $||e||$ is equal to $w(H)/2$;

(**) e is directed from the negative hyperplane towards the positive hyperplane.

Evidently, the conditions (*) and (**) uniquely determine the vector e, so one can state that e is canonically associated with the strip H whose boundary hyperplanes Γ_1 and Γ_2 are treated as positive and negative, respectively.

It is clear that if we consider Γ_1 as a negative hyperplane and Γ_2 as a positive hyperplane, then the vector canonically associated with H is equal to $-e$. In other words, the orientation (Γ_2, Γ_1) of H implies the replacement of e by $-e$. Formally, we may write $-H$ for the strip H equipped with the orientation (Γ_2, Γ_1).

Lemma 4. *If H is an oriented strip in the space $\mathbf{R}^m$, then the equation of its positive hyperplane is*

$$\langle z, e\rangle + c = \langle e, e\rangle = e^2$$

and the equation of its negative hyperplane is

$$\langle z, e\rangle + c = -\langle e, e\rangle = -e^2,$$

where z is a variable point of a hyperplane, e is the vector canonically associated with H, and c is a real constant uniquely determined by the oriented strip H.

Conversely, let (e, c) be any pair, where e is a nonzero vector in $\mathbf{R}^m$ and c is a real constant. Then this pair uniquely determines in $\mathbf{R}^m$ two parallel affine hyperplanes Γ_1 and Γ_2 given by the equations

$$\langle z, e\rangle + c = e^2, \qquad \langle z, e\rangle + c = -e^2,$$

respectively, and so (e, c) determines the corresponding oriented strip H whose width is equal to $2||e||$.

The easy analytic proof of this lemma is left to the reader.

Thus, according to Lemma 4, the oriented strip H produces the uniquely determined pair (e, c) and it is not hard to see that $-H$ produces $(-e, -c)$.

Naturally, we shall say in the sequel that the pair (e, c) is canonically associated with H.

It also follows from the above lemma that a point z of the space $\mathbf{R}^m$ belongs to H if and only if

$$\langle z, e\rangle + c \leq e^2, \qquad \langle z, e\rangle + c \geq -e^2.$$

It is important for our further purposes to observe that this condition of the membership of z to H does not depend on the orientation of H (more precisely, the condition is invariant under the changing an orientation of H). So, a point z of $\mathbf{R}^m$ does not belong to the oriented strip H if and only if

$$\langle z, \varepsilon e\rangle + \varepsilon c > (\varepsilon e)^2$$

for a certain $\varepsilon \in \{-1, 1\}$ depending on z.

Let now $H_1, H_2, \ldots, H_n$ be finitely many oriented strips in the Euclidean space $\mathbf{R}^m$ and let

$$(e_1, c_1), \quad (e_2, c_2), \quad \ldots, \quad (e_n, c_n)$$

be the pairs canonically associated with $H_1, H_2, \ldots, H_n$, respectively. Obviously, all these strips produce a finite decomposition $\mathcal{P}$ of the entire space $\mathbf{R}^m$ into m-dimensional convex polyhedral sets (among them there are unbounded ones). Each facet of any such polyhedral set is contained in the boundary hyperplane of some strip H_i, where $i \in \{1, 2, ..., n\}$.

Further, for every n-tuple

$$\varepsilon = (\varepsilon_1, \varepsilon_2, ..., \varepsilon_n) \in \{-1, 1\}^n$$

and for every vector e_i ($1 \leq i \leq n$), we stipulate

$$D(\varepsilon, i) = \{z \in \mathbf{R}^m : \langle z, \varepsilon_i e_i \rangle + \varepsilon_i c_i \geq (\varepsilon_i e_i)^2\}.$$

In view of the remarks made above, it can readily be seen that $D(\varepsilon, i)$ is a closed half-space in $\mathbf{R}^m$ which does not contain the strip H_i and the boundary hyperplane of which coincides with the positive boundary hyperplane of $\varepsilon_i H_i$. In particular, we have

$$D(\varepsilon, i) \cap D(-\varepsilon, i) = \emptyset.$$

In connection with the half-spaces of the above-mentioned type, it is useful to observe that a point $z \in \mathbf{R}^m$ does not belong to the interior of the set

$$H_1 \cup H_2 \cup ... \cup H_n$$

if and only if $z \in P_\varepsilon$ for some $\varepsilon \in \{-1, 1\}^n$, where

$$P(\varepsilon) = D(\varepsilon, 1) \cap D(\varepsilon, 2) \cap ... \cap D(\varepsilon, n).$$

In other words, we have the following statement.

Lemma 5. *The family of polyhedral sets*

$$\{P(\varepsilon) : \varepsilon \in \{-1, 1\}^n\}$$

consists of all those members of the decomposition $\mathcal{P}$ of $\mathbf{R}^m$, which do not have common points with the interior of $H_1 \cup H_2 \cup ... \cup H_n$.

A detailed proof of Lemma 5 is also left to the reader.

For every n-tuple $\delta = (\delta_1, \delta_2, ..., \delta_n) \in \mathbf{R}^n$, let us introduce one more notation. Consider the pair

$$(\delta_1 e_1 + \delta_2 e_2 + ... + \delta_n e_n, \delta_1 c_1 + \delta_2 c_2 + ... + \delta_n c_n) = (e, c).$$

If $e \neq 0$, then we have the oriented strip $H(\delta)$ canonically associated with this pair. In such a case, let $D(\delta)$ denote the closed half-space of $\mathbf{R}^m$ which does not contain the strip $H(\delta)$ and whose boundary hyperplane coincides with the positive boundary hyperplane of $H(\delta)$.

In our considerations below we always assume that

$$e = \delta_1 e_1 + \delta_2 e_2 + ... + \delta_n e_n \neq 0$$

whenever

$$\delta = (\delta_1, \delta_2, ..., \delta_n) \in \{-1, 0, 1\}^n \ \& \ \delta \neq (0, 0, ..., 0).$$

So we have the corresponding oriented strip $H(\delta)$ for any such δ.

The reader will see that this additional assumption does not restrict the generality of our argument in the process of solving Tarski's problem (see Exercise 7 of this chapter).

Further, let us denote

$$Q(\varepsilon) = \cap\{D((\varepsilon - \varepsilon')/2) : \varepsilon' \in \{-1, 1\}^n\}$$

for each $\varepsilon \in \{-1, 1\}^n$ (here, by definition, $D(0, 0, ..., 0) = \mathbf{R}^m$).

Under the above notation, the next statement holds true.

Lemma 6. *For any $\varepsilon \in \{-1, 1\}^n$, the inclusion $Q(\varepsilon) \subset P(\varepsilon)$ is valid.*

Proof. Take $\varepsilon \in \{-1, 1\}^n$ and pick arbitrarily a natural index $i \in [1, n]$. Consider $\varepsilon' \in \{-1, 1\}^n$ defined by

$$\varepsilon' = (\varepsilon'_1, \varepsilon'_2, ..., \varepsilon'_n) = (\varepsilon_1, \varepsilon_2, ..., \varepsilon_{i-1}, -\varepsilon_i, \varepsilon_{i+1}, ..., \varepsilon_n).$$

Then it is easy to see that

$$D((\varepsilon - \varepsilon')/2) = D(\varepsilon, i).$$

Therefore, ranging i over $\{1, 2, ..., n\}$, we may write

$$Q(\varepsilon) \subset D(\varepsilon, 1) \cap D(\varepsilon, 2) \cap ... \cap D(\varepsilon, n) = P(\varepsilon),$$

which completes the proof of Lemma 6. □

Intuitively it is clear that the union of all polyhedral sets of the form $P(\varepsilon)$ is much bigger, than the union of all given strips $H_1, H_2, \ldots, H_n$.

Speaking more precisely, we now are able to formulate and prove the following key auxiliary proposition.

Lemma 7. *For the polyhedral sets $P(\varepsilon)$ $(\varepsilon \in \{-1, 1\}^n)$, the equality*

$$\mathbf{R}^m = \cup\{P(\varepsilon) - \varepsilon_1 e_1 - \varepsilon_2 e_2 - ... - \varepsilon_n e_n : \varepsilon \in \{-1, 1\}^n\}$$

holds true.

Proof. By virtue of Lemma 6, it suffices to demonstrate that

$$\mathbf{R}^m = \cup\{Q(\varepsilon) - \varepsilon_1 e_1 - \varepsilon_2 e_2 - ... - \varepsilon_n e_n : \varepsilon \in \{-1, 1\}^n\}.$$

In other words, for any point $z \in \mathbf{R}^m$, we must find $\varepsilon \in \{-1, 1\}^n$ such that

$$z + \varepsilon_1 e_1 + \varepsilon_2 e_2 + ... + \varepsilon_n e_n \in Q(\varepsilon).$$

Remembering the definition of $Q(\varepsilon)$, the last relation is equivalent to

$$z + \varepsilon_1 e_1 + \varepsilon_2 e_2 + ... + \varepsilon_n e_n \in D((\varepsilon - \varepsilon')/2)$$

for all $\varepsilon' \in \{-1,1\}^n$. By definition of $D((\varepsilon-\varepsilon')/2)$, this means that

$$\langle z+\varepsilon_1 e_1+\varepsilon_2 e_2+\ldots+\varepsilon_n e_n, (\varepsilon_1-\varepsilon_1')e_1/2+(\varepsilon_2-\varepsilon_2')e_2/2+\ldots+(\varepsilon_n-\varepsilon_n')e_n/2\rangle$$
$$+((\varepsilon_1-\varepsilon_1')c_1/2+(\varepsilon_2-\varepsilon_2')c_2/2+\ldots+(\varepsilon_n-\varepsilon_n')c_n/2)$$
$$\geq ((\varepsilon_1-\varepsilon_1')e_1/2+(\varepsilon_2-\varepsilon_2')e_2/2+\ldots+(\varepsilon_n-\varepsilon_n')e_n/2)^2.$$

After straightforward computation, one can see that the last relation is equivalent to the following

$$2\langle z, \varepsilon_1 e_1+\ldots+\varepsilon_n e_n\rangle+2(\varepsilon_1 c_1+\ldots+\varepsilon_n c_n)+(\varepsilon_1 e_1+\ldots+\varepsilon_n e_n)^2 \geq$$
$$2\langle z, \varepsilon_1' e_1+\ldots+\varepsilon_n' e_n\rangle+2(\varepsilon_1' c_1+\ldots+\varepsilon_n' c_n)+(\varepsilon_1' e_1+\ldots+\varepsilon_n' e_n)^2.$$

Consequently, the required ε always exists and, actually, is identical to one of those n-tuples ε' for which the expression

$$2\langle z, \varepsilon_1' e_1+\ldots+\varepsilon_n' e_n\rangle+2(\varepsilon_1' c_1+\ldots+\varepsilon_n' c_n)+(\varepsilon_1' e_1+\ldots+\varepsilon_n' e_n)^2$$

attains maximum value.

Lemma 7 has thus been proved. □

Summarizing all the stated above, we get the positive solution of Tarski's problem, first given by Bang [14] and Fenchel [95].

Theorem 1. *Let C be a compact convex body in $\mathbf{R}^m$ covered by finitely many strips H_1, H_2, . . ., H_n whose widths are, respectively,*

$$w_1, \quad w_2, \quad \ldots, \quad w_n.$$

Then, for the width of C, the inequality

$$w(C) \leq w_1+w_2+\ldots+w_n$$

holds true.

Proof. Associate with H_1, H_2, . . . , H_n their canonical pairs

$$(e_1,c_1), \quad (e_2,c_2), \quad \ldots, \quad (e_n,c_n).$$

By definition of these pairs, we have

$$w(H_1)=2||e_1||, \quad w(H_2)=2||e_2||, \quad \ldots, \quad w(H_n)=2||e_n||.$$

Suppose to the contrary that

$$w(C)>w_1+w_2+\ldots+w_n=2(||e_1||+||e_2||+\ldots+||e_n||)$$

and consider the set

$$K=C\setminus \operatorname{int}(H_1\cup H_2\cup\ldots\cup H_n).$$

According to the assumption of the theorem, K is small in the sense that its dimension is strictly less than m (because K is contained in the union of a finite family of affine hyperplanes of $\mathbf{R}^m$). At the same time, in view of Lemma 5, we have

$$K = \cup\{C \cap P(\varepsilon) : \varepsilon \in \{-1, 1\}^n\},$$

where the convex polyhedral sets $P(\varepsilon)$ were described earlier. Each set $C \cap P(\varepsilon)$ is small, too, as well as its translate of the form

$$(C \cap P(\varepsilon)) - \varepsilon_1 e_1 - \varepsilon_2 e_2 - ... - \varepsilon_n e_n.$$

In our situation, the union of a finite family of small sets is also small, so

$$K' = \cup\{(C \cap P(\varepsilon)) - \varepsilon_1 e_1 - \varepsilon_2 e_2 - ... - \varepsilon_n e_n : \varepsilon \in \{-1, 1\}^n\}$$

is a small set. Let us denote

$$K'' = \cap\{C - \varepsilon_1 e_1 - \varepsilon_2 e_2 - ... - \varepsilon_n e_n : \varepsilon \in \{-1, 1\}^n\}.$$

Then, by virtue of Lemma 7, we get

$$K'' = \cup\{K'' \cap (P(\varepsilon) - \varepsilon_1 e_1 - \varepsilon_2 e_2 - ... - \varepsilon_n e_n) : \varepsilon \in \{-1, 1\}^n\} \subset K'.$$

Consequently, the set K'' being a subset of the small set K' must be small, too.

However, according to Lemma 3, the same set

$$K'' = \cap\{C - \varepsilon_1 e_1 - \varepsilon_2 e_2 - ... - \varepsilon_n e_n : \varepsilon \in \{-1, 1\}^n\}$$

is a convex body whose width is strictly positive. Namely, $w(K'')$ is greater than or equal to

$$w(C) - 2(||e_1|| + ||e_2|| + ... + ||e_n||) > 0.$$

The obtained contradiction ends the proof of Theorem 1. □

As a straightforward consequence of the above theorem we get the following statement.

Theorem 2. *If a compact convex body C in the space $\mathbf{R}^m$ is decomposed into finitely many convex parts $C_1, C_2, \ldots, C_n$, then the least 1-dimensional orthogonal projection of C does not exceed the sum of the least 1-dimensional orthogonal projections of $C_1, C_2, \ldots, C_n$.*

Remark 1. Theorem 1 was extended and generalized in several directions. However, we do not touch upon those results here and refer the reader to [12], [22], and [123].

EXERCISES

1. Let C be a compact convex set in the space $\mathbf{R}^m$, where $m \geq 1$, and let d denote the diameter of C (suppose that $d > 0$).

   Verify that:

   (a) there exists a line segment $[z_1, z_2] \subset C$ such that $||z_1 - z_2|| = d$;

   (b) $[z_1, z_2]$ is an affine diameter of C.

   More precisely, check that the affine hyperplane Γ_1 (respectively, the affine hyperplane Γ_2) passing through the point z_1 (respectively, through the point z_2) and perpendicular to the line segment $[z_1, z_2]$ is a supporting hyperplane of C. Moreover, the equalities

   $$\Gamma_1 \cap C = \{z_1\}, \quad \Gamma_2 \cap C = \{z_2\}$$

   hold true.

2. Let C be a compact convex body in the space $\mathbf{R}^m$, where $m \geq 2$, and let e be any nonzero vector in $\mathbf{R}^m$.

   Show that there exist two points $z_1 \in \mathrm{bd}(C)$ and $z_2 \in \mathrm{bd}(C)$ such that the vector $z_1 - z_2$ is parallel to e and the chord $[z_1, z_2]$ of C is an affine diameter of C.

   For this purpose, argue as follows. First of all, by using the compactness of C, infer that there exists a chord of C parallel to e and having the maximal length. Let z_1 and z_2 be the endpoints of such a chord. Further, consider the body $C + (z_1 - z_2)$ and observe that:

   (a) z_1 is a common point of C and $C + (z_1 - z_2)$;

   (b) the bodies C and $C + (z_1 - z_2)$ have no common internal points.
   According to one of the separations theorems for disjoint open convex sets (see Appendix 1), there is an affine hyperplane Γ in $\mathbf{R}^m$ separating the interior of C and the interior of $C + (z_1 - z_2)$. Check that:

   (c) Γ passes through z_1 and is a supporting hyperplane of C;

   (d) the affine hyperplane $\Gamma + (z_2 - z_1)$ passes through z_2 and also is a supporting hyperplane of C.
   Conclude that the chord $[z_1, z_2]$ of C is an affine diameter of C.

3. Let C be a closed disc in the plane $\mathbf{R}^2$ and suppose that the strips H_1, H_2, . . . , H_n in $\mathbf{R}^2$ collectively cover this C.

   Give a direct argument for proving the inequality

   $$w(C) \leq w_1 + w_2 + ... + w_n,$$

   where w_1, w_2, . . . , w_n are, respectively, the widths of H_1, H_2, . . . , H_n.

For this purpose, consider in the space $\mathbf{R}^3 = \mathbf{R}^2 \times \mathbf{R}$ the sphere S whose center coincides with the center of C and whose radius equals the radius of C. Also, for each natural index $i \in [1, n]$, consider the strip H'_i in $\mathbf{R}^3$ whose orthogonal projection on $\mathbf{R}^2$ is identical with H_i. Check that

$$S = (H'_1 \cap S) \cup (H'_2 \cap S) \cup ... \cup (H'_n \cap S)$$

and, taking into account that the area of each spherical zone $H'_i \cap S$ does not exceed $\pi w(C) w_i$, obtain the required result.

4. Give a detailed proof of Lemma 4.

5. Give a detailed proof of Lemma 5.

6*. Let X be a nonempty compact subset of the Euclidean space $\mathbf{R}^m$, where $m \geq 2$.

Check that there exist two parallel affine hyperplanes G_1 and G_2 in $\mathbf{R}^m$ such that X is contained in the strip $H(G_1, G_2)$ determined by G_1 and G_2, and the width of $H(G_1, G_2)$ is minimal (in some cases it may happen that G_1 and G_2 coincide). Obviously,

$$G_1 \cap X \neq \emptyset, \quad G_2 \cap X \neq \emptyset.$$

The width of $H(G_1, G_2)$ is usually called the width of a given set X and is denoted by $w(X)$ (obviously, this definition is compatible with the definition of the width of a compact convex body in $\mathbf{R}^m$).

Show that:

(a) $w(X) = w(\text{conv}(X))$ for any nonempty compact set $X \subset \mathbf{R}^m$;

(b) if $n \geq 2$ and $X_1, X_2, \ldots, X_n$ are nonempty compact subsets of the space $\mathbf{R}^m$, then

$$w(X_1 + X_2 + ... + X_n) \geq w(X_1) + w(X_2) + ... + w(X_n);$$

(c) there are many nonempty compact sets $Y_1, Y_2, \ldots, Y_m$ in $\mathbf{R}^m$ such that

$$w(Y_1) = w(Y_2) = ... = w(Y_m) = 0, \quad w(Y_1 \cup Y_2 \cup ... \cup Y_m) > 0.$$

For (b), first consider the case $n = 2$ and then use induction on n.

For (c), take e.g. all those edges of the cube $[0, 1]^m$ which are incident to a certain vertex of $[0, 1]^m$.

Remark 2. Assertion (c) of Exercise 6 shows that, in general, the analog of Tarski's plank problem for a non-convex compact set $X \subset \mathbf{R}^m$ has a negative solution.

7. Let H_1, H_2, . . . , H_n be a finite family of strips in the space $\mathbf{R}^m$ with the corresponding pairs (e_1, c_1), (e_2, c_2), . . . , (e_n, c_n) and let, for a convex compact body $C \subset \mathbf{R}^m$, the inclusion

$$C \subset H_1 \cup H_2 \cup ... \cup H_n$$

be satisfied.

Demonstrate that, for every real $r > 0$, there exist strips H'_1, H'_2, . . . , H'_n in $\mathbf{R}^m$ with the corresponding pairs (e'_1, c'_1), (e'_2, c'_2), . . . , (e'_n, c'_n) such that:

(i) $C \subset H'_1 \cup H'_2 \cup ... \cup H'_n$;

(ii) for each $i \in \{1, 2, ..., n\}$, one has $w(H'_i) \leq w(H_i) + r$;

(iii) $\delta_1 e'_1 + \delta_2 e'_2 + ... + \delta_n e'_n \neq 0$ whenever the relation

$$\delta = (\delta_1, \delta_2, ..., \delta_n) \in \{-1, 0, 1\}^n \ \ \& \ \ \delta \neq (0, 0, ..., 0)$$

holds true.

Remark 3. In view of Exercise 7, to solve positively Tarski's plank problem, it suffices to restrict considerations to the case where the given strips

$$H_1, \ \ H_2, \ . \ . \ . \ , \ \ H_n$$

in $\mathbf{R}^m$, which collectively cover a compact convex body $C \subset \mathbf{R}^m$, satisfy the analog of condition (iii).

CHAPTER 13

Borsuk's conjecture

For a long period of time, one of the main and best known open questions in combinatorial (discrete, convex) geometry was Borsuk's famous problem. It is formulated as follows (see [40]).

Borsuk's Problem. Is it true that in the Euclidean space $\mathbf{R}^m$ every bounded set of diameter $d > 0$ admits a decomposition into $m + 1$ many subsets, all of which have diameters strictly less than d?

For lower dimensions $m = 1, 2, 3$ the just formulated question was solved positively. However, even for $m = 3$ the solution is very nontrivial (in this connection, see e.g. [34], [36], [157], [287], and references given therein; cf. also some exercises of the present chapter).

Borsuk's conjecture means that the answer to the above problem is positive for all natural numbers m.

This conjecture was extremely attractive for many people working in the area of discrete and combinatorial geometry, and there were a lot of attempts to confirm the conjecture. However, it turned out that, for higher dimension m, the answer is negative. So this beautiful conjecture was finally rejected (see [154]).

In fact, the diameter of any bounded subset X of $\mathbf{R}^m$ is equal to the diameter of $\text{conv}(X)$. Therefore, without loss of generality, it suffices to study Borsuk's problem (and other questions of similar type) only for bounded convex sets in $\mathbf{R}^m$.

Here we wish to discuss the negative solution to Borsuk's conjecture. We follow the most simplified version of an argument leading to the desired result (see [7]). We will demonstrate in the sequel that some purely combinatorial properties of subsets of a multi-dimensional Euclidean cube play a key role in the process of solving Borsuk's problem (in this context, cf. Chapter 3).

Let us begin with several auxiliary notions and relatively simple facts about these notions.

Our starting object is the standard cube $[-1, 1]^m$ in the Euclidean space $\mathbf{R}^m$, where $m \geq 1$.

DOI: 10.1201/9781003458708-13

We denote by V_m the set of all vertices of $[-1,1]^m$. So each vector $v \in V_m$ has the form

$$v = (v_1, v_2, ..., v_m) \qquad (v_k \in \{-1, 1\},\ k = 1, 2, ..., m).$$

Lemma 1. *For any two vertices $v \in V_m$ and $u \in V_m$, one has*

$$||v - u||^2 = 2m - 2\langle v, u\rangle,$$

where $\langle v, u\rangle$ denotes, as usual, the inner (scalar) product of the vectors v and u.

This lemma is trivial, so we do not give its proof here.

Further, we may consider the elements of V_{m^2} as $(m \times m)$-matrices, all terms of which are either 1 or -1. Let us define a mapping $f : V_m \to V_{m^2}$ by the formula

$$f(v) = (v_i \cdot v_j)_{(i,j)\in\{1,2,...,m\}^2} \qquad (v \in V_m).$$

The mapping f has the following simply verified properties.

Lemma 2. *$f(v) = f(-v)$ for any $v \in V_m$.*

Lemma 3. *If $f(v) = f(u)$, then $v = u$ or $v = -u$.*

Lemma 2 is obvious and Lemma 3 can easily be proved by induction on m.

Now, let us denote

$$V_{m-1} = \{v \in V_m : v_1 = 1\}.$$

Clearly, V_{m-1} is the set of all vertices of the special facet of $[-1,1]^m$.

Lemma 4. *The restriction of f to V_{m-1} is injective, i.e., for any two distinct vertices $v \in V_{m-1}$ and $u \in V_{m-1}$, one has $f(v) \neq f(u)$.*

Proof. The injectivity of the restriction of f to V_{m-1} immediately follows from Lemma 3, because V_{m-1} contains no two points which are symmetric with respect to the origin of $\mathbf{R}^m$. □

Lemma 5. *If $f(v) = \{z_{ij} : (i,j) \in \{1, 2, ..., m\}^2\}$, then*

$$z_{ii} = 1, \quad z_{ij} = z_{ji} \qquad ((i,j) \in \{1, 2, ..., m\}^2).$$

In other words, the $(m \times m)$-matrix $\{z_{ij} : (i,j) \in \{1, 2, ..., m\}^2\}$ is symmetric and all terms of the main diagonal of this matrix are equal to 1.

The above lemma is also trivial.

Lemma 6. *For any vectors $v \in V_m$ and $u \in V_m$, the equality*

$$\langle f(v), f(u)\rangle \ = \ \langle v, u\rangle^2$$

holds true.

Proof. The required result immediately follows from the formula

$$\sum\{(v_iv_j)(u_iu_j) : (i,j) \in \{1,2,...,m\}^2\} = (\sum\{v_iu_i : 1 \leq i \leq m\})^2,$$

which is readily verified. □

The next statement is less trivial.

Lemma 7. *Let the dimension m be divisible by 4. Let $v \in V_m$ and $u \in V_m$ be such that the number of terms $v_i = -1$ is even and the number of terms $u_i = -1$ is also even.*

Then the integer $\langle v, u\rangle$ is divisible by 4.

Proof. It is convenient to denote

$$I = \{i \in \{1,2,...,m\} : v_i = -1\}, \quad J = \{i \in \{1,2,...,m\} : u_i = -1\},$$

$$K = I \cap J.$$

Under this notation, it is not hard to see that

$$\langle v, u\rangle = \text{card}(\{1,2,...,m\} \setminus (I \cup J)) + \text{card}(K) - \text{card}(I \setminus K) - \text{card}(J \setminus K)$$

$$= m - \text{card}(I \cup J) - \text{card}(I) - \text{card}(J) + 3\text{card}(K) =$$

$$m - (\text{card}(I) + \text{card}(J) - \text{card}(K)) - \text{card}(I) - \text{card}(J) + 3\text{card}(K)$$

$$= m - 2(\text{card}(I) + \text{card}(J)) + 4\text{card}(K).$$

Keeping in mind the fact that both $\text{card}(I)$ and $\text{card}(J)$ are even natural numbers, we obtain the required result.

Lemma 7 has thus been proved. □

Further, let us introduce the set

$$B = \{v \in V_{m-1} : \text{ the number of } v_i = -1 \text{ is even}\}.$$

It is easy to compute the cardinality of this B.

Lemma 8. *If $m \geq 2$ is even, then the equality* $\text{card}(B) = 2^{m-2}$ *holds true.*

Proof. Let $m \geq 2$ be an even natural number. Then $m - 1$ is odd. Consider any element

$$v = (1, v_2, v_3, ..., v_m) \in B.$$

Evidently, the element $(1, -v_2, -v_3, ..., -v_m)$ belongs to the set $V_{m-1} \setminus B$. In other words, the mapping

$$(1, v_2, v_3, ..., v_m) \to (1, -v_2, -v_3, ..., -v_m)$$

establishes a one-to-one correspondence between the sets B and $V_{m-1} \setminus B$. Since we have $\text{card}(V_{m-1}) = 2^{m-1}$, we finally get $\text{card}(B) = 2^{m-2}$, which completes the proof. □

Taking into account Lemma 6, we also obtain

Lemma 9. *For any two vectors $v \in V_m$ and $u \in V_m$, the equality*

$$||f(v) - f(u)||^2 = 2m^2 - 2(\langle v, u\rangle)^2$$

is valid and, consequently, one has the following two relations:

(1) $||f(v) - f(u)|| \leq 2^{1/2}m$;

(2) $||f(v) - f(u)|| = 2^{1/2}m \Leftrightarrow \langle v, u\rangle = 0$.

From Lemma 9 one can deduce the next auxiliary statement.

Lemma 10. *If $m \geq 4$ and m is divisible by 4, then the set B contains a pair of orthogonal vectors (points).*

So the diameter of the set $f(B)$ is equal to $2^{1/2}m$.

Proof. Suppose first that $m = 4$. Then the vectors

$$(1, -1 - 1, 1), \qquad (1, 1 - 1 - 1)$$

belong to B and are orthogonal. Now, it directly follows from this fact that if m is divisible by 4, then B necessarily contains two orthogonal vectors. Keeping in mind Lemma 9, we conclude that the diameter of $f(B)$ is equal to $2^{1/2}m$, which ends the proof. □

Lemma 11. *If the set $f(B)$ is arbitrarily decomposed into finitely many subsets Y_1, Y_2, ... , Y_k, then the set B is decomposed into its subsets*

$$X_1 = f^{-1}(Y_1), \quad X_2 = f^{-1}(Y_2), \quad \ldots, \quad X_k = f^{-1}(Y_k)$$

and $\mathrm{card}(X_i) = \mathrm{card}(Y_i)$ *for each index* $i \in \{1, 2, ..., k\}$.

Proof. Indeed, in view of the inclusion $B \subset V_{m-1}$, this statement directly follows from Lemma 4. □

Lemma 12. *Under the notation of Lemma 11, if the diameters of all sets Y_i are strictly less than $2^{1/2}m$, then no set X_i contains a pair of orthogonal vectors.*

Consequently, if $k = m^2 + 1$ and the inequality

$$k(\max\{\mathrm{card}(X_i) : 1 \leq i \leq k\}) < 2^{m-2}$$

holds true for every decomposition $\{Y_1, Y_2, ..., Y_k\}$ of the set $f(B) \subset \mathbf{R}^{m^2}$, then Borsuk's conjecture is false for $f(B)$.

This statement trivially follows from Lemma 11.

Lemma 13. *Let $m = 4p$, where p is an odd prime number. Let v and u be any two distinct points from the set B such that $\langle v, u\rangle \neq 0$.*

Then the integer $\langle v, u\rangle$ is not divisible by p.

Proof. Indeed, supposing for a moment otherwise, we would have

$$0 < |\langle v, u\rangle| < m = 4p$$

and, therefore,

$$\langle v, u\rangle \in \{p, -p, 2p, -2p, 3p, -3p\}.$$

On the other hand, by virtue of Lemma 7, $\langle v, u\rangle$ should be divisible by 4, and we obtain a contradiction which completes the proof. □

Lemma 14. *Let p be a prime number. Consider the following polynomial of one real variable t:*

$$P(t) = (t-1)(t-2)...(t-(p-1)).$$

For any integer t, the integer $P(t)$ is divisible by p if and only if t is not divisible by p.

The proof of Lemma 14 is almost trivial and is left to the reader.

Now, we have to introduce some polynomials of several variables.

Take the above-mentioned polynomial $P(t)$ and, for every point

$$b = (1, b_2, b_3, ..., b_m) \in B,$$

introduce the polynomial

$$F_b(x_2, x_3, ..., x_m) = P(\langle b, x\rangle) = P(1 + b_2x_2 + b_3x_3 + ... + b_mx_m),$$

where

$$x = (1, x_2, x_3, ..., x_m).$$

Obviously, $F_b(x_2, ..., x_m)$ is a finite sum of monomials of the form

$$c_{i_2 i_3 ... i_m} x_2^{i_2} x_3^{i_3} ... x_m^{i_m},$$

where $c_{i_2 i_3 ... i_m}$ is some integer and

$$i_2 = 2r_2 + j_2, \quad i_3 = 2r_3 + j_3, \quad . \; . \; . \; , \quad i_m = 2r_m + j_m,$$

$$\{r_2, r_3, ..., r_m\} \subset \mathbf{N},$$

$$\{j_2, j_3, ..., j_m\} \subset \{0, 1\}.$$

Replace each expression $c_{i_2 i_3 ... i_m} x_2^{i_2} x_3^{i_3} ... x_m^{i_m}$ by the expression of the form

$$c_{i_2 i_3 ... i_m} x_2^{j_2} x_3^{j_3} ... x_m^{j_m}$$

and the obtained in this way new polynomial denote by $F_b^*(x_2, x_3, ..., x_m)$.

Clearly, the degree of $F_b^*(x_2, x_3, ..., x_m)$ does not exceed $p-1$ (because the degree of $P(t)$ is $p-1$).

Lemma 15. *If $m = 4p$ and a set $B' \subset B$ contains no pair of orthogonal vectors, then the polynomials*

$$F_b^*(x_2, x_3, ..., x_m) \qquad (b \in B')$$

are linearly independent over the field $\mathbf{Q}$ *of all rational numbers.*

Proof. Let $B' = \{b^{(1)}, b^{(2)}, ..., b^{(k)}\}$ be an injective enumeration of a set B' which contains no pair of orthogonal vectors, and suppose on the contrary that

$$q_1 F_{b^{(1)}}^*(x_2, x_3, ..., x_m) + q_2 F_{b^{(2)}}^*(x_2, x_3, ..., x_m) + ... + q_k F_{b^{(k)}}^*(x_2, x_3, ..., x_m) = 0,$$

where all q_j $(1 \leq j \leq k)$ are rational numbers and at least one of them differs from zero. We may assume, without loss of generality, that all q_j are integers. Furthermore, we may assume that at least one of q_j is not divisible by p. Specifically, let us stipulate that the integer q_1 is not divisible by p.

Taking in the above equality

$$x_2 = b_2^{(1)}, \quad x_3 = b_3^{(1)}, \quad . . . , \quad x_m = b_m^{(1)},$$

we get the value

$$F_{b^{(1)}}^*(x_2, x_3, ..., x_m) = P(\langle b^{(1)}, b^{(1)} \rangle) = P(m) = P(4p),$$

which is not divisible by p (see Lemma 14). Consequently, $q_1 F_{b^{(1)}}^*(x_2, x_3, ..., x_m)$ is not divisible by p. On the other hand, all the expressions

$$F_{b^{(2)}}^*(x_2, x_3, ..., x_m) = P(\langle b^{(1)}, b^{(2)} \rangle), \ . . . , \ F_{b^{(k)}}^*(x_2, x_3, ..., x_m) = P(\langle b^{(1)}, b^{(k)} \rangle)$$

are divisible by p (see Lemmas 13 and 14). Therefore, the sum

$$q_1 F_{b^{(1)}}^*(x_2, x_3, ..., x_m) + q_2 F_{b^{(2)}}^*(x_2, x_3, ..., x_m) + ... + q_k F_{b^{(k)}}^*(x_2, x_3, ..., x_m)$$

is not divisible by p and, clearly, cannot be equal to zero.

The obtained contradiction ends the proof of Lemma 15. □

Below, for two natural numbers m and n, where $n \leq m$, we will use the notation

$$C_m^n = \frac{m!}{n!(m-n)!}.$$

As is well known, C_m^n is equal to the number of all n-element subsets of an m-element set.

Under this notation, we have the following auxiliary proposition.

Lemma 16. *Let p be an odd prime, $m = 4p$, and let $x_2, x_3, . . . , x_m$ be independent real variables. Consider the monomials of the form $x_{i_1} x_{i_2} ... x_{i_s}$, where $i_1, i_2, . . . , i_s$ are distinct indices from $\{2, 3, ..., m\}$ and s ranges over $\{0, 1, ..., p-1\}$.*

These monomials are linearly independent over $\mathbf{Q}$ *and their total number is less than or equal to $(m/4) C_{m-1}^{m/4-1}$.*

Proof. The linear independence over **Q** of the indicated monomials can readily be obtained by induction on m, and we leave the corresponding details to the reader. Notice also that the total number of the above-mentioned monomials is equal to

$$C^0_{m-1} + C^1_{m-1} + ... + C^{p-1}_{m-1}$$

and this number does not exceed pC^{p-1}_{m-1} which equals $(m/4)C^{m/4-1}_{m-1}$.

Lemma 16 has thus been proved. □

Now, let us formulate and prove the crucial auxiliary statement.

Lemma 17. *Suppose that $p > 2$ is a sufficiently large prime number and put $m = 4p$. Let Z be a subset of B not containing a pair of orthogonal vectors.*

Then the relation

$$\mathrm{card}(Z) \leq (m/4)C^{m/4-1}_{m-1} < 2^{m-2}/(m^2+1)$$

holds true.

Proof. First of all, observe that the second inequality

$$(m/4)C^{m/4-1}_{m-1} < 2^{m-2}/(m^2+1)$$

is an easy consequence of Stirling's well-known asymptotic formula:

$$m! = (m/\mathrm{e})^m(2\pi m)^{1/2}(1+\alpha_m),$$

where e is the Napier constant and α_m tends to zero as m tends to infinity.

Now, consider the polynomials

$$F^*_z(x_2, x_3, ..., x_m) \qquad (z \in Z).$$

As was shown earlier, these polynomials are linearly independent over **Q**. At the same time, these polynomials can be linearly expressed (over the same **Q**) by the monomials indicated in Lemma 16.

Therefore, the total number of polynomials $F^*_z(x_2, x_3, ..., x_m)$ or, equivalently, the cardinality of Z does not exceed the total number of monomials of Lemma 16, i.e., $\mathrm{card}(Z)$ does not exceed $(m/4)C^{m/4-1}_{m-1}$.

This completes the proof of Lemma 17. □

Summarizing all the stated above, we can formulate the following main result (see [7], [154]).

Theorem 1. *Let $p > 2$ be a sufficiently large prime number and let m be defined as before, i.e., $m = 4p$.*

Then there exists no partition of the set $f(B) \subset \mathbf{R}^{m^2}$ into m^2+1 many subsets of strictly smaller diameter.

Proof. Denote $k = m^2 + 1$. According to Lemmas 9 and 10, the diameter of $f(B)$ is equal to $2^{1/2}m$. Suppose to the contrary that there exists a partition of $f(B)$ into sets

$$Y_1, \; Y_2, \; \ldots, \; Y_k,$$

all of which have diameters strictly less than $2^{1/2}m$, and consider the pre-images

$$X_1 = f^{-1}(Y_1), \quad X_2 = f^{-1}(Y_2), \; \ldots, \quad X_k = f^{-1}(Y_k).$$

By virtue of Lemma 12, no set X_i $(1 \leq i \leq k)$ contains a pair of orthogonal vectors. So, remembering that p is sufficiently large and using Lemma 17, we must have

$$\mathrm{card}(X_i) < 2^{m-2}/(m^2+1) \qquad (1 \leq i \leq k).$$

Consequently,

$$2^{m-2} = \mathrm{card}(B) = \mathrm{card}(\cup\{X_i : 1 \leq i \leq k\}) < k2^{m-2}/(m^2+1) = 2^{m-2},$$

which yields a contradiction.

The obtained contradiction ends the proof. □

Remark 1. For any natural number $m > 0$, denote by $b(m)$ the least natural number having the property that every bounded subset Z of $\mathbf{R}^m$ with $\mathrm{card}(Z) \geq 2$ admits a decomposition into $b(m)$ parts, all of which have strictly smaller diameters than the diameter of Z (the existence of such $b(m)$ is not difficult to show). As known, $b(m) = m + 1$ for $m = 1, 2, 3$ and, according to Theorem 1 proved above, $b(m) > m + 1$ for sufficiently large m. The reader probably could see that the argument given above does not enable one to make any conclusion on the growth of $b(m)$ with respect to m. One may conjecture that, although $b(m) > m + 1$ for higher dimensions m, the growth of $b(m)$ is polynomial with respect to m. However, this much weaker hypothesis is false, too. Indeed, as was demonstrated in [154], the growth of $b(m)$ turns out to be non-polynomial with respect to m (for more details, see [36], [154], [157]).

Recently, the following interesting result connected with Borsuk's problem was obtained in [39].

Theorem 2. *For some sufficiently large natural number m, there exists a 2-distance subset X of the space $\mathbf{R}^m$ such that X cannot be decomposed into $m + 1$ parts of strictly smaller diameters.*

Notice that any 2-distance subset of $\mathbf{R}^m$ is necessarily finite (cf. Chapter 4). For the more precise value of m in Theorem 2 and for further details concerning the above result, we refer the reader to [39].

EXERCISES

1. Let Z be a bounded subset of the plane $\mathbf{R}^2$ and let z_1, z_2, z_3, z_4 be some points of Z such that both values $||z_1 - z_2||$ and $||z_3 - z_4||$ coincide with the diameter of Z.

Check that the line segments $[z_1, z_2]$ and $[z_3, z_4]$ have a common point.

Give an example of a 4-element subset Z of $\mathbf{R}^3$ for which the analogous assertion is not true.

2. For every finite set Z in the space $\mathbf{R}^m$, denote by $d(Z)$ the total number of subsets $\{z, z'\}$ of Z such that the value $||z - z'||$ coincides with the diameter of Z.

Prove that if $Z \subset \mathbf{R}^2$ contains exactly $n \geq 1$ elements, then

$$\text{card}(d(Z)) \leq n,$$

and the above estimation is precise.

For this purpose, use induction on n and consider the two possible cases.

(a) For any point $z \in Z$, there are at most two points z_1 and z_2 of Z such that both $||z - z_1||$ and $||z - z_2||$ are diameters of Z.
In this case, $d(Z)$ trivially does not exceed n.

(b) There exists a point $z \in Z$ for which there are at least three distinct points z_1, z_2, z_3 of Z such that $||z - z_1||$, $||z - z_2||$, $||z - z_3||$ are diameters of Z.
In this case, taking into account Exercise 1, verify that one of the points z_1, z_2, z_3, say z_i, is such that z is a unique point z' in Z for which $||z_i - z'||$ coincides with the diameter of Z. Consider the set $Z \setminus \{z_i\}$ and use the inductive assumption to this set.

3*. Preserving the same notation as in the previous exercise, show that if a finite set $Z \subset \mathbf{R}^3$ contains exactly $n \geq 2$ points, then

$$\text{card}(d(Z)) \leq 2(n-1),$$

and the above estimation is precise.

For this purpose, follow the argument presented by K. J. Swanepoel, which is based on several auxiliary statements.

(a) Let $x_1, x_2, \ldots, x_k, x_{k+1}$ be some unit vectors in a real pre-Hilbert space $(H, \langle \cdot, \cdot \rangle)$ and let

$$x_{k+1} = \sum \{t_i x_i : i = 1, 2, ..., k\},$$

where all coefficients t_i are non-negative real numbers. Suppose also that a vector $y \in H$ satisfies the relations

$$||y - x_i|| \leq 1 \qquad (i = 1, 2, ..., k).$$

Then the inequality $||y - x_{k+1}|| \leq 1$ takes place.
Indeed, one may write

$$1 = ||\sum\{t_i x_i : i = 1, 2, ..., n\}|| \leq \sum\{t_i : i = 1, 2, ..., n\}.$$

Since the relation $||y - x_i|| \leq 1$ is equivalent to $-2\langle x_i, y\rangle \leq -||y||^2$, one obtains

$$||y - \sum\{t_i x_i : i = 1, 2, ..., n\}||^2$$
$$= ||y||^2 - 2\sum\{\langle t_i x_i, y\rangle : i = 1, 2, ..., n\} + 1 \leq$$
$$(1 - \sum\{t_i : i = 1, 2, ..., n\})||y||^2 + 1 \leq 1,$$

as required.

(b) Assume, without loss of generality, that the diameter of the given set Z is equal to 1, and, for each point $z \in Z$, there are at least two points x and y in Z such that

$$||z - x|| = ||z - y|| = 1.$$

Otherwise, an easy induction on $n = \text{card}(Z)$ gives $\text{card}(d(Z)) \leq 2(n - 1)$.
Further, for $z \in Z$, let $S(z)$ be the intersection of $\mathbf{S}_2$ with the convex cone in $\mathbf{R}^3$ whose vertex is 0 and which is generated by the set

$$\{p - z : p \in Z,\ ||p - z|| = 1\}.$$

Actually, $S(z)$ is a convex spherical polygon on $\mathbf{S}_2$ (and it may happen that this polygon is degenerate).
It turns out that, for any two distinct points z and z' from Z, one has $S(z) \cap S(z') = \emptyset$.
To prove this fact, denote by $\{x_i : i \in I\}$ the family of all those points of Z which satisfy

$$||z - x_i|| = 1 \qquad (i \in I)$$

and, analogously, denote by $\{y_j : j \in J\}$ the family of all those points of Z which satisfy

$$||z' - y_j|| = 1 \qquad (j \in J).$$

Also, define

$$x_i^* = x_i - z, \quad y_j^* = y_j - z' \qquad (i \in I,\ j \in J).$$

Suppose for a moment that

$$\sum\{t_i x_i^* : i \in I\} = \sum\{r_j y_j^* : j \in J\} \in S(z) \cap S(z')$$

for some real numbers $t_i \geq 0$ $(i \in I)$ and $r_j \geq 0$ $(j \in J)$.
Then, for all $i \in I$ and for all $j \in J$, one has

$$||z + x_i^* - z'|| \leq 1, \quad ||z - y_j^* - z'|| \leq 1.$$

In view of (a), one can infer

$$||z + \sum\{t_i x_i^* : i \in I\} - z'|| \leq 1, \quad ||z - \sum\{r_j y_j^* : j \in J\} - z'|| \leq 1.$$

At the same time,

$$2 = ||(z + \sum\{t_i x_i^* : i \in I\} - z') - (z - \sum\{r_j y_j^* : j \in J\} - z')|| \leq$$

$$||z + \sum\{t_i x_i^* : i \in I\} - z'|| + ||z - \sum\{r_j y_j^* : j \in J\} - z'|| \leq 2,$$

whence it readily follows that $z - z' = 0$, i.e., $z = z'$.

(c) If for some two points $z \in Z$ and $z' \in Z$ the relation

$$S(z) \cap (-S(z')) \neq \emptyset$$

is valid, then

$$||z - z'|| = 1, \quad S(z) \cap (-S(z')) = \{z' - z\}.$$

To check this fact, first observe that

$$S(z) \cap (-S(z)) = \emptyset$$

for any point $z \in Z$, so $S(z) \cap (-S(z')) \neq \emptyset$ implies $z \neq z'$.
Preserving the notation of (b), suppose that

$$\sum\{t_i x_i^* : i \in I\} = -\sum\{r_j y_j^* : j \in J\} \in S(z) \cap (-S(z')).$$

If an index $j \in J$ is fixed, then

$$||z + x_i^* - z' - y_j^*|| \leq 1 \qquad (i \in I),$$

so by virtue of (a), one gets

$$||z + \sum\{t_i x_i^* : i \in I\} - z' - y_j^*|| \leq 1.$$

Ranging j over J and using repeatedly (a), one obtains

$$||z + \sum\{t_i x_i^* : i \in I\} - z' - \sum\{r_j y_j^* : j \in J\}|| \leq 1.$$

Therefore,

$$2 = ||(z + \sum\{t_i x_i^* : i \in I\} - z' - \sum\{r_j y_j^* : j \in J\}) - (z - z')|| \leq$$

$$||z + \sum\{t_i x_i^* : i \in I\} - z' - \sum\{r_j y_j^* : j \in J\}|| + ||z - z'|| \leq 2,$$

which gives

$$||z' - z|| = 1, \quad z' = z + \sum\{t_i x_i^* : i \in I\},$$

$$S(z) \cap (-S(z')) = \{z' - z\},$$

as required.

(d) For each point $z \in Z$, choose an internal point p_z of $S(z)$ (if $S(z)$ is degenerate i.e., is an arc on the sphere $\mathbf{S}_2$, then let p_z be an internal point of this arc). Also, let q_z be the antipodal point for p_z, so $||p_z - q_z|| = 2$. One may assume that

$$p_z \neq q_{z'} \quad (z \in Z,\ z' \in Z).$$

Construct a certain 2-chromatic graph G on $\mathbf{S}_2$ whose vertices are

$$p_z, \quad q_z \quad (z \in Z)$$

and an edge of G is defined as a curve on $\mathbf{S}_2$ whose endpoints are p_z and $q_{z'}$, where $z \in Z$ and $z' \in Z$ satisfy $||z - z'|| = 1$.
This means that if $\{z, z'\} \subset Z$ is such that $||z - z'|| = 1$, then $\{z, z'\}$ corresponds to exactly two edges of G.
Taking into account (b) and (c), verify that G can be realized as a planar graph. Clearly, the number of vertices of G is equal to

$$2\text{card}(Z) = 2n.$$

According to Exercise 26 from Appendix 5, the number of edges of G does not exceed $2(2n - 2)$. Consequently, $\text{card}(d(Z)) \leq 2(n - 1)$.

4. Using the results of Exercises 2 and 3, demonstrate that:

(i) any finite subset Z of the plane $\mathbf{R}^2$ with $\text{card}(Z) \geq 2$ can be decomposed into three parts each of which has strictly smaller diameter;

(ii) any finite subset Z of the space $\mathbf{R}^3$ with $\text{card}(Z) \geq 2$ can be decomposed into four parts each of which has strictly smaller diameter.

In both cases argue by induction on $\text{card}(Z)$. In case (i) keep in mind the result of Exercise 2, and in case (ii) keep in mind the result of Exercise 3.

Remark 2. The assertions (i) and (ii) of Exercise 4 show that Borsuk's conjecture is valid for all those finite subsets of $\mathbf{R}^2$ and of $\mathbf{R}^3$, respectively, which contain at least two distinct points. Unfortunately, it is impossible to infer from (i) and (ii) that the same conjecture is valid for all bounded subsets of $\mathbf{R}^2$ and of $\mathbf{R}^3$, containing at least two distinct points. For this purpose, another more delicate topological argument is needed (see, e.g., [34], [36], [287]). It should be mentioned that the case of $\mathbf{R}^2$ is much easier than the case of $\mathbf{R}^3$.

5. Starting with Stirling's classical formula

$$m! = (2\pi m)^{1/2}(m/\mathrm{e})^m(1 + \alpha_m),$$

where e is Napier's constant and α_m tends to zero as m tends to infinity, check that the growth of the expression $2^m/C_m^{m/4}$ is exponential with respect to m, and deduce from this fact that

$$(m/4)C_{m-1}^{m/4-1} < 2^{m-2}/(m^2 + 1)$$

for sufficiently large natural numbers m.

6. Give a simple direct proof of the existence of $b(m)$ for every natural number $m > 0$.

For this purpose, argue as follows. First of all, using induction on m show that, for any bounded subset X of the space $\mathbf{R}^m$ with strictly positive diameter d, there exists an m-dimensional cube $C = C_X$ in $\mathbf{R}^m$ such that:

(a) C contains X;

(b) the length of an edge of C is equal to d.

Then consider the decomposition of C into k^m pairwise congruent cubes, where $k \in \mathbf{N}$ satisfies the inequalities

$$m^{1/2} < k \leq m^{1/2} + 1,$$

and verify that

$$b(m) \leq (m^{1/2} + 1)^m,$$

which trivially yields the desired result.

7*. Show that there exists a real constant $\alpha > 0$ such that

$$b(m^2) \geq (1 + \alpha)^{m^2},$$

where m is a sufficiently large natural number as in Theorem 1.

For this purpose, take into account the hint of Exercise 5.

8*. The celebrated fixed point theorem of Brouwer states that, for any continuous mapping $f : \mathbf{B}_m \to \mathbf{B}_m$, there exists at least one point x in $\mathbf{B}_m$ such that $f(x) = x$.

Recall that the Borsuk–Ulam theorem on antipodes states that if

$$g : \mathbf{S}_m \to \mathbf{R}^m$$

is an arbitrary continuous mapping, then there exist two points $x \in \mathbf{S}_m$ and $y \in \mathbf{S}^m$ which satisfy the relations

$$||x - y|| = 2, \qquad g(x) = g(y).$$

Demonstrate that it is possible to infer Brouwer's fixed point theorem from the Borsuk–Ulam theorem on antipodes.

For this purpose, argue as follows.

Suppose that the Borsuk–Ulam theorem has already been proved and consider any continuous mapping $f : \mathbf{B}_m \to \mathbf{B}_m$.

To show the existence of a fixed point for f, take the topological product of $\mathbf{B}_m$ and $[-1, 1]$ and define a new continuous mapping

$$g : \mathbf{B}_m \times [-1, 1] \to \mathbf{R}^m$$

by the formula

$$g(z, t) = (t + 1)(f(z) - z) \qquad ((z, t) \in \mathbf{B}_m \times [-1, 1]).$$

Since the cylinder $\mathbf{B}_m \times [-1, 1]$ is homeomorphic to $\mathbf{B}_{m+1}$ with preserving central symmetry and antipodes, one can apply the Borsuk–Ulam theorem to the restriction of g to $\mathrm{bd}(\mathbf{B}_m \times [-1, 1])$.

So there exists a point $(z, t) \in \mathrm{bd}(\mathbf{B}_m \times [-1, 1])$ such that

$$(t + 1)(f(z) - z) = g(z, t) = g(-z, -t) = (-t + 1)(f(-z) - (-z)).$$

Here only three cases are possible.

(a) $t = 1$. In this case, $2(f(z) - z) = 0$ and z is a fixed point for f.

(b) $t = -1$. In this case, $2(f(-z) - (-z)) = 0$ and $-z$ is a fixed point for f.

(c) $-1 < t < 1$. In this case, z belongs to $\mathrm{bd}(\mathbf{B}_m) = \mathbf{S}_{m-1}$ and both coefficients $t+1$ and $-t+1$ are strictly positive. Consider the affine hyperplane Γ in $\mathbf{R}^m$ tangent to $\mathbf{S}_{m-1}$ at the point z and assume for a moment that none of the two equal vectors

$$(t + 1)(f(z) - z), \qquad (-t + 1)(f(-z) - (-z))$$

is zero. Then it is clear that the point $z + (t + 1)(f(z) - z)$ lies in one open half-space of $\mathbf{R}^m$ determined by Γ, while the point $z + (-t + 1)(f(-z) - (-z))$ lies in the other open half-space of $\mathbf{R}^m$ determined by the same Γ. But the latter is impossible, because these two points coincide. The obtained contradiction yields the desired result.

Remark 3. The above argument is due to A. Yu. Volovikov [354].

9*. Prove that among all two-dimensional compact convex subsets of $\mathbf{R}^2$ having a fixed diameter $d > 0$, the sets of constant width d are of maximal perimeter equal to πd.

For this purpose, argue step by step as follows.

(i) First, using the compactness argument, show that there exists a two-dimensional compact convex set $Z \subset \mathbf{R}^2$ of diameter d and of maximal perimeter p.

(ii) Then show that there exists a compact convex set $Z' \subset \mathbf{R}^2$ of constant width d, such that $Z \subset Z'$.

(iii) Keeping in mind that the diameter of Z' is the same d and that $Z \subset Z'$, infer the equality $Z = Z'$.

(iv) Finally, taking into account that the perimeter of any two-dimensional compact convex set of constant width d is equal to πd, obtain that $p = \pi d$.

10. Let $n \geq 3$ be a natural number and let $e_1, e_2, \ldots, e_n$ be nonzero pairwise distinct coplanar vectors such that $e_1 + e_2 + ... + e_n = 0$.

Prove that there exists an oriented convex n-gon P in $\mathbf{R}^2$ (possibly, degenerate), the set of all oriented sides of which coincides with the set $\{e_1, e_2, ..., e_n\}$.

Here we do not exclude the cases when some sides of P are collinear.

11*. Let $\{e_i : i \in I\}$ be a finite family of nonzero vectors in the plane $\mathbf{R}^2$ and let $\sum\{||e_i|| : i \in I\} > \pi$.

Verify that there exists a subfamily $\{e_i : i \in J\}$ of $\{e_i : i \in I\}$ such that

$$||\sum\{e_i : i \in J\}|| \geq 1.$$

For this purpose, consider the vector e defined by the formula

$$e = -\sum\{e_i : i \in I\}$$

and assume, without loss of generality, that the system of vectors

$$\{e\} \cup \{e_i : i \in I\}$$

satisfies the conditions of Exercise 10. According to the result formulated in Exercise 10, there is a convex polygon P in $\mathbf{R}^2$, all oriented sides of which constitute the family $\{e\} \cup \{e_i : i \in I\}$. Obviously, the perimeter of P is strictly greater than π. As well known, the diameter of P coincides with the length of one of the diagonals of P. Supposing that the length of the above-mentioned diagonal is strictly less than 1, obtain a contradiction with the result of Exercise 9. Consequently, the length of this diagonal is greater than or equal to 1, so the subfamily of $\{e_i : i \in I\}$ produced by the diagonal is as required.

Remark 4. It is easy to see that the constant π in Exercise 11 is precise in the sense that it cannot be replaced by a strictly less constant with preserving the result of Exercise 11.

12. Let F be a compact strictly convex body in the Euclidean space $\mathbf{R}^m$ and let $e \in \mathbf{R}^m$ be a nonzero vector.

Show that the set $F \cap (F + e)$ is contained in the interior of $F + e/2$.

Deduce from this fact that, for any point $z_0 \in \mathrm{int}(F + e/2)$, there exists $t \in]0, 1[$ such that

$$F \cap (F + e) \subset \{z_0 + t(z - z_0) : z \in F + e/2\}.$$

Also, check that the assumption of the strict convexity of F cannot be omitted here.

13. Let Z be a nonempty bounded subset of the space $\mathbf{R}^m$ and let F be a compact strictly convex body in $\mathbf{R}^m$.

Prove that there exists a unique smallest (in size) positive-homothetic image of F entirely containing Z.

For this purpose, use the result of Exercise 12.

Remark 5. Let F be a compact strictly convex body in the space $\mathbf{R}^m$ which is centrally symmetric with respect to the origin of $\mathbf{R}^m$ (in other words, F may be treated as the unit ball in Minkowski's geometry). Let Z be any nonempty bounded subset of $\mathbf{R}^m$. It follows from the result of Exercise 13 that there exist a least coefficient $t \geq 0$ and a uniquely determined vector $z_0 \in \mathbf{R}^m$ which satisfy the inclusion $Z \subset z_0 + tF$.

The point z_0 is usually called Chebyshev's center of the given set Z (in Minkowski's geometry produced by F).

CHAPTER 14

Piecewise affine approximations of continuous functions of several variables and Caratheodory–Gale polyhedra

In this chapter we will be dealing with some geometric phenomena which unexpectedly occur when approximations of continuous functions of many variables are studied. Namely, it will be shown below that the structure of such approximations cannot be simple from the computational point of view.

As is well known, continuous real-valued functions defined on compact subsets of the Euclidean space $\mathbf{R}^m$ play an important role in many questions of mathematical analysis and its applications. Since every compact subset of $\mathbf{R}^m$ can be (more or less successfully) approximated by polyhedra, it makes sense to restrict our further consideration to continuous real-valued functions defined on polyhedra whose dimension does not exceed m.

So, let P be an m-dimensional polyhedron in the space $\mathbf{R}^m$ and suppose that this polyhedron is homogeneous, i.e., any point $x \in P$ belongs to some m-dimensional simplex entirely contained in P (this restriction is convenient, but is not essential for our purposes). Let now $\phi : P \to [0, 1]$ be a continuous function. Any triangulation of P enables us to construct a certain piecewise affine continuous approximation $\psi : P \to [0, 1]$ of ϕ. Indeed, consider a trian-

DOI: 10.1201/9781003458708-14

gulation $\{T_i : i \in I\}$ of P into m-dimensional simplices. If x is an arbitrary point of P, then there exists a unique simplex T_i such that $x \in T_i$. Let $x_{i,0}$, $x_{i,1}$,..., $x_{i,m}$ denote the vertices of T_i. Clearly, x admits a unique representation in the form

$$x = \alpha_0 x_{i,0} + \alpha_1 x_{i,1} + ... + \alpha_m x_{i,m},$$

where all α_i $(i = 0, 1, ..., m)$ are non-negative real numbers and

$$\alpha_0 + \alpha_1 + \ . \ . \ . \ + \alpha_m = 1.$$

Putting now

$$\psi(x) = \alpha_0 \phi(x_{i,0}) + \alpha_1 \phi(x_{i,1}) + ... + \alpha_m \phi(x_{i,m}),$$

we get the desired continuous piecewise affine approximation ψ of ϕ.

The correctness of the definition of the function ψ is guaranteed, because $\{T_i : i \in I\}$ is a triangulation of P. In other words, the approximation ψ is well-defined, because the intersection of any two simplices from $\{T_i : i \in I\}$ is necessarily one of their common faces.

We thus see that the obtained approximating function ψ can be represented, e.g., in the form

$$\psi = \max\{\psi_i : i \in I\},$$

where, for each index $i \in I$, the function $\psi_i : P \to [0, 1]$ is affine on T_i and is identically equal to zero on the set $P \setminus T_i$.

In such a case, all functions ψ_i $(i \in I)$ may be treated as "affine pieces" of the approximating function ψ and it is natural to ask about the minimal possible value of the number of these pieces (or, equivalently, about the minimum of card(I)).

Let us first consider the most simple situation when a convex polygon P with $v = v(P)$ vertices is given in the plane $\mathbf{R}^2$ (obviously, $v \geq 3$). Suppose that this polygon is arbitrarily decomposed (dissected) into finitely many triangles, i.e., the relation

$$P = \cup\{T_i : 1 \leq i \leq n\}$$

holds true, where n is some natural number and all triangles T_i have pairwise disjoint interiors. Then we can assert that $n \geq v - 2$. Indeed, the sum of all interior angles of the triangles T_i $(1 \leq i \leq n)$ is equal to $\pi \cdot n$. As is well known, the sum of all interior angles of P is equal to $\pi(v - 2)$ (see, e.g., Exercise 2). But the latter sum is a part of the first one. So we may write

$$\pi(v - 2) \leq \pi n, \quad v - 2 \leq n.$$

Moreover, it is easy to show that there exists a triangulation of P consisting of exactly $v - 2$ triangles whose vertices belong to the set of all vertices of P

(note that the same result remains valid for any simple polygon in $\mathbf{R}^2$ with v vertices).

The above-mentioned facts directly lead to the following conclusion.

For any convex polygon $P \subset \mathbf{R}^2$, denote by $s(P)$ the minimal cardinality of a dissection of P into triangles. Then we have the equality

$$s(P) = v - 2,$$

where $v = v(P)$ is again the number of all vertices (equivalently, the number of all sides) of P.

In particular, the value $s(P)$ is completely determined by a canonical parameter associated with P. The role of such a parameter is played by the number of vertices (sides) of P.

Remark 1. For non-convex polygons, this conclusion fails to be true. For instance, it is not difficult to give an example of a simple polygon $Q \subset \mathbf{R}^2$ with 6 vertices, such that $s(Q) = 2$. On the other hand, let $P \subset \mathbf{R}^2$ be a simple polygon with $v = v(P)$ vertices and let

$$x_1, \ x_2, \ \ldots, \ x_w$$

be some points lying in the interior of P. Consider an arbitrary triangulation $\{T_i : 1 \leq i \leq n\}$ of P such that the set of all vertices of this triangulation contains all vertices of P and all the points x_1, x_2, ..., x_w. Then the relation

$$v + 2w - 2 \leq n.$$

holds true. Moreover, the relation

$$v + 2w - 2 = n$$

is valid if and only if the set of all vertices of $\{T_i : 1 \leq i \leq n\}$ is equal to the union of $\{x_1, x_2, ..., x_w\}$ with the set of all vertices of P.

If we turn our attention to the Euclidean space $\mathbf{R}^3$, then a rather surprising circumstance occurs, namely, the number $v = v(P)$ of vertices of a convex three-dimensional polyhedron $P \subset \mathbf{R}^3$ does not allow anyone to determine uniquely the analogous value $s(P)$, where:

$s(P) =$ the minimal cardinality of a dissection of P into three-dimensional simplices (i.e., tetrahedra).

This fact can readily be derived from the following example.

Example 1. Let P_1 be a prism whose base is a triangle and let P_2 be an octahedron. Obviously, the number of vertices of P_1 coincides with the number of vertices of P_2, and both of them are equal to 6. An easy argument shows that $s(P_1) = 3$ and $s(P_2) = 4$, so we conclude that

$$s(P_1) \neq s(P_2).$$

More generally, let P be a convex bipyramid in $\mathbf{R}^3$ with $4n$ facets, where $n \geq 2$ (i.e., the number of vertices of P is equal to $2n + 2$). It is not hard to prove that $s(P) \geq 2n$. Namely, it suffices to observe that there is a family of $2n$ facets of P possessing the following property: any two distinct facets of this family either have no common points or have only one common vertex. From the above-mentioned fact one immediately obtains that if Q is any octahedron in $\mathbf{R}^3$, then $s(Q) = 4$.

Example 1 shows us that if the number $v = v(P)$ of all vertices of a convex three-dimensional polyhedron $P \subset \mathbf{R}^3$ is given, then we only can speak of some estimates for $s(P)$. Of course, it is reasonable to try to describe such estimates in terms of v. More precisely, taking into account that P is picked arbitrarily, it is natural to try to establish the validity of two inequalities of the following type:

$$g_1(v) \leq s(P) \leq g_2(v),$$

where both functions g_1 and g_2 have the same order of growth as v tends to infinity. In other words, we would like to have the inequalities

$$0 < \text{liminf}_{v \to \infty}(g_1(v)/g_2(v)) \leq \text{limsup}_{v \to \infty}(g_1(v)/g_2(v)) < +\infty.$$

In particular, if both functions g_1 and g_2 are polynomials (of a variable v) whose degrees coincide, then the situation may be regarded as sufficiently nice from the point of view of general approximation theory.

It turns out that, in the case of $\mathbf{R}^3$, the required estimating functions g_1 and g_2 do exist and they can be chosen to be polynomials of degree 1, i.e., g_1 and g_2 are affine functions of v. To demonstrate this circumstance, we need the classical Euler formula for an arbitrary convex three-dimensional polyhedron $P \subset \mathbf{R}^3$. Namely, recall Euler's seminal equality

$$v - e + f = 2,$$

where the symbol $v = v(P)$ denotes the number of all vertices of P, the symbol $e = e(P)$ denotes the number of all edges of P, and $f = f(P)$ stands for the number of all facets of P (see also Appendix 3 where a multi-dimensional generalization of this remarkable formula is discussed).

Theorem 1. *For any convex three-dimensional polyhedron $P \subset \mathbf{R}^3$ with $v = v(P)$ vertices, the inequality $v(P) - 3 \leq s(P)$ holds true.*

Proof. Suppose to the contrary that the inequality of this theorem fails to be satisfied for some convex three-dimensional polyhedra $Q \subset \mathbf{R}^3$. Obviously, in such a case we may choose a convex three-dimensional polyhedron P for which $s(P) + 3 < v(P)$ and the value $s(P)$ is minimal. Consider a dissection $\{T_i : 1 \leq i \leq s(P)\}$ of P into tetrahedra. Only two cases are possible.

Case 1. For some natural index $j \in \{1, 2, ..., s(P)\}$, the tetrahedron T_j has three facets which lie in the corresponding three facets of P.

It can easily be seen that, in this case, the vertex of T_j incident to the above-mentioned three facets of T_j is simultaneously a vertex of P. Moreover, the polyhedron

$$P' = \cup\{T_i : i \in \{1, 2, ..., s(P)\} \setminus \{j\}\}$$

must be convex, three-dimensional and satisfying the relations

$$v(P') \geq v - 1, \quad s(P') \leq s(P) - 1 < s(P).$$

We thus obtain

$$s(P') + 3 \leq (s(P) - 1) + 3 = (s(P) + 3) - 1 < v(P) - 1 \leq v(P'),$$

which contradicts the choice of P.

Case 2. Every tetrahedron from the family $\{T_i : 1 \leq i \leq s(P)\}$ has at most two facets lying in the corresponding facets of P.

Let $\{D_1, D_2, ..., D_f\}$ be the family of all facets of P (clearly, here $f = f(P)$). Consider any facet D from this family. The decomposition $\{T_i : 1 \leq i \leq s(P)\}$ induces the corresponding decomposition of D into some triangles. Let $t(D)$ denote the total number of those triangles. We already know that the inequality

$$v(D) - 2 \leq t(D)$$

holds true, where $v(D)$ stands for the number of vertices (equivalently, sides) of the facet D. Now, let us put

$$w = t(D_1) + t(D_2) + \ldots + t(D_f).$$

Our assumption readily implies that $w \leq 2s(P)$. On the other hand, keeping in mind the Euler formula, we may write

$$2(v(P) - 2) = 2(e(P) - f(P)) = 2e(P) - 2f(P) =$$

$$(v(D_1) - 2) + (v(D_2) - 2) + ... + (v(D_f) - 2) \leq$$

$$t(D_1) + t(D_2) + ... + t(D_f) = w \leq 2s(P),$$

whence it immediately follows that $v(P) - 2 \leq s(P)$, and we again come to a contradiction, because $s(P) + 3 < v(P)$.

The obtained contradiction shows that neither Case 1 nor Case 2 is possible, so the inequality $v(P) - 3 \leq s(P)$ must be true for all three-dimensional convex polyhedra $P \subset \mathbf{R}^3$.

Theorem 1 has thus been proved. □

The above theorem enables us to conclude that the function $g_1(v) = v - 3$ is a lower estimate for $s(P)$, where P is an arbitrary convex three-dimensional polyhedron with exactly $v = v(P)$ vertices. In fact, this is a precise lower estimate. To explain the situation in more details, let us introduce one definition.

Let Q be a convex three-dimensional polyhedron in the space $\mathbf{R}^3$.

We shall say that a convex polyhedron Q' is a primitive extension of Q if there exist a triangular facet D of Q and a tetrahedron T in $\mathbf{R}^3$ such that D is also a facet of T and

(*) $T \cap Q = D$,

(**) the set of all vertices of Q' coincides with the union of the sets of all vertices of Q and T.

In particular, Q' can be obtained by adding to Q some tetrahedron whose base is one of the facets of Q.

Let $\{Q_1, Q_2, ..., Q_k\}$ be a finite sequence of convex three-dimensional polyhedra in $\mathbf{R}^3$. We say that this sequence is primitive if Q_1 is a tetrahedron and, for each natural index $i \in [1, k-1]$, the polyhedron Q_{i+1} is a primitive extension of Q_i.

Finally, a convex three-dimensional polyhedron $Q \subset \mathbf{R}^3$ is called primitive if $Q = Q_k$ for some primitive sequence $\{Q_1, Q_2, ..., Q_k\}$ of convex three-dimensional polyhedra in $\mathbf{R}^3$.

Actually, the argument used in the proof of Theorem 1 yields that, for any convex three-dimensional polyhedron $P \subset \mathbf{R}^3$, the following two assertions are equivalent:

(a) $s(P) = v(P) - 3$;

(b) P is a primitive polyhedron.

Remark 2. It is not hard to show that the three-dimensional unit cube in the space $\mathbf{R}^3$ is a primitive polyhedron and the same is true for any three-dimensional parallelepiped in $\mathbf{R}^3$.

In the analogous manner, the notion of a convex m-dimensional primitive polyhedron can be introduced for the space $\mathbf{R}^m$, where $m \geq 4$. If P is such a polyhedron, then we obviously have

$$s(P) \leq v(P) - m.$$

It turns out that, for $m \geq 4$, the m-dimensional unit cube in $\mathbf{R}^m$ is not a primitive polyhedron (see [205], [220]).

Let us return to the case of the Euclidean space $\mathbf{R}^3$ and let us try to find an appropriate affine function g_2 which will play the role of an upper estimate for $s(P)$, when P ranges over the family of all convex three-dimensional polyhedra in $\mathbf{R}^3$.

In this direction, we have the following statement.

Theorem 2. *For any convex three-dimensional polyhedron $P \subset \mathbf{R}^3$ with $v(P)$ vertices, the inequality*

$$s(P) \leq 2(v(P) - 2)$$

holds true, so the function $g_2(v) = 2(v - 2)$ may be regarded as an upper estimate for $s(P)$.

Proof. The argument is also substantially based on the Euler formula

$$e(P) + 2 = v(P) + f(P).$$

Indeed, consider again the family $\{D_1, D_2, ..., D_f\}$ of all facets of P and let D be any member of this family. Obviously, we can triangulate D into $v(D) - 2$ triangles (without adding new vertices). Proceeding in this manner for all facets of P, we will come to a triangulation of the boundary $\mathrm{bd}(P)$ of P. The total number of all these triangles is equal to

$$(v(D_1) - 2) + (v(D_2) - 2) + ... + (v(D_f) - 2) = 2(e(P) - f(P)) = 2(v(P) - 2).$$

Now, take any point z in the interior of P and utilize the standard conical construction. Namely, associate to each triangle T of the obtained triangulation of $\mathrm{bd}(P)$ the tetrahedron whose vertex is z and whose base is T. As a result, we get a decomposition of P into finitely many tetrahedra. Clearly, the total number of those tetrahedra is equal to the total number of triangles of the above-mentioned triangulation, i.e., is equal to $2(v(P) - 2)$.

This completes the proof of Theorem 2. □

We thus conclude that, in the case of the three-dimensional Euclidean space $\mathbf{R}^3$, both estimating functions g_1 and g_2 can be chosen to be affine.

Recall that, for $\mathbf{R}^2$, we have a much simpler situation, namely, we may take

$$g_1(v) = g_2(v) = v - 2.$$

Briefly speaking, if the dimension m of $\mathbf{R}^m$ does not exceed 3, then the estimating functions g_1 and g_2 do exist and both of them are affine.

Dealing with the four-dimensional Euclidean space $\mathbf{R}^4$, we encounter a surprise concerning the existence of natural analogues of the estimating functions g_1 and g_2. To explain this extraordinary situation, we need the notion of Carathéodory–Gale polyhedra (see [104]). This type of polyhedra was first indicated by Carathéodory in 1907, but his result was not widely recognized. In 1956, Gale rediscovered these polyhedra and considered them in more details.

In $\mathbf{R}^3$ every convex three-dimensional polyhedron with at least five vertices necessarily has two vertices such that the line segment determined by them is not an edge of the polyhedron. It was demonstrated by Carathéodory and, much later, by Gale that in the space $\mathbf{R}^4$ there exists a convex four-dimensional polyhedron G which has arbitrarily many vertices and possesses the property that any two distinct vertices of G are the endpoints of one of its edges. The construction of such a polyhedron G is very clever and intriguing. We would like to present it here.

Example 2. In the space $\mathbf{R}^4$ take a finite sequence of points

$$(t_1, t_1^2, t_1^3, t_1^4), \ (t_2, t_2^2, t_2^3, t_2^4), \ \ldots, \ (t_v, t_v^2, t_v^3, t_v^4),$$

where $v \geq 5$ is a natural number and $t_1 < t_2 < ... < t_v$ are some strictly positive real numbers. It is easy to verify that these points are in general position, i.e., no five of them belong to an affine hyperplane of $\mathbf{R}^4$ (the reader may check it by utilizing appropriate Vandermonde determinants). Denote by G the convex hull of the family of these points and let us verify that G has the above-mentioned property, namely, any two distinct vertices of G are the endpoints of an edge of G.

For this purpose, fix two distinct t_i and t_j and consider the polynomial

$$(t - t_i)^2(t - t_j)^2 = a_0 + a_1 t + a_2 t^2 + a_3 t^3 + t^4.$$

Evidently, we may associate to this polynomial the affine hyperplane Γ in $\mathbf{R}^4$, defined as follows:

$$\Gamma = \{(x_1, x_2, x_3, x_4) \in \mathbf{R}^4 : a_0 + a_1 x_1 + a_2 x_2 + a_3 x_3 + x_4 = 0\}.$$

The definition of this hyperplane directly implies that:

(1) both points $(t_i, t_i^2, t_i^3, t_i^4)$ and $(t_j, t_j^2, t_j^3, t_j^4)$ lie in Γ;

(2) all other points $(t_k, t_k^2, t_k^3, t_k^4)$ $(k \neq i,\ k \neq j)$ lie in one of the open half-spaces of $\mathbf{R}^4$ determined by Γ.

The relations (1) and (2) immediately give us that all v points described above are convexly independent (so they are vertices of G) and any two of them determine an edge of G.

Starting with Example 2 and using an easy induction on $m \geq 4$, it can be shown that, for every natural number $v \geq m + 1$, there exists a convex m-dimensional polyhedron $G \subset \mathbf{R}^m$ with v vertices such that any two distinct vertices of G are the endpoints of some edge of G. In this manner, we get many convex m-dimensional polyhedra of Carathéodory–Gale type in the space $\mathbf{R}^m$, where $m \geq 4$.

Let P be a convex four-dimensional polyhedron in the space $\mathbf{R}^4$. Similarly to the 2-dimensional and 3-dimensional cases, we may introduce the following number:

$s(P)$ = the minimal cardinality of a dissection of P into four-dimensional simplices.

Keeping in mind the above-mentioned property of Carathéodory–Gale polyhedra, it is not difficult to deduce the next statement.

Theorem 3. *If G is an arbitrary Carathéodory-Gale polyhedron in $\mathbf{R}^4$ with $v(G)$ vertices, then the inequality*

$$v(G)(v(G) - 1)/20 \leq s(G)$$

holds true.

Proof. Consider any decomposition $\{T_i : 1 \leq i \leq k\}$ of G into finitely many four-dimensional simplices. Let τ be an arbitrary edge of G and let an index

$$i = i(\tau) \in \{1, 2, ..., k\}$$

be such that the simplex $T_{i(\tau)}$ has an edge entirely contained in τ (the existence of $i(\tau)$ is evident). In this manner, we define a mapping

$$\theta : E \to \{1, 2, ..., k\},$$

where $E = E(G)$ denotes the set of all edges of G and $\theta(\tau) = i(\tau)$ for each $\tau \in E$. Since any four-dimensional simplex has 10 edges, we see that

$$\mathrm{card}(\theta^{-1}(i)) \leq 10$$

for every number $i \in \{1, 2, ..., k\}$. At the same time, since G is a Carathéodory–Gale polyhedron, we have

$$\mathrm{card}(E) = v(G)(v(G) - 1)/2.$$

From the above facts the inequality $v(G)(v(G) - 1)/2 \leq 10k$ readily follows, which trivially yields the desired estimation

$$v(G)(v(G) - 1)/20 \leq s(G).$$

This completes the proof of Theorem 3. □

Let us indicate some nontrivial corollaries of Theorem 3. Indeed, this theorem shows that the situations in $\mathbf{R}^3$ and $\mathbf{R}^4$ substantially differ from each other.

Namely, it is natural to introduce the class $\mathcal{P}_m$ of all those convex m-dimensional polyhedra Q in $\mathbf{R}^m$ $(m \geq 2)$ for which the inequality

$$s(Q) \leq v(Q) - m$$

holds true. As has already been indicated in Remark 2, all primitive convex m-dimensional polyhedra are in the class $\mathcal{P}_m$, so there exist polyhedra $Q \in \mathcal{P}_m$ for which the corresponding numbers $v(Q)$ are arbitrarily large.

For $m = 2$, the class $\mathcal{P}_m$ coincides with the family of all two-dimensional convex polyhedra in $\mathbf{R}^m$.

For $m \geq 3$, the class $\mathcal{P}_m$ is a proper (and rather poor) subfamily of the family of all m-dimensional convex polyhedra in $\mathbf{R}^m$.

For $m \geq 4$, the existence of polyhedra belonging to $\mathcal{P}_m$ and the simultaneous existence of Carathéodory-Gale polyhedra in $\mathbf{R}^m$ imply that there are no functions g_1 and g_2 of one variable, satisfying the inequalities

$$g_1(v(P)) \leq s(P) \leq g_2(v(P))$$

and such that

$$0 < \text{liminf}_{v\to\infty}(g_1(v)/g_2(v)) \leq \text{limsup}_{v\to\infty}(g_1(v)/g_2(v)) < +\infty.$$

A more thorough consideration leads to a somewhat deeper result.

Let $m > 2$ be an even natural number, i.e., $m = 2n$, where $n \in \{2, 3, ...\}$. For any natural number $v \geq m + 1$, we may take in the Euclidean space $\mathbf{R}^m$ a sequence of points:

$$(t_1, t_1^2, ..., t_1^m),\ (t_2, t_2^2, ..., t_2^m),\ \ldots,\ (t_v, t_v^2, ..., t_v^m),$$

where again real numbers $t_1, t_2, \ldots, t_v$ satisfy the inequalities

$$0 < t_1 < t_2 < ... < t_v.$$

Considering, as in Example 2, the polynomials

$$(t - t_{i_1})^2(t - t_{i_2})^2...(t - t_{i_n})^2 = a_0 + a_1 t + ... + a_{m-1}t^{m-1} + t^m,$$

where $i_1, i_2, ..., i_n$ are pairwise distinct indices from the set $\{1, 2, ..., v\}$, we conclude that all these points are in general position and, simultaneously, they are convexly independent. The convex hull G of them is an m-dimensional polyhedron possessing the following property:

Every n-element subset of the set of all vertices of G coincides with the set of all vertices of an $(n-1)$-dimensional simplex which is a face of G.

The existence of such a G yields an interesting consequence. Namely, by using an argument similar to the proof of Theorem 3, one can deduce the following statement.

Theorem 4. *There exists no upper estimate for $s(P)$ having the polynomial form with respect to both variables m and $v = v(P)$, when P ranges over the class of all convex m-dimensional polyhedra in the space $\mathbf{R}^m$.*

More precisely, there exists no polynomial $h(m, v)$ of two variables m and v, such that

$$s(P) \leq h(m, v(P))$$

for every convex m-dimensional polyhedron $P \subset \mathbf{R}^m$ with $v = v(P)$ vertices.

Proof. Suppose to the contrary that such a polynomial $h(m, v)$ does exist. Let $m = 2n$ be an even natural number satisfying the relation $n > \deg(h)$, where $\deg(h)$ stands, as usual, for the degree of h. In the space $\mathbf{R}^m$ consider an m-dimensional convex polyhedron P having $v = v(P)$ vertices and possessing the following property:

Every n-element subset of the set of all vertices of P coincides with the set of all vertices of an $(n-1)$-dimensional simplex which is a face of P.

As we already know, the number $v = v(P)$ of vertices of such a polyhedron P can be taken arbitrarily large.

Using an argument analogous to the proof of Theorem 3, we come to the estimation

$$C_v^n / C_{m+1}^n \leq s(P) \leq h(m, v),$$

where C_v^n and C_{m+1}^n denote the binomial coefficients:

$$C_v^n = v!/(n!(v-n)!), \quad C_{m+1}^n = (m+1)!/(n!(m+1-n)!).$$

From the above estimation we infer that the inequality

$$v(v-1)...(v-n+1) \leq a v^{\deg(h)}$$

must be valid for some constant $a > 0$ (depending on n) and for all large values v. On the other hand, it is easy to see that the latter inequality is impossible for sufficiently large v. Indeed, in view of the choice of m (recall that $n = m/2 > \deg(h)$), we have

$$\lim_{v \to \infty} \frac{v(v-1)...(v-n+1)}{a v^{\deg(h)}} = +\infty.$$

The obtained contradiction ends the proof of Theorem 4. □

Let us return to a continuous function $\phi : P \to [0,1]$ defined on an m-dimensional polyhedron $P \subset \mathbf{R}^m$ and to its continuous piecewise affine approximation $\psi : P \to [0,1]$ described at the beginning of this chapter. Recall that ψ admits an expression in the form

$$\psi = \max\{\psi_i : i \in I\},$$

where ψ_i $(i \in I)$ are the "affine pieces" of ψ. By virtue of Theorem 4, we may assert that, in general, the number card(I) of the "affine pieces" of such an approximation ψ is very large in comparison with the number m of variables of ϕ and the number $v(P)$ of vertices of P.

Moreover, let $x_1, x_2, ..., x_w$ be some points lying in the interior of P and let $\{T_i : i \in I\}$ be a triangulation of P into simplices, corresponding to the family $\{\psi_i : i \in I\}$ and having the property that all points $x_1, x_2, ..., x_w$ are among the vertices of this triangulation. Supposing that $w = w(m, v)$ is a function of two arguments m and v, we again can assert that, in general, the growth of card(I) is faster than the polynomial one.

Indeed, only two cases are possible.

Case 1. The growth of $w = w(m, v)$ is non-polynomial.

In this case, we use the simple inequality

$$(w(m, v) + v)/(m+1) \leq \mathrm{card}(I)$$

and readily deduce that the growth of card(I) must be non-polynomial, too.

Case 2. The growth of $w = w(m, v)$ is of polynomial character, i.e.,

$$w(m, v) \leq p(m, v)$$

for some fixed polynomial $p = p(m, v)$ with strictly positive coefficients.

In this case, suppose for a while that there is a polynomial $h_0 = h_0(m, v + w)$ such that

$$\mathrm{card}(I) \leq h_0(m, v + w).$$

Assuming, without loss of generality, that all coefficients of h_0 are strictly positive, we readily get

$$\mathrm{card}(I) \leq h_0(m, v + w) \leq h_0(m, v + p(m, v)) = h(m, v),$$

where $h(m, v)$ is also a polynomial. But this relation contradicts Theorem 4.

Some other nontrivial and interesting results should be mentioned in connection with the existence of Carathéodory–Gale polyhedra in the Euclidean space $\mathbf{R}^m$, where $m \geq 4$.

We present them in the next two examples.

Example 3. Here the concept of duality for convex polyhedra is needed (see, e.g., [5], [20], [123], [126], [249], [364]). By applying this duality to four-dimensional Carathéodory–Gale polyhedra in the space $\mathbf{R}^4$, one can easily show that there exists a convex four-dimensional polyhedron P in $\mathbf{R}^4$ having arbitrarily many facets and possessing the following property:

Any two facets of P have a common two-dimensional face of P.

It follows from this result that, for every natural number k, there exists a family $\{P_j : j \in \{1, 2, ..., k\}\}$ of convex three-dimensional polyhedra in the space $\mathbf{R}^3$ such that the intersection of any two distinct members of this family is their common facet (hence is a convex polygon).

It is useful to compare this fact with the circumstance that in $\mathbf{R}^2$ there are no five simple polygons any two of which have one-dimensional intersection. The latter circumstance follows directly from the non-planarity of Kuratowski's graph K_5 (which is a full graph with five vertices). On the other hand, there is a simple configuration of four triangles in $\mathbf{R}^2$, any two of which have one-dimensional intersection (we suggest the reader think a little about such a configuration).

In connection with Example 3, see also [123], [126], [344]; some interesting results in this direction are presented in [361].

Example 4. Let (V, E) be a graph (the symbol V denotes, as usual, the set of all vertices and the symbol E stands for the set of all edges of this graph).

Recall that an injective family $\{X_v : v \in V\}$ of nonempty sets is a set-theoretical representation (realization) of (V, E) if, for any two distinct vertices $v \in V$ and $u \in V$, the relation

$$X_v \cap X_u \neq \emptyset \Leftrightarrow \{v, u\} \in E$$

holds true. It turns out that any graph (V, E) admits a set-theoretic representation (this is Marczewski's theorem; see Exercise 6 from Appendix 5).

By using the result of Example 3, one can readily derive that, for every finite graph (V, E), there exists its set-theoretic representation consisting of three-dimensional convex polyhedra in $\mathbf{R}^3$. Notice, by the way, that convex polygons in the plane $\mathbf{R}^2$ are not sufficient for set-theoretic realizations of all finite graphs.

We would like to finish this chapter with one, more or less trivial, observation. Returning to the question of approximation of functions of several variables, it is natural to ask: why only decompositions into simplices are discussed here? As is widely known, finite decompositions of polyhedra into multi-dimensional cubes are often used in various topics of applied mathematics, where different approximation problems arise permanently. The answer is quite simple. Indeed, any m-dimensional cube C in the space $\mathbf{R}^m$ has precisely 2^m vertices

$$v_1, \; v_2, \; v_3, \; \ldots, \; v_{2^m}.$$

So, if we are given a real-valued function ϕ defined on an m-dimensional domain $Q \subset \mathbf{R}^m$ and C is a member of some finite decomposition of Q, then even the calculation of the values $\phi(v_1), \; \phi(v_2), \; \ldots, \; \phi(v_{2^m})$ needs exponential (with respect to m) number of elementary operations.

Remark 3. It would be interesting to describe a class of polyhedra, which guarantees the minimal growth of the number of elementary operations for calculating suitable continuous piecewise affine approximations of continuous real-valued functions of m real variables.

EXERCISES

1. Check in detail that a continuous piecewise affine approximation ψ introduced at the beginning of this chapter is well-defined.

2. Let P be a simple polygon in $\mathbf{R}^2$ with $v = v(P)$ vertices.

Show that the sum of all internal angles of P is equal to $\pi(v - 2)$.

Deduce this fact from the existence of a triangulation $\{T_i : 1 \leq i \leq v-2\}$ of P such that the vertices of each T_i belong to the set of all vertices of P.

3. Give a proof of the assertion formulated in Remark 1.

4. Let P be a three-dimensional convex polyhedron in the space $\mathbf{R}^3$ with $v(P) \geq 5$.

Demonstrate that there are two distinct vertices x and y of P such that the line segment $[x, y]$ differs from all edges of P.

Now, let $n \geq 4$ be a natural number.

Check that there exists a convex three-dimensional polyhedron $Q \subset \mathbf{R}^3$ such that $v(Q) = n$ and any two distinct vertices x and y of Q determine the line segment $[x, y]$ entirely contained in the boundary of Q.

5. Let $m \geq 4$ and $k \geq m+1$ be two natural numbers.

Show that there exists an m-dimensional convex polyhedron $P \subset \mathbf{R}^m$ such that $v(P) \geq k$ and any two distinct vertices of P are the endpoints of some edge of P.

For this purpose, use the standard conical construction and induction on m.

6. Prove the facts indicated in Example 3.

7. Prove the facts indicated in Example 4.

8. Taking into account the non-planarity of Kuratowski's graph K_5, infer that there exist no five simple polygons in $\mathbf{R}^2$ such that the intersection of any two of them is a one-dimensional set.

9. Show that there are $8 = 2^3$ tetrahedra in the space $\mathbf{R}^3$, such that the intersection of any two of them is a non-degenerate convex polygon (for this purpose, keep in mind the text after Example 3).

Try to generalize this result to the case of $\mathbf{R}^m$.

10. Let P be a convex n-gonal prism in the space $\mathbf{R}^3$, where $n \geq 5$, and let T be a tetrahedron contained in P. Denote by $\mathrm{vol}(P)$ the volume of P and, respectively, denote by $\mathrm{vol}(T)$ the volume of T.

Demonstrate that $\mathrm{vol}(T) < (1/3)\mathrm{vol}(P)$ and check that this inequality is precise in the sense that the coefficient $1/3$ cannot be replaced by any strictly smaller coefficient.

11*. Prove that if $n \geq 5$, then no convex n-gonal prism P in the space $\mathbf{R}^3$ is a primitive polyhedron.

For this purpose, it suffices to verify that $s(P) > 2n - 3$, where $s(P)$ is the least number of tetrahedra into which P can be decomposed.

Let $\{T_i : i \in I\}$ be any finite decomposition (dissection) of P into tetrahedra and let $J \subset I$ be defined as follows:

$$j \in J \Leftrightarrow T_j \text{ has a facet lying in a base of } P.$$

Infer that, for the set J, the relations

$$\mathrm{card}(J) \geq 2(n-2), \qquad \mathrm{vol}(\cup\{T_j : j \in J\}) \leq (2/3)\mathrm{vol}(P)$$

hold true and then use Exercise 10 for obtaining the desired result.

12*. Let (V, E) be an arbitrary finite planar graph.

Show that:

(a) there exists a set-theoretical realization $\{X_v : v \in V\}$ of (V, E) such that the sets X_v ($v \in V$) are simple polygons in the plane $\mathbf{R}^2$;

(b) there exists a realization of (V, E) in $\mathbf{R}^2$, all edges of which are (non-degenerate) line segments.

For (b), reduce the problem to a special case when all regions produced by (V, E) in $\mathbf{R}^2$ are triangular, and then argue by induction on $\operatorname{card}(V)$.

CHAPTER 15

Dissecting a square into triangles of equal areas

Undoubtedly, problems concerning dissections of given geometric figures into finitely many parts with prescribed properties are typical for combinatorial, discrete, and convex geometry (see, e.g., [2], [11], [29], [31], [34], [46], [47], [56], [58], [130], [153], [155], [185], [193], [283], [347], [355], [358], [359]).

In the present chapter, we would like to consider one problem of this kind. The reader will see in the sequel that, although the formulation of the problem is quite elementary and understandable even for school students, its solution needs substantially advanced mathematical techniques, such as valuations with values in linearly ordered groups, Sperner's combinatorial lemma (see Exercise 19 from Appendix 1), and the Kuratowski–Zorn lemma which is equivalent to the Axiom of Choice (**AC**).

For this reason, before speaking of the problem itself we have to discuss some auxiliary facts which are necessary for our further considerations.

A group $(G, \cdot)$ is called a partially ordered group if a partial order $\leq$ is given on G that is compatible with the group operation $\cdot$, i.e., for any two elements x and y from G, one has

$$x \leq y \Leftrightarrow (\forall z \in G)(xz \leq yz \ \& \ zx \leq zy).$$

Throughout the chapter, we will be dealing with commutative partially ordered groups $(G, \cdot)$, so in our case it suffices to require

$$x \leq y \Leftrightarrow (\forall z \in G)(xz \leq yz)$$

instead of the above condition for non-commutative groups.

Moreover, all groups $(G, \cdot)$ under our consideration are assumed to be linearly ordered by $\leq$ and, of course, we always suppose that $\leq$ is compatible with the operation $\cdot$ in G.

The neutral element of G will be denoted by e. The reader will see below the principal reason that explains why throughout this chapter we prefer to

DOI: 10.1201/9781003458708-15

use the multiplicative notation $\cdot$ instead of $+$ and, accordingly, why we prefer to write e instead of 0.

For the sake of convenience, we first expand G by adding to it a new element not belonging to G and denoted by 0. Further, we extend both $\cdot$ and $\leq$ by putting

$$(\forall x \in G)(0 < x \ \& \ 0x = x0 = 0).$$

The new structure obtained in such a manner will be denoted by G^*. Clearly, G^* is a linearly ordered set with the least element 0.

Example 1. The subset $G =]0, +\infty[$ of the real line $\mathbf{R}$ is a trivial example of a linearly ordered group, with respect to the standard ordering and multiplication operation in $\mathbf{R}$. In this case, we obviously may take

$$G^* = G \cup \{0\},$$

where 0 is zero in $\mathbf{R}$.

Now, the crucial definition follows.

Let K be a field in the usual algebraic sense. The canonical binary operations in K are denoted by $+$ and $\cdot$, and the constants 0 and 1 in K have their standard meaning.

A non-Archimedean valuation (with values in G^*) is an arbitrary mapping

$$v : K \to G^*$$

satisfying the following three conditions:

(i) for each $x \in K$, the equality $v(x) = 0$ is equivalent to $x = 0$;

(ii) $v(xy) = v(x)v(y)$ for all x and y from K;

(iii) $v(x + y) \leq \max(v(x), v(y))$ for all x and y from K.

Example 2. If we put $v(0) = 0$ and $v(x) = e$ for any $x \in K \setminus \{0\}$, then we come to a valuation on K which is called trivial one. In many cases the existence of nontrivial valuations on K can also be established, but needs a more delicate argument (cf. Example 3 below).

Lemma 1. *For any valuation $v : K \to G^*$, these five assertions hold true:*

(1) $v(1) = e$, where 1 is the unit of K;

(2) if $x \in K \setminus \{0\}$, then $v(x^{-1}) = v(x)^{-1}$;

(3) if $x \in K$, then $v(x) = v(-x)$;

(4) if $x \in K$ and $y \in K$ are such that $v(x) \neq v(y)$, then

$$v(x + y) = \max(v(x), v(y));$$

(5) if $x \in K \setminus \{0\}$ and $v(x) > e$, then $v(x^{-1}) < e$.

Proof. The assertions (1) and (2) trivially follow from (ii).

To see the validity of (3), write

$$e = v(1) = v((-1)(-1)) = v(-1)v(-1).$$

Putting $t = v(-1)$, we get $t^2 = e$, from which one can easily infer that $t = e$. Indeed, supposing for a moment that $t < e$, we come to the relation

$$e = t^2 \leq te = t < e,$$

which is impossible. The analogous contradiction is obtained when we assume that $t > e$, because in this case.

$$e = t^2 \geq te = t > e.$$

So $v(-1) = e$ and, consequently,

$$v(-x) = v((-1)x) = v(-1)v(x) = ev(x) = v(x)$$

for every element $x \in K$.

To see the validity of (4), suppose that $x \in K$, $y \in K$ and $v(x) < v(y)$. Obviously, we may write

$$v(y) = v(x + y + (-x)) \leq \max(v(x + y), v(-x)) \leq \max(v(x), v(y)) = v(y),$$

whence it immediately follows that

$$v(x + y) = v(y).$$

To see the validity of (5), observe that $v(x^{-1}) = v(x)^{-1}$ in view of (2). Supposing for a moment that $v(x) > e$ and $v(x^{-1}) \geq e$, we get

$$e = v(x)v(x^{-1}) > e \cdot e = e,$$

which yields a contradiction (cf. Exercise 1 of this chapter).

The obtained contradiction completes the proof. □

Lemma 2. *If $v : K \to G^*$ is an arbitrary valuation on a field K and*

$$x_1, \quad x_2, \quad \dots, \quad x_n, \quad y$$

are any elements of K such that

$$v(x_i) < v(y) \qquad (1 \leq i \leq n),$$

then one has

$$v(y) = v(y + \varepsilon_1 x_1 + \varepsilon_2 x_2 + \dots + \varepsilon_n x_n),$$

where $\varepsilon_i \in \{-1, 1\} \subset K$ for all natural indices $i \in [1, n]$.

This statement can readily be deduced from (4) of Lemma 1 by using induction on n (the details are left to the reader).

Lemma 3. *Let K be a field and let $v : K \to G^*$ be a valuation. Let us introduce the notation*

$$R = \{x \in K : v(x) \leq e\},$$

$$U = \{x \in K : v(x) = e\}.$$

The following three assertions are valid:

(1) R is a subring of the additive ring of K and $1 \in R$;

(2) U is a subgroup of the multiplicative group of K and $U \subset R$;

(3) $K = R \cup R^{-1}$, where $R^{-1} = \{1/x : x \neq 0 \ \& \ x \in R\}$.

Proof. To demonstrate (1), first observe that since $v(1) = e$, one has $1 \in R$. Further, if $x \in R$ and $y \in R$, then

$$v(x + y) \leq \max(v(x), v(y)) \leq e,$$

so $x + y \in R$. Similarly, we can write

$$v(xy) = v(x)v(y) \leq e \cdot e = e,$$

which also implies $xy \in R$ and proves (1).

The validity of (2) is checked completely analogously (the inclusion $U \subset R$ is trivial).

To establish (3), consider any element $x \in K$. If $v(x) \leq e$, then $x \in R$ by definition of R. If $v(x) > e$, then obviously $x \neq 0$ and, by virtue of Lemma 1, $v(x^{-1}) < e$, whence it follows that $x \in R^{-1}$.

Lemma 3 has thus been proved. □

The next auxiliary statement is, in a certain sense, converse to the previous lemma.

Lemma 4. *Let K be a field and let R be a subring of the additive ring of K such that*

$$K = R \cup R^{-1}.$$

Then there exist a commutative linearly ordered group $(G, \cdot, \leq)$ and a certain valuation

$$v : K \to G^*$$

canonically associated with G.

Proof. Consider the multiplicative group $K \setminus \{0\}$ of K and denote by U the set of all those nonzero elements $x \in R$ which satisfy $x^{-1} \in R$. A straightforward verification shows that U is a subgroup of the multiplicative group $K \setminus \{0\}$.

Let us introduce the quotient group

$$G = (K \setminus \{0\})/U$$

and define a binary relation $<$ on G by putting

$$xU < yU \Leftrightarrow xy^{-1} \in R \setminus U,$$

where x and y range over all nonzero elements of K.

We are going to verify in detail that $<$ is a strong linear ordering of G compatible with the group structure of G.

We will do it step by step as follows.

First, observe that if $xU = x'U$ and $yU = y'U$, then

$$xy^{-1} \in R \setminus U \Leftrightarrow x'(y')^{-1} \in R \setminus U.$$

Consequently, the binary relation $<$ is well-defined.

Let $xU < yU$ and $yU < zU$. This means that $xy^{-1} \in R \setminus U$ and $yz^{-1} \in R \setminus U$. The latter two formulas imply that

$$yx^{-1} \notin R, \quad zy^{-1} \notin R, \quad xz^{-1} = (xy^{-1})(yz^{-1}) \in R.$$

Supposing for a while that $xz^{-1} \in U \subset R$, we get $zx^{-1} \in R$ and

$$yx^{-1} = (yz^{-1})(zx^{-1}) \in R,$$

which contradicts $yx^{-1} \notin R$. Therefore, $xz^{-1} \in R \setminus U$, and the transitivity of $<$ has thus been established.

Observe now that if $xU < yU$ and $yU < xU$, then $xU < xU$, so $e \in R \setminus U$ which is impossible. Thus, $<$ is indeed a strong partial ordering of G.

Now, take any x and y from $K \setminus \{0\}$ and consider xy^{-1}. In view of the equality

$$K = R \cup R^{-1},$$

either $xy^{-1} \in R$ or $yx^{-1} \in R$. Without loss of generality, we may assume that $xy^{-1} \in R$.

If $xy^{-1} \in U$, then $yx^{-1} \in U$ and, by virtue of the commutativity of the operation $\cdot$ in K, it is clear that

$$x \in yU, \quad y \in xU, \quad xU = yU,$$

which shows us that the elements xU and yU of G are identical. On the other hand, if $xy^{-1} \notin U$, then by definition $xU < yU$. So, $<$ turns out to be a strong linear ordering of G.

For checking the compatibility of $<$ with the group structure of G, it suffices to demonstrate that

$$xU \leq yU \Leftrightarrow (xU) \cdot (zU) \leq (yU) \cdot (zU)$$

for any nonzero elements x, y and z from K. Indeed, the relation $xU < yU$ means that $xy^{-1} \in R \setminus U$. In view of the commutativity of G, the relation

$$(xU) \cdot (zU) < (yU) \cdot (zU)$$

is equivalent to the relation $xzU < yzU$ which, in its turn, is equivalent to the same $xy^{-1} \in R \setminus U$.

Further, we expand the group G in the manner described earlier and obtain

$$G^* = G \cup \{0\}.$$

The required canonical valuation $v : K \to G^*$ is now defined as follows:

$v(0) = 0$ and $v(x) = xU$ for all elements $x \in K \setminus \{0\}$.

It remains to verify that v is indeed a certain valuation on K, i.e., the conditions (i), (ii) and (iii) are valid for this v.

Condition (i) is trivially fulfilled by definition of v.

Condition (ii) is implied by the relation

$$(xU) \cdot (yU) = xyU,$$

because G is a commutative group.

Further, for proving (iii), take any two nonzero elements x and y from K.

Suppose first that $xU = yU$ or, equivalently, $yx^{-1} \in U \subset R$. In this case, let us show that

$$(x + y)U \leq xU = yU.$$

Assume on the contrary that $xU < (x + y)U$, which means that

$$x(x + y)^{-1} \in R \setminus U.$$

This implies that $(x + y)x^{-1} \notin R$, so $1 + yx^{-1} \notin R$ and $yx^{-1} \notin R$. We thus come to a contradiction.

Let now $xU \neq yU$. Since G is linearly ordered by $\leq$, we may assume without loss of generality that $xU < yU$ or, equivalently, $xy^{-1} \in R \setminus U$. Supposing for a moment that $yU < (x + y)U$, we get

$$y(x + y)^{-1} \in R \setminus U, \quad (x + y)y^{-1} \notin R, \quad xy^{-1} \notin R,$$

which yields again a contradiction.

Therefore, in all possible cases we obtain

$$v(x + y) \leq \max(v(x), v(y)) \quad (x \in K,\ y \in K),$$

so the validity of condition (iii) is also established.

Lemma 4 has thus been proved. □

Lemma 5. *Let the relation $2 = 1 + 1 \neq 0$ be satisfied for a given field K and let R be a maximal (by inclusion) subring of K containing 1 and not containing the element $1/2 \in K$.*

Then the equality $K = R \cup R^{-1}$ holds true and, consequently, K admits a valuation $v : K \to G^*$ canonically associated with some commutative linearly ordered group G.

Proof. The existence of a subring R mentioned in the formulation of Lemma 5 is an easy consequence of the Kuratowski–Zorn lemma. So, keeping in mind Lemma 4, we only have to show that

$$K = R \cup R^{-1}.$$

Suppose otherwise, i.e., there is an element $c \in K$ such that $c \notin R \cup R^{-1}$. We denote by $R(c)$ and $R(1/c)$, respectively, the ring generated by $R \cup \{c\}$ and the ring generated by $R \cup \{1/c\}$. In view of the maximality of R, we must have $1/2 \in R(c)$ and $1/2 \in R(1/c)$. More precisely,

$$1 = 2a_0 + 2a_1 c + 2a_2 c^2 + ... + 2a_m c^m,$$

$$1 = 2b_0 + 2b_1(1/c) + 2b_2(1/c)^2 + ... + 2b_n(1/c)^n,$$

where m and n are some nonzero natural numbers and

$$\{a_0,\ a_1,\ a_2,\ \ldots,\ a_m,\ b_0,\ b_1,\ b_2,\ \ldots,\ b_n\} \subset R.$$

We may assume, without loss of generality, that in the above representations of $1 \in K$ the degrees m and n are as small as possible and $m \geq n$. Further, write

$$c^n(1 - 2b_0) = 2b_1 c^{n-1} + 2b_2 c^{n-2} + ... + 2b_{n-1}c + 2b_n,$$

$$1 - 2b_0 = 2a_0(1 - 2b_0) + 2a_1(1 - 2b_0)c + ... + 2a_m(1 - 2b_0)c^m,$$

$$1 = 2b_0 + 2a_0(1 - 2b_0) + 2a_1(1 - 2b_0)c + ... + 2a_m(1 - 2b_0)c^m,$$

$$2a_m(1 - 2b_0)c^m = 2a_m((1 - 2b_0)c^n)c^{m-n}.$$

Finally, replacing the term $2a_m(1 - 2b_0)c^m$ by

$$2a_m(2b_1 c^{n-1} + 2b_2 c^{n-2} + ... + 2b_{n-1}c + 2b_n)c^{m-n},$$

we come to a representation of $1 \in K$ in the form of an element from $2R(c)$ in which the degree of c is strictly less than m. But this contradicts the choice of m.

The obtained contradiction ends the proof of Lemma 5. □

Example 3. Take as K the standard field $\mathbf{R}$ of all real numbers and let R be a maximal subring of $\mathbf{R}$ containing 1 and not containing $1/2$. Denote by G^* the expanded structure produced by R and let

$$v : \mathbf{R} \to G^*$$

be the valuation canonically associated with R (see Lemma 4 and its proof). Since the relation

$$1 \cdot (1/2)^{-1} = 2 \in R \setminus U$$

trivially holds in our case, we infer $1 \cdot U < (1/2) \cdot U$, which is equivalent to $v(1/2) > e$. Consequently, $v(2) < e$ (see (5) of Lemma 1). Further,

$$v(2+2) \leq \max(v(2), v(2)) < e$$

and, using induction on $k \in \mathbf{N}$, we easily deduce that $v(2k) < e$ for every $k \in \mathbf{N}$. Simultaneously, taking into account (4) of Lemma 1 and the relation $v(2k) \neq e = v(1)$, we have

$$v(2k+1) = \max(v(2k), v(1)) = e,$$

whence it immediately follows that $v(1/(2k+1)) = e$. These simple facts will be essentially used in the sequel.

We have finished the preliminary purely algebraic considerations which are necessary for solving one concrete problem from combinatorial geometry. This problem is formulated as follows.

Give a characterization of all those natural numbers $n \geq 2$ which have the property that the unit square $[0,1]^2$ in $\mathbf{R}^2$ can be dissected into n triangles of equal areas.

It is easy to see that for any even $n \geq 2$ the square $[0,1]^2$ admits a dissection into n many pairwise congruent triangles (see Exercise 2). So, every such n has the above-mentioned property. In the work [263] it was proved that no odd natural number satisfies this property. We are going to present below the argument of [263] which leads to the desired result. For this purpose, we will be dealing with the valuation

$$v : \mathbf{R} \to G^*$$

described in Example 3 and, starting with this v, we will define a special 3-coloring of the plane $\mathbf{R}^2$.

It will be convenient to treat the colors used in the sequel as blue, green, and red (instead of 1, 2, and 3; cf. Chapter 8). Let (x, y) be an arbitrary point of $\mathbf{R}^2 = \mathbf{R} \times \mathbf{R}$. We stipulate that:

(x, y) is blue if and only if $v(x) \geq v(y)$ and $v(x) \geq e$;

(x, y) is green if and only if $v(x) < v(y)$ and $v(y) \geq e$;

(x, y) is red if and only if $v(x) < e$ and $v(y) < e$.

A straightforward verification (left to the reader) shows that in this manner a certain 3-coloring of $\mathbf{R}^2$ is well-defined.

We shall say in the sequel that a triangle $[A, B, C]$ in $\mathbf{R}^2$ (possibly, degenerate) is rainbow if its vertices A, B and C are of pairwise different colors (cf. again Chapter 8).

Lemma 6. *Let (x_b, y_b) be a blue point in $\mathbf{R}^2$, let (x_g, y_g) be a green point in $\mathbf{R}^2$, and let (x_r, y_r) be a red point in $\mathbf{R}^2$. Denote*

$$\delta = x_b y_g + x_g y_r + y_b x_r - x_r y_g - x_g y_b - y_r x_b.$$

Then the inequality $v(\delta) \geq e$ holds true.

Proof. First of all, we have

$$v(x_b y_g) = v(x_b)v(y_g) \geq e \cdot e = e.$$

Further, we also may write

$$v(x_g y_r) = v(x_g)v(y_r) < v(y_g)v(x_b) = v(x_b y_g),$$

$$v(y_b x_r) = v(y_b)v(x_r) < v(x_b)v(y_g) = v(x_b y_g),$$
$$v(x_r y_g) = v(x_r)v(y_g) < v(x_b)v(y_g) = v(x_b y_g),$$
$$v(x_g y_b) = v(x_g)v(y_b) < v(y_g)v(x_b) = v(x_b y_g),$$
$$v(y_r x_b) = v(y_r)v(x_b) < v(y_g)v(x_b) = v(x_b y_g).$$

Now, according to Lemma 2, we get $v(\delta) = v(x_b y_g) \geq e$ which completes the proof of Lemma 6. □

Lemma 7. *The following three assertions are valid:*

(1) every straight line in the plane $\mathbf{R}^2$ carries at most two colors;

(2) no rainbow triangle in $\mathbf{R}^2$ can be degenerate;

(3) the area of an arbitrary rainbow triangle in $\mathbf{R}^2$ differs from $1/n$, where n is any odd natural number.

Proof. Actually, the assertions (1) and (2) are equivalent. Lemma 6 states that $v(\delta) \geq e$ for any rainbow triangle with vertices

$$(x_b, y_b), \quad (x_g, y_g), \quad (x_r, y_r).$$

But the area of this triangle is equal to $(1/2)|\delta|$, where δ is as in Lemma 6, and

$$v(|\delta|/2) = v(1/2)v(\delta) > e,$$

which shows us that $|\delta|/2 \neq 0$, and so the triangle is not degenerate.

Further, supposing for a moment that

$$(1/2)|\delta| = 1/n$$

for some odd natural number n, we get $|\delta| = 2/n$, whence it follows that

$$v(\delta) = v(2/n) = v(2)v(1/n) < e$$

in view of $v(2) < e$ and $v(1/n) = e$ (see Example 3). But the latter contradicts Lemma 6.

The obtained contradiction ends the proof. □

Lemma 8. *If* $\{T_i : i \in I\}$ *is any dissection of the square* $[0,1]^2$ *into finitely many triangles, then among those triangles there are rainbow ones.*

More precisely, the total number of rainbow triangles in $\{T_i : i \in I\}$ *is odd.*

Proof. Obviously, all sides of triangles from the family $\{T_i : i \in I\}$ constitute some system of polygonal lines in $[0,1]^2$ which are subdivided by the vertices of T_i $(i \in I)$ into non-degenerate line segments (minimal by the standard inclusion relation). We only are interested in the line segments whose endpoints carry two different colors. Each of those segments can be of one of the following three types:

red-blue, i.e, one of its endpoints is red and the other is blue;

blue-green, i.e., one of its endpoints is blue and the other is green;

green-red, i.e, one of its endpoints is green and the other is red.

Notice that the point $(0,0)$ is red, the point $(1,0)$ is blue, the point $(0,1)$ is green, and the point $(1,1)$ is blue. Also, no point of the line segment $\{1\}\times[0,1]$ can be red and, by virtue of Lemma 7, all those vertices of triangles from $\{T_i : i \in I\}$ which belong to $[0,1] \times \{0\}$ are either red or blue. A trivial inductive argument gives us that the total number of red-blue segments on $[0,1] \times \{0\}$ is necessarily odd.

Now, take any triangle T from the family $\{T_i : i \in I\}$. It is easy to verify that if the vertices of T carry at most two colors, then the boundary $\mathrm{bd}(T)$ of T contains an even number of red-blue segments. On the other hand, if the vertices of T carry all three colors, then $\mathrm{bd}(T)$ contains an odd number of red-blue segments.

Consider the set P of all pairs (s,T), where s is a red-blue segment and T is a triangle from $\{T_i : i \in I\}$ such that $s \subset \mathrm{bd}(T)$. Computing $\mathrm{card}(P)$ by means of all T_i, we may write

$$\mathrm{card}(P) = 2k + p_1 + p_2 + ... + p_r,$$

where $k \in \mathbf{N}$, $p_1, p_2, \ldots, p_r$ are some odd numbers, and r is the total number of rainbow triangles from $\{T_i : i \in I\}$. Computing $\mathrm{card}(P)$ by means of red-blue segments and keeping in mind that any red-blue segment intersecting the interior of $[0,1]^2$ is contained in exactly two triangles from $\{T_i : i \in I\}$, while any red-blue segment on $[0,1] \times \{0\}$ is contained in exactly one triangle from $\{T_i : i \in I\}$, we get

$$\mathrm{card}(P) = 2n + p,$$

where $n \in \mathbf{N}$ and p denotes the total number of red-blue segments on $[0,1] \times \{0\}$. Therefore,

$$2k + p_1 + p_2 + ... + p_r = 2n + p.$$

But we already know that p is an odd number, whence it follows that r is also an odd number.

Lemma 8 has thus been proved. □

Remark 1. It is useful to compare the above argument with the standard proof of Sperner's combinatorial lemma, which has many applications in topology, optimization theory, game theory, and other fields of modern mathematics (see, e.g., [4], [5], [82], [240]; cf. also Exercise 19 from Appendix 1). By the way, Sperner's lemma is used in one of the proofs of Helly's celebrated theorem on intersections of convex sets in the space $\mathbf{R}^m$).

We now are ready to formulate and prove the main result of this chapter, due to P. Monsky [263].

Theorem 1. *The square* $[0,1]^2$ *cannot be dissected into an odd number of triangles of equal areas.*

Proof. Suppose for a moment otherwise, i.e., there exists a dissection

$$\{T_i : i \in I\}$$

of $[0,1]^2$ into triangles of equal area, where $\mathrm{card}(I)$ is an odd natural number. According to Lemma 8, a certain triangle T_i of this dissection is rainbow one. Evidently, the area of T_i must be equal to $1/\mathrm{card}(I)$. But this contradicts (3) of Lemma 7.

The obtained contradiction completes the proof. □

Remark 2. Surprisingly, in the book [2] Monsky's theorem is included in the section of Analysis. But it is evident that no analytic technique is used in proving this remarkable result. Since the formulation of the theorem is due to elementary geometry, it would be interesting to try to find a relatively elementary proof of it or to show that such a proof is impossible. In particular, it is interesting to specify the role of the Axiom of Choice in establishing this result.

Monsky's theorem is very elegant and impressive from the general mathematical viewpoint and, naturally, some further results extending this theorem were afterwards published. For instance, there were considered dissections of certain kinds of polygons into triangles of equal areas (in this connection, see especially [160], [326]). Also, the analogous problem was considered for dissections of an m-dimensional cube in the space $\mathbf{R}^m$ into n many m-dimensional simplices of equal volumes. A result in this direction is as follows.

Theorem 2. *A dissection of an* m*-dimensional cube in* $\mathbf{R}^m$ *into* n *many* m*-dimensional simplices of equal volumes is possible if and only if the number* $m!$ *is a divisor of the number* n.

For a proof, see [255]. However, it should be mentioned that the obtained result is more complicated only from the technical point of view, but does not involve principally new ideas.

EXERCISES

1. Let $(G, \cdot, \leq)$ be an arbitrary partially ordered group.

Show that if a, b, c, d are any four elements of G such that $a \leq c$ and $b \leq d$, then $ab \leq cd$.

Moreover, show that if $a < c$ and $b \leq d$, then $ab < cd$.

2. Demonstrate that, for any even natural number $n \geq 2$, the square $[0,1]^2$ admits a dissection into n many pairwise congruent triangles.

For this purpose, first dissect $[0,1]^2$ into $n/2$ pairwise congruent rectangles.

3. Give a direct elementary proof of the fact that $[0,1]^2$ cannot be dissected into three triangles of equal areas.

4. Give a direct elementary proof of the fact that $[0,1]^2$ cannot be dissected into five triangles of equal areas.

5. Check that the 3-coloring of $\mathbf{R}^2$ described after Example 3 is well-defined.

6. Preserving the notation of Lemma 8 and of its proof, verify in detail that if a triangle in the plane $\mathbf{R}^2$ is not rainbow, then the total number of red-blue segments contained in its boundary is necessarily even.

7*. Let k be an arbitrary odd natural number.

Demonstrate that there exists a real $\varepsilon = \varepsilon(k) > 0$ such that, for any dissection

$$\{T_i : 1 \leq i \leq k\}$$

of $[0,1]^2$ into triangles, there is at least one triangle T_i satisfying the inequality

$$|a(T_i) - 1/k| \geq \varepsilon,$$

where $a(T_i)$ denotes the area of T_i.

For this purpose, suppose to the contrary that ε with the indicated property does not exist and take a sequence $\{\varepsilon_n : n \in \mathbf{N}\}$ of strictly positive real numbers tending to zero. Associate with each ε_n some dissection

$$\{T_{n,i} : 1 \leq i \leq k\}$$

of $[0,1]^2$ into triangles such that

$$|a(T_{n,i}) - 1/k| < \varepsilon_n \qquad (1 \leq i \leq k).$$

Further, verify that the sequence of dissections $\{\{T_{n,i} : 1 \leq i \leq k\} : n \in \mathbf{N}\}$ contains a subsequence converging (in the natural sense) to some dissection $\{T_i : 1 \leq i \leq k\}$ of $[0,1]^2$ into k triangles of equal areas, which contradicts Monsky's theorem.

The obtained contradiction yields the desired result.

Remark 3. In the book [2] it is a picture of seven triangles which dissect $[0,1]^2$, and the authors of [2] state that the areas of those triangles are nearly equal. In view of Exercise 7, it becomes clear that the statement is misleading.

In this context, see the recent work [244] containing some estimates for the value $\delta(n)$ in terms of the growth of n, where $\delta(n)$ is defined as follows.

Let $a_1, a_2, \ldots, a_n$ denote the areas of n many triangles $T_1, T_2, \ldots, T_n$, into which the unit square $[0,1]^2$ is dissected. Denote

$$a(n) = \max\{|a_i - a_j| : 1 \le i \le n,\ 1 \le j \le n\},$$

$$\delta(n) = \min\{a(n) : \{T_1, T_2, ..., T_n\} \text{ is a dissection of } [0,1]^2\}.$$

Actually, in [244] the value $\delta(n)$ is estimated from below and from above in terms of $1/n$ for a nonzero n.

8. Consider the field $\mathbf{Q}$ of all rational numbers and fix a prime number p. Every nonzero element $x \in \mathbf{Q}$ can be represented in the form $x = p^k(r/s)$, where r and s are two integers, both co-prime with p, and k is some integer uniquely determined by x.

The p-norm $|\cdot|_p : \mathbf{Q} \to [0, +\infty[$ is defined as follows:

$|0|_p = 0$ and $|x|_p = p^{-k}$ for $x \in \mathbf{Q} \setminus \{0\}$.

Verify that, for any $x \in \mathbf{Q}$ and $y \in \mathbf{Q}$, these four formulas are true:

(a) $|-x|_p = |x|_p$;

(b) $|xy|_p = |x|_p|y|_p$;

(c) $|x+y|_p \le \max(|x|_p, |y|_p)$;

(d) if $|x|_p \ne |y|_p$, then $|x+y|_p = \max(|x|_p, |y|_p)$.
In particular, check that $|1/n|_2 = 1$ for every odd natural number n.

9. Work in **ZF** set theory and prove that there exists no dissection of the square $[0,1]^2$ into an odd number of triangles of equal areas, under the assumption that all vertices of those triangles have rational coordinates.

For this purpose, using the result of Exercise 8, define an appropriate three-coloring of the rational plane $\mathbf{Q}^2$ and use an argument similar to the proof of Theorem 1.

10*. Deduce Brouwer's fixed point theorem from one of the consequences of Sperner's lemma (namely, from the result of Exercise 21 of Appendix 1).

For this purpose, it suffices to show that if $S = [x_0, x_1, ..., x_m]$ is an m-dimensional simplex in the space $\mathbf{R}^m$ and f is a continuous mapping of S into itself, then there exists a point $z \in S$ such that $f(z) = z$.

For every natural index $i \in [0, m]$, denote by Γ_i the affine hyperplane in $\mathbf{R}^m$ carrying the facet S_i of S opposite the vertex x_i, and put

$$P_i = \{x \in S : \mathrm{dist}(x, \Gamma_i) \leq \mathrm{dist}(f(x), \Gamma_i)\}.$$

Assuming that f does not possess any fixed points, verify that these three assertions are valid:

(a) the sets P_i $(0 \leq i \leq m)$ are closed in S;

(b) $S = P_0 \cup P_1 \cup ... \cup P_m$ and $x_i \notin P_i$ for each vertex x_i of S;

(c) $S_i \subset P_i$ for each facet S_i of S.

By virtue of Exercise 21 of Appendix 1, there exists some point

$$z \in P_0 \cap P_1 \cap ... \cap P_m.$$

Conclude that z is a fixed point for f, contradicting the made above assumption.

CHAPTER 16

Geometric realizations of finite and infinite families of sets

Geometric models (realizations) of various families of sets play an important role in many fields of contemporary mathematics and its applications. The best known examples of such realizations are Euler–Venn diagrams (see, for instance, [9], [124], [226], [227], [243], [279], [293], [333], and the original work by Venn [352]).

As a rule, any course of set theory or mathematical logic oriented to beginners contains some information on Euler–Venn diagrams. These diagrams help students to see visually various kinds of sets, inclusion relations between them, and the results of standard set-theoretical operations applied to them, such as: the union, intersection, difference, and symmetric difference of two given sets. At first sight, the material about Euler–Venn diagrams seems to be easy and not problematic. However, it turns out that these diagrams have interesting connections with purely combinatorial (sometimes, rather difficult) problems, with discrete geometry, with the theory of knots, and many other topics. In the present chapter, we would like to discuss a few questions concerning Euler–Venn diagrams.

Of course, at the beginning we have to rigorously determine what is meant by Euler–Venn diagrams in our further considerations. Surprisingly, it should be mentioned that a precise definition of Euler–Venn diagrams is absent in many books and papers devoted to the subject. In those works the diagrams are introduced at intuitive or visual level. Therefore, it is reasonable to start in this chapter with a variant of the precise definition of Euler–Venn diagrams. We do not think that only one way is possible to formulate such a definition. But, in the sequel, only a concrete version of an Euler–Venn diagram for a given finite (or infinite, or even uncountable) family of subsets of a universal

 DOI: 10.1201/9781003458708-16

set will be considered. By starting with this version, some geometrical properties of such diagrams will be discussed and close connections with purely combinatorial problems and the theory of convex sets (e.g., Helly's type theorems and related topics) will be indicated. In particular, some geometric realizations of uncountable independent families of sets will be touched upon.

Let U and V be two nonempty ground sets and let

$$\{X_1, X_2, ..., X_n\} \subset \mathcal{P}(U), \quad \{Y_1, Y_2, ..., Y_n\} \subset \mathcal{P}(V)$$

be two finite families of sets.

We shall say that these two families are combinatorially isomorphic (or combinatorially equivalent) if, for each subset I of $\{1, 2, ..., n\}$ and for any function

$$f : I \to \{0, 1\},$$

one has the relation

$$\cap\{X_i^{f(i)} : i \in I\} \neq \emptyset \Leftrightarrow \cap\{Y_i^{f(i)} : i \in I\} \neq \emptyset,$$

where $X_i^{f(i)}$ (respectively, $Y_i^{f(i)}$) coincides with X_i (respectively, with Y_i) if $f(i) = 0$, and coincides with $U \setminus X_i$ (respectively, with $V \setminus Y_i$) if $f(i) = 1$.

It is not hard to see that two families of sets

$$\{X_1, X_2, ..., X_n\} \subset \mathcal{P}(U), \quad \{Y_1, Y_2, ..., Y_n\} \subset \mathcal{P}(V)$$

are combinatorially isomorphic if and only if

$$\cap\{X_i^{f(i)} : i \in \{1, 2, ..., n\}\} \neq \emptyset \Leftrightarrow \cap\{Y_i^{f(i)} : i \in \{1, 2, ..., n\}\} \neq \emptyset$$

for every function

$$f : \{1, 2, ..., n\} \to \{0, 1\},$$

where the sets $X_i^{f(i)}$ and $Y_i^{f(i)}$ are defined in the same manner as above.

Remark 1. Obviously, the notion of the combinatorial isomorphism (or combinatorial equivalence) can be introduced for any two (not necessarily finite) families $\{X_j : j \in J\}$ and $\{Y_j : j \in J\}$ of subsets of ground sets U and V respectively. In this substantially more general case, it is natural to require that

$$\cap\{X_i^{f(i)} : i \in I\} \neq \emptyset \Leftrightarrow \cap\{Y_i^{f(i)} : i \in I\} \neq \emptyset$$

for each subset I of J and for every function $f : I \to \{0, 1\}$.

But in some works the notion of a combinatorial isomorphism is introduced under much weaker assumption.

Namely, two families $\{X_j : j \in J\}$ and $\{Y_j : j \in J\}$ of subsets of ground sets U and V, respectively, are called weakly combinatorially isomorphic (or weakly combinatorially equivalent) if

$$\cap\{X_i^{f(i)} : i \in I\} \neq \emptyset \Leftrightarrow \cap\{Y_i^{f(i)} : i \in I\} \neq \emptyset$$

for each finite subset I of J and for every function $f : I \to \{0, 1\}$.

Let D $(= V)$ be a geometric figure in the Euclidean plane $\mathbf{R}^2$ and let

$$D_1 = Y_1, \quad D_2 = Y_2, \quad . \; . \; . \; , \quad D_n = Y_n$$

be geometric figures, all of which are contained in D.

We shall say that $\{D_1, D_2, ..., D_n\}$ is an Euler–Venn diagram of a family $\{X_1, X_2, ..., X_n\}$ of subsets of a ground set U if $\{D_1, D_2, ..., D_n\}$ is combinatorially isomorphic to $\{X_1, X_2, ..., X_n\}$.

In this case, we shall also say that the family $\{D_1, D_2, ..., D_n\}$ is a geometric realization of $\{X_1, X_2, ..., X_n\}$ by means of figures $D_1, D_2, ..., D_n$.

In connection with the introduced notion, the following question arises:

What kind of figures can be used in geometric realizations of various finite families of subsets of a ground set U?

Simple (and, probably, well-known) examples show that, in general, one cannot guarantee the existence of an Euler–Venn diagram whose figures satisfy natural geometric assumptions.

For instance, if one has any four subsets X_1, X_2, X_3, X_4 of U such that

$$X_1 \cap X_2 \cap X_3 \neq \emptyset,$$

$$X_1 \cap X_3 \cap X_4 \neq \emptyset,$$

$$X_1 \cap X_2 \cap X_4 \neq \emptyset,$$

$$X_2 \cap X_3 \cap X_4 \neq \emptyset,$$

$$X_1 \cap X_2 \cap X_3 \cap X_4 = \emptyset,$$

then, using an easy geometric argument, one can infer that there exists no Euler–Venn diagram of $\{X_1, X_2, X_3, X_4\}$, consisting of convex subsets D_1, D_2, D_3, D_4 of the plane $\mathbf{R}^2$.

Actually, this fact readily follows from the widely known Helly theorem on intersections of convex sets (see, for instance, [68], [131], [132], or Exercises 18 and 23 from Appendix 1), but the same result can also be obtained directly, without referring to the above-mentioned important theorem of convex geometry.

Every set representable in the form $[a, b[\times[c, d[$, where $\{a, b, c, d\} \subset \mathbf{R}$, is called a (half-open) rectangle in the plane $\mathbf{R}^2$.

A set $X \subset \mathbf{R}^2$ is called rectangular if X is representable as the union of a finite family of rectangles. Observe that this family can always be chosen to be disjoint.

Notice that if D is a nonempty rectangular set, then the family of all rectangular subsets of D is an algebra of sets (in D).

The following statement is valid.

Theorem 1. *Let D be a nonempty rectangular set in the plane $\mathbf{R}^2$.*

For any finite family $\{X_1, X_2, ..., X_n\}$ of subsets of U, there exists an Euler–Venn diagram $\{D_1, D_2, ..., D_n\}$ of $\{X_1, X_2, ..., X_n\}$ such that all D_i $(1 \leq i \leq n)$ are rectangular subsets of D.

Proof. We use induction on n. For $n = 1$ the statement is trivial. Suppose that it is true for n and consider an arbitrary family $\{X_1, X_2, ..., X_n, X_{n+1}\}$ of subsets of U. By inductive assumption, there exists an Euler–Venn diagram $\{D_1, D_2, ..., D_n\}$ of $\{X_1, X_2, ..., X_n\}$, where all D_i $(1 \leq i \leq n)$ are some rectangular subsets of D. Let $Z_1, Z_2, ..., Z_k$ (respectively, $C_1, C_2, ..., C_k$) denote the family of all nonempty sets of the form $X_1^{f(1)} \cap X_2^{f(2)} \cap ... \cap X_n^{f(n)}$ (respectively, of the form $D_1^{f(1)} \cap D_2^{f(2)} \cap ... \cap D_n^{f(n)}$), where f ranges over the family of all mappings acting from $\{1, 2, ..., n\}$ into $\{0, 1\}$. Here the sets $X_i^{f(i)}$ and $D_i^{f(i)}$ $(1 \leq i \leq n)$ are defined as earlier (see the text preceding Remark 1).

Clearly, we have the inequality $k \leq 2^n$.

Moreover, all the sets $Z_1, Z_2, ..., Z_k$ (respectively, all the sets $C_1, C_2, ..., C_k$) are disjoint and their union coincides with U (respectively, with D). Besides, all C_i $(1 \leq i \leq k)$ are rectangular subsets of the plane $\mathbf{R}^2$. We may assume, without loss of generality, that

$$\{Z_1, Z_2, ..., Z_k\} = \{P_1, P_2, ..., P_r\} \cup \{Q_1, Q_2, ..., Q_m\} \cup \{S_1, S_2, ..., S_l\},$$

where

$$P_j \cap X_{n+1} \neq \emptyset, \quad P_j \cap (U \setminus X_{n+1}) = \emptyset \quad (1 \leq j \leq r),$$
$$Q_j \cap X_{n+1} = \emptyset, \quad Q_j \cap (U \setminus X_{n+1}) \neq \emptyset \quad (1 \leq j \leq m),$$
$$S_j \cap X_{n+1} \neq \emptyset, \quad S_j \cap (U \setminus X_{n+1}) \neq \emptyset \quad (1 \leq j \leq l).$$

Let us represent the family $\{C_1, C_2, ..., C_k\}$ in a similar manner:

$$\{C_1, C_2, ..., C_k\} = \{T_1, T_2, ..., T_r\} \cup \{F_1, F_2, ..., F_m\} \cup \{H_1, H_2, ..., H_l\},$$

where all T_j correspond to P_j $(1 \leq j \leq r)$, all F_j correspond to Q_j $(1 \leq j \leq m)$ and all H_j correspond to S_j $(1 \leq j \leq l)$.

Further, since each rectangular set H_j $(1 \leq j \leq l)$ is nonempty, we can define two nonempty disjoint rectangular sets H'_j and H''_j such that

$$H_j = H'_j \cup H''_j.$$

Now, we put

$$D_{n+1} = (\cup\{T_j : 1 \leq j \leq r\}) \cup (\cup\{H'_j : 1 \leq j \leq l\})$$

and then readily verify that

$$D \setminus D_{n+1} = (\cup\{F_j : 1 \leq j \leq m\}) \cup (\{H''_j : 1 \leq j \leq l\}).$$

It directly follows from the last two equalities that $\{D_1, D_2, ..., D_n, D_{n+1}\}$ is an Euler–Venn diagram of the family of sets $\{X_1, X_2, ..., X_n, X_{n+1}\}$.

This completes the proof of Theorem 1. □

Remark 2. Actually, the argument of the proof presented above does not depend on the dimension of $\mathbf{R}^2$. The same argument works for Euclidean space of any nonzero dimension. In particular, in the case of the real line $\mathbf{R} = \mathbf{R}^1$, we can assert that an arbitrary finite family $\{X_1, X_2, ..., X_n\}$ of subsets of U admits a geometric realization $\{D_1, D_2, ..., D_n\}$ such that each set D_i $(1 \leq i \leq n)$ is representable as the union of a finite family of half-open (from the right) subintervals of D, where $D = [a, b[$ is a fixed nonempty half-open interval in $\mathbf{R}$. Of course, for certainty one may take the standard half-open interval $D = [0, 1[$.

In the sequel, we need one important notion from abstract set theory.

A family $\{X_1, X_2, ..., X_n\}$ of subsets of U is independent (in the purely set-theoretical sense) if

$$X_1^{f(1)} \cap X_2^{f(2)} \cap ... \cap X_n^{f(n)} \neq \emptyset$$

for all functions

$$f : \{1, 2, ..., n\} \to \{0, 1\}.$$

In such a case, $X_1, X_2, ..., X_n$ are also called mutually independent subsets of U.

The nonempty sets of the form $X_1^{f(1)} \cap X_2^{f(2)} \cap ... \cap X_n^{f(n)}$ are sometimes called atomic components of $\{X_1, X_2, ..., X_n\}$ or constituents of $\{X_1, X_2, ..., X_n\}$ (see, e.g., [241]). Clearly, the number of all atomic components does not exceed 2^n and is equal to 2^n only in the case of an independent family $\{X_1, X_2, ..., X_n\}$.

Let us observe that if each set X_r $(r = 1, 2, ..., n)$ is represented in the form

$$X_r = \cup\{X_{i,r} : i = 1, 2, ..., k(r)\}$$

or in the form

$$X_r = \cap\{X_{i,r} : i = 1, 2, ..., k(r)\},$$

then the number of all atomic components of $\{X_1, X_2, ..., X_n\}$ does not exceed the number of all atomic components of the double family

$$\{X_{i,r} : i = 1, 2, ..., k(r);\ r = 1, 2, ..., n\}.$$

In the general case, when an arbitrary family $\{X_j : j \in J\}$ of subsets of U is given, we say that $\{X_j : j \in J\}$ is independent if every finite subfamily of this family is independent in the above-mentioned sense (see again [241]).

Independent families of sets play a significant role in many fields of modern mathematics, especially, in general topology and measure theory (see, for instance, [82], [141], [186], [241]). So, natural questions arise about appropriate geometric realizations of independent families of sets.

It is well known that there are no four discs in the plane $\mathbf{R}^2$, which yield an Euler–Venn diagram of an independent family $\{X_1, X_2, X_3, X_4\}$ of subsets of U. A more precise result will be established in our further considerations.

First, let us notice that the number of all open connected pairwise disjoint regions (domains) which are produced by the union of n circles in $\mathbf{R}^2$ does not exceed $n(n-1)+2$. To prove this fact, it suffices to apply induction on n.

In particular, if $n = 4$, then we have at most 14 ($< 16 = 2^4$) such regions.

But, for obtaining the required result, only this fact is not sufficient. Indeed, we cannot exclude the situation where certain atomic components of an Euler–Venn diagram of a given family $\{X_1, X_2, X_3, X_4\}$ differ from all above-mentioned regions. For instance, an atomic component may have empty interior (as simple examples show). Therefore, a more delicate argument is necessary here.

For a given figure $D \subset \mathbf{R}^2$, let the symbol $\mathrm{bd}(D)$ denote, as usual, the boundary of D. In particular, if D is a disc, then $\mathrm{bd}(D)$ stands for the boundary circle of D.

Lemma 1. *Let $D_1, D_2, ..., D_n$ be discs in the plane $\mathbf{R}^2$, such that some atomic component of the family $\{D_1, D_2, ..., D_n\}$ has empty interior.*

Then this atomic component is of the form $\{x\}$, where $x \in \mathbf{R}^2$, and the disjunction of the following two assertions is fulfilled:

(1) there are two tangent discs D_i and D_j from the family $\{D_1, D_2, ..., D_n\}$, for which the equality $D_i \cap D_j = \{x\}$ holds true;

(2) there are three discs D_i, D_j, D_k from this family, such that

$$\mathrm{bd}(D_i) \cap \mathrm{bd}(D_j) \cap \mathrm{bd}(D_k) = \{x\}.$$

Lemma 2. *Let D_1, D_2, D_3, D_4 be four discs in the plane $\mathbf{R}^2$.*

The following two assertions hold true:

(1) if two of these discs are tangent to each other, then the number of atomic components of $\{D_1, D_2, D_3, D_4\}$ does not exceed 14;

(2) if the sets $\mathrm{bd}(D_1) \cap \mathrm{bd}(D_2) \cap \mathrm{bd}(D_3)$ and $\mathrm{bd}(D_1) \cap \mathrm{bd}(D_2) \cap \mathrm{bd}(D_4)$ are two distinct atomic components of $\{D_1, D_2, D_3, D_4\}$, then

$$D_1 \cap D_3 \cap D_4 = \emptyset \quad \vee \quad D_2 \cap D_3 \cap D_4 = \emptyset.$$

Consequently, the number of atomic components of $\{D_1, D_2, D_3, D_4\}$ does not exceed 14.

We omit simple geometric proofs of Lemmas 1 and 2 and leave them to the reader.

By using these lemmas, one easily obtains that, for any four discs

$$D_1 \subset \mathbf{R}^2, \quad D_2 \subset \mathbf{R}^2, \quad D_3 \subset \mathbf{R}^2, \quad D_4 \subset \mathbf{R}^2,$$

the number of atomic components of the family $\{D_1, D_2, D_3, D_4\}$ does not exceed 14.

In particular, no four discs in the plane are mutually independent.

Remark 3. It is not hard to see that in $\mathbf{R}^2$ there exist three mutually independent discs D_1, D_2, D_3 such that $\mathrm{bd}(D_1) \cap \mathrm{bd}(D_2) \cap \mathrm{bd}(D_3)$ is a singleton and, simultaneously, is an atomic component of $\{D_1, D_2, D_3\}$.

Remark 4. For the real line $\mathbf{R}$, we have a result similar to the case of the Euclidean plane $\mathbf{R}^2$. Namely, no three non-degenerate closed bounded intervals in $\mathbf{R}$ are mutually independent (this fact is almost trivial).

More generally, if $m+1$ points of the Euclidean space $\mathbf{R}^m$ are not in general position, then no $m+1$ balls with centers in these points are mutually independent (see [9] and [293]). This result for $m = 3$ immediately yields the non-existence of four mutually independent discs in the plane $\mathbf{R}^2$.

Lemma 3. *Let $C_1, C_2, ..., C_n, C_{n+1}$ be pairwise distinct circles in the plane $\mathbf{R}^2$ and let*

$$X = C_{n+1} \cap (C_1 \cup C_2 \cup ... \cup C_n).$$

Denote:

X_0 = the set of all points x from X such that $\{x\} = C_{n+1} \cap C_i$ for some $i \leq n$ and x does not belong to $C_1 \cup C_2 \cup ... \cup C_{i-1} \cup C_{i+1} \cup ... \cup C_n$;

X_1 = the set of all points x from X such that $\{x\}$ is a proper subset of $C_{n+1} \cap C_i$ for some $i \leq n$ and x does not belong to the above-mentioned set $C_1 \cup C_2 \cup ... \cup C_{i-1} \cup C_{i+1} \cup ... \cup C_n$;

$X_2 = X \setminus (X_0 \cup X_1)$;

$n_0 = \mathrm{card}(X_0), \quad n_1 = \mathrm{card}(X_1), \quad n_2 = \mathrm{card}(X_2)$.

Then the relation $2n_0 + n_1 + 2n_2 \leq 2n$ is satisfied.

The proof of Lemma 3 is based on a simple purely combinatorial argument, which is omitted here and is left to the reader as a useful exercise.

Theorem 2. *For any discs $D_1, D_2, ..., D_n$ in the plane $\mathbf{R}^2$, the number of all atomic components of $\{D_1, D_2, ..., D_n\}$ does not exceed $n(n-1)+2$.*

Proof. We argue by induction on n.

Actually, the cases $n = 1, 2$ are trivial and the cases $n = 3, 4$ have already been discussed earlier.

Suppose that the assertion of the theorem is valid for any n discs in the plane $\mathbf{R}^2$ and consider a family of discs D_1, D_2, ..., D_n, D_{n+1} in $\mathbf{R}^2$. Taking into account Lemma 1 and using the notation of Lemma 3, we deduce that the number of all atomic components of the family $\{D_1, D_2, ..., D_n, D_{n+1}\}$ cannot exceed

$$(n(n-1)+2)+(n_0+n_1+n_2)+(n_0+n_2) = (n(n-1)+2)+(2n_0+n_1+2n_2).$$

According to Lemma 3, we have the inequality

$$2n_0 + n_1 + 2n_2 \leq 2n,$$

which immediately yields that the number of all atomic components of the family $\{D_1, D_2, ..., D_n, D_{n+1}\}$ does not exceed

$$n(n-1) + 2 + 2n = (n+1)((n+1)-1) + 2.$$

This completes the proof of Theorem 2. □

Now, let us consider finite independent families of polygons in the plane $\mathbf{R}^2$. It is well known that, for every natural number $n \geq 1$, there exists an independent family of Jordan (i.e., simple) polygons in $\mathbf{R}^2$ containing exactly n members (see, e.g., Exercise 4 of this chapter).

There are further and much stronger versions of the above fact. Actually, its direct analogue is true for convex polygons in $\mathbf{R}^2$. For the sake of completeness, we formulate and prove this analogue below.

Lemma 4. *For any natural number $n \geq 1$, there exist n convex polygons in the plane $\mathbf{R}^2$, which form an independent family of sets.*

Proof. As in the proofs of Theorems 1 and 2, we argue by induction on n.

The trivial cases $n = 1$ and $n = 2$ are omitted.

Suppose now that the following much stronger assertion is established for a natural number $n \geq 2$:

(*) There exist mutually independent convex polygons $W_1, W_2, ..., W_n$ in $\mathbf{R}^2$ such that there are 2^n convexly independent points $x_1, x_2, ..., x_{2^n}$ which lie in the interiors of pairwise distinct atomic components of the family $\{W_1, W_2, ..., W_n\}$.

For each natural index $i \in [1, 2^n]$, choose two distinct points y_i and z_i satisfying the following relations:

(a) x_i, y_i and z_i lie in the interior of some atomic component of the family $\{W_1, W_2, ..., W_n\}$;

(b) x_i coincides with the mid-point of the line segment $[y_i, z_i]$;

(c) $[y_i, z_i] \cap \mathrm{conv}\{x_1, x_2, ..., x_{2^n}\} = \{x_i\}$;

(d) the set $\cup\{\{y_i, z_i\} : i \in \{1, 2, ..., 2^n\}\}$ is convexly independent.

It is not difficult to see that the above-mentioned choice of the points y_i and z_i $(i = 1, 2, ..., 2^n)$ can always be realized. Further, let us put

$$W_{n+1} = \mathrm{conv}(\{y_1, y_2, ..., y_{2^n}, z_1, z_2, ..., z_{2^n}\}).$$

A straightforward verification shows that $W_1, W_2, ..., W_n, W_{n+1}$ are mutually independent convex polygons in the plane $\mathbf{R}^2$. Moreover, we can pick some convexly independent points

$$y'_1, y'_2, ..., y'_{2^n}, z'_1, z'_2, ..., z'_{2^n}$$

which all lie in the interiors of pairwise distinct atomic components of the family $W_1, W_2, ..., W_n, W_{n+1}$. Indeed, for this purpose, we may take as y'_i (respectively, as z'_i) any point belonging to a sufficiently small neighborhood of y_i (respectively, of z_i) and lying inside of W_{n+1} (respectively, outside of W_{n+1}).

Thus, we see that the direct analogue of assertion (*) holds true for the family of points

$$\{y'_1, y'_2, ..., y'_{2^n}, z'_1, z'_2, ..., z'_{2^n}\},$$

and our recursive construction can work at the next step as well as before.

This ends the proof of Lemma 4. □

Theorem 3. *There exists an infinite family of mutually independent convex polygons in the plane* $\mathbf{R}^2$.

Proof. Actually, the argument used in the proof of Lemma 4 shows that, at each finite step, the described above recursive construction can be continued and a new convex polygon can be added to the previous finite family of convex polygons with preserving the property of mutual independence.

Proceeding in this manner, we come to a countably infinite family of mutually independent convex polygons in $\mathbf{R}^2$.

Theorem 3 has thus been proved. □

Remark 5. It follows from Theorem 3 that, for any natural number $m \geq 3$, there exists an infinite independent family of convex m-dimensional polyhedra in the Euclidean space $\mathbf{R}^m$ (it suffices to apply induction on m).

In view of Theorem 3, it is natural to ask whether there exists an uncountable independent family of convex polygons in the plane $\mathbf{R}^2$. We shall see below that the answer to this question is negative (even without the assumption of convexity of polygons).

For any simple polygon W, denote by $s(W)$ the number of all sides (or, equivalently, the number of all vertices) of W. If $\{W_1, W_2, ..., W_n\}$ is a finite family of polygons, then we denote

$$s(W_1, W_2, ..., W_n) = \max(s(W_1), s(W_2), ..., s(W_n)).$$

Let us establish that if $\{W_1, W_2, ..., W_n\}$ is an independent family of polygons in $\mathbf{R}^2$ and n tends to infinity, then $s(W_1, W_2, ..., W_n)$ tends to infinity, too.

For this purpose, we need several auxiliary propositions.

Lemma 5. *Let* $\{[a_i, b_i] : i \in \{1, 2, ..., n\}\}$ *be an arbitrary finite family of closed bounded subintervals of* $\mathbf{R}$, *where* $n \geq 2$. *Then the number of atomic components of this family does not exceed* $2n + 1$.

Proof. Suppose first that some half-lines $L_1, L_2, ..., L_m$ are given in $\mathbf{R}$. Then an easy induction on m shows that the number of all atomic components of the family $\{L_1, L_2, ..., L_m\}$ is less than or equal to $m + 1$. Since every segment $[a_i, b_i]$ is representable as the intersection of the two half-lines

$$[a_i, +\infty[, \qquad]-\infty, b_i],$$

we immediately obtain the desired result by putting $m = 2n$, which completes the proof. □

Lemma 6. *Let $W_1, W_2, ..., W_n$ be any triangles in the plane $\mathbf{R}^2$.*

Then the number of all atomic components of the family $\{W_1, W_2, ..., W_n\}$ does not exceed $18n^2 + 2$.

Proof. Similarly to the proof of Lemma 5, first consider an arbitrary nonempty finite family $\{H_1, H_2, ..., H_m\}$ of closed half-planes in $\mathbf{R}^2$. Any atomic component of this family is either a two-dimensional convex polygonal region whose boundary is contained in the set $\cup\{\mathrm{bd}(H_j) : j = 1, 2, ..., m\}$ or is a nonempty convex subset of some $\mathrm{bd}(H_j)$.

As is well known (and can easily be shown by induction on m), the number of the above-mentioned regions does not exceed $m(m+1)/2 + 1$. In view of the obvious inequality

$$m(m+1)/2 + 1 \leq m(m-1) + 2,$$

we deduce that the number of the produced regions does not exceed $m(m-1) + 2$.

Further, the number of those atomic components of $\{H_1, H_2, ..., H_m\}$ which are contained in $\cup\{\mathrm{bd}(H_j) : j = 1, 2, ..., m\}$ is less than $m(m+1)$ (see the proof of Lemma 5).

So we get the upper estimate $2m^2 + 2$ for the total number of atomic components of $\{H_1, H_2, ..., H_m\}$.

Now, every given triangle W_i $(i = 1, 2, ..., n)$ can be represented in the form

$$W_i = H_{1,i} \cap H_{2,i} \cap H_{3,i},$$

where $H_{1,i}, H_{2,i}, H_{3,i}$ are some closed half-planes in $\mathbf{R}^2$. Consequently, the number of all atomic components of $\{W_1, W_2, ..., W_n\}$ does not exceed the number of all atomic components of the family

$$\{H_{k,i} \ : \ k = 1, 2, 3; \ i = 1, 2, ..., n\},$$

i.e., is less than or equal to $18n^2 + 2$.

This completes the proof of Lemma 6. □

Lemma 7. *Let $\{W_1, W_2, ..., W_n\}$ be an independent family of simple polygons in the plane $\mathbf{R}^2$ and let $s(W_1, W_2, ..., W_n) = p$.*

Then the inequality

$$(p-2)^2 \geq (2^{n-1} - 1)/9n^2$$

is fulfilled.

Proof. Recall that, for each natural number $r \in [1, n]$, there exists a decomposition $\{T_{i,r} : 1 \leq i \leq k(r)\}$ of the polygon W_r into triangles, where

$$k(r) = s(W_r) - 2 \leq p - 2.$$

So we get the family of triangles

$$\{T_{i,r} : 1 \le i \le k(r), 1 \le r \le n\},$$

whose cardinality does not exceed $(p-2)n$. Again, the number of all atomic components of $\{W_1, W_2, ..., W_n\}$ is less than or equal to the number of all atomic components of $\{T_{i,r} : 1 \le i \le k(r),\ 1 \le r \le n\}$. In view of Lemma 6, the latter number does not exceed $18(p-2)^2 n^2 + 2$. Taking into account the independence of $\{W_1, W_2, ..., W_n\}$, we can write

$$2^n \le 18(p-2)^2 n^2 + 2$$

and, therefore,

$$(p-2)^2 \ge (2^{n-1} - 1)/9n^2.$$

Lemma 7 has thus been proved. □

It is clear that both estimates given in Lemmas 6 and 7 are far from the precise values which are not interesting for us at the moment. Nevertheless, even these rough estimates are completely sufficient for establishing Theorems 4 and 6 below.

Theorem 4. *Let $\{W_1, W_2, ..., W_n\}$ be an independent family of simple polygons in $\mathbf{R}^2$.*

Then the formula $\lim_{n\to\infty} s(W_1, W_2, ..., W_n) = +\infty$ holds true.

Proof. Suppose otherwise, i.e., suppose that $s(W_1, W_2, ...W_n) \le p$ for some natural number $p \ge 3$. According to the preceding lemma, the inequality

$$(p-2)^2 \ge (2^{n-1} - 1)/9n^2$$

must be satisfied for all natural numbers $n \ge 1$. But this is absolutely impossible for sufficiently large n.

The contradiction obtained yields the required result. □

Remark 6. The direct analogue of Theorem 4 can be established for all finite independent families of simple m-dimensional polyhedra in the Euclidean space $\mathbf{R}^m$. Naturally, this analogue is formulated in terms of the value

$$f(W_1, W_2, ..., W_n) = \max(f(W_1), f(W_2), ..., f(W_n)),$$

where $f(W_r)$ $(r = 1, 2, ..., n)$ denotes the number of all facets of the polyhedron W_r. To obtain the corresponding result, one may restrict his/her consideration to the case where all W_r $(r = 1, 2, ..., n)$ are convex m-dimensional polyhedra in $\mathbf{R}^m$ and then argue similarly to the proof of Theorem 4.

Theorem 5. *Let $D_1, D_2, ..., D_n$ be figures in $\mathbf{R}^2$ whose boundaries are nondegenerate irreducible algebraic curves of a degree $\le m$, where $m > 0$ is a fixed natural number.*

Then there exists a natural number $n(m)$ such that, for any $n > n(m)$, the family $\{D_1, D_2, ..., D_n\}$ is not independent.

The proof of Theorem 5 is similar to the previous argument and is based on the purely algebraic fact stating that the set $\mathrm{bd}(D_i) \cap \mathrm{bd}(D_j)$ contains at most m^2 points for any two distinct figures D_i and D_j.

Recall that this fact is an immediate consequence of the classical Bezout theorem in elementary theory of plane algebraic curves.

Theorem 6. *There is no uncountable independent family of simple polygons in the plane* $\mathbf{R}^2$*.*

Proof. Suppose otherwise, and let $\{W_j : j \in J\}$ be an uncountable family of mutually independent simple polygons in $\mathbf{R}^2$. Then there exist a natural number p and an uncountable set $I \subset J$ such that every polygon of the partial family $\{W_i : i \in I\}$ has at most p sides. Clearly, $\{W_i : i \in I\}$ is also a family of mutually independent simple polygons. So, we come to a contradiction with Theorem 4, which ends the proof. □

Arguing in the analogous manner, one can establish that there exists no uncountable independent family of simple m-dimensional polyhedra in the Euclidean space $\mathbf{R}^m$ (cf. Remark 6).

On the other hand, the following statement is valid.

Theorem 7. *There exists an uncountable independent family* $\{P_j : j \in J\}$ *of two-dimensional compact convex figures in the plane* $\mathbf{R}^2$*. Moreover, one can suppose that the boundary of each figure* P_j *is representable as the union of countably many non-degenerate line segments.*

The proof of this theorem is essentially non-elementary. It appeals to the method of transfinite induction up to the first uncountable ordinal number ω_1. Notice also that the argument is based on some modified versions of Lemma 4 and Theorem 3.

Speaking more precisely, Theorem 7 can be reduced to one statement of mathematical analysis, which is formulated in terms of certain families of convex functions (cf. [207]). Let us introduce the definition of such families.

Let X be a set and let $\{f_j : j \in J\}$ be a family of functions acting from X into $\mathbf{R}$.

We say that this family has the interference property if, for any two nonempty finite disjoint sets $J_1 \subset J$ and $J_2 \subset J$, there exists at least one point $x \in X$ such that

$$\max\{f_j(x) : j \in J_1\} < \min\{f_j(x) : j \in J_2\}.$$

Actually, the above definition states that, for any two nonempty finite disjoint sets $J_1 \subset J$ and $J_2 \subset J$, the growth of the function $\max\{f_j : j \in J_1\}$ is controlled by the function $\min\{f_j : j \in J_2\}$ at some point $x \in X$ (of course, this point depends on the choice of J_1 and J_2).

We have the next auxiliary proposition (see [207]).

Lemma 8. *Suppose that a countable family of convex strictly increasing functions*

$$f_n : [0,1] \to [0,1] \quad (n \in \mathbf{N})$$

is given satisfying the following condition:

For any two nonempty finite disjoint sets $J_1 \subset \mathbf{N}$ and $J_2 \subset \mathbf{N}$ and for every nonzero number $k \in \mathbf{N}$, there exists a point $(x,y) \in [0,1]^2$ such that

$$1 - 1/k < x < 1, \quad 1 - 1/k < y < 1, \quad (1-y)/(1-x) > k,$$

$$\max\{f_j(x) : j \in J_1\} < y < \min\{f_j(x) : j \in J_2\}.$$

Then there exists a convex strictly increasing function

$$f : [0,1] \to [0,1]$$

such that the same condition is preserved for the extended family of functions $\{f\} \cup \{f_n : n \in \mathbf{N}\}$. Moreover, one can assume that $f(0) = 0$ and that the graph of f is representable as the union of countably many non-degenerate line segments with the singleton $\{(1,1)\}$.

The proof of Lemma 8 is left to the reader as a useful exercise.

This lemma allows one to obtain (with the aid of transfinite induction) the following statement.

Theorem 8. *There exists a family of convex strictly increasing functions*

$$f_j : [0,1] \to [0,1] \quad (j \in J)$$

satisfying the following three conditions:

(1) *$card(J) = \omega_1$;*

(2) *for each $j \in J$, the equality $f_j(0) = 0$ holds and the graph of f_j is representable as the union of countably many non-degenerate line segments with the singleton $\{(1,1)\}$;*

(3) *the family $\{f_j : j \in J\}$ has the interference property.*

Observe that Theorem 8 directly implies Theorem 7.

Finally, notice that a rich additional material concerning various properties of Euler–Venn diagrams and their applications can be found in many textbooks, manuals, and articles (see, for instance, [124], [243], and [333]).

EXERCISES

1. Let U and V be any two nonempty ground (base) sets and let

$$\{X_1, X_2, ..., X_n\} \subset \mathcal{P}(U), \qquad \{Y_1, Y_2, ..., Y_n\} \subset \mathcal{P}(V)$$

be two finite families of sets.

Show that these families are combinatorially isomorphic if and only if

$$\cap\{X_i^{f(i)} : i \in \{1, 2, ..., n\}\} \neq \emptyset \Leftrightarrow \cap\{Y_i^{f(i)} : i \in \{1, 2, ..., n\}\} \neq \emptyset$$

for every function

$$f : \{1, 2, ..., n\} \to \{0, 1\}$$

(where the sets $X_i^{f(i)}$ and $Y_i^{f(i)}$ are defined at the beginning of this chapter).

2. As was indicated in Remark 1, for any two infinite families of sets, at least two approaches are possible if one wishes to define the concept of a combinatorial isomorphism (combinatorial equivalence) of these families.

Recall that two families $\{X_j : j \in J\}$ and $\{Y_j : j \in J\}$ of subsets of nonempty ground sets U and V, respectively, are combinatorially isomorphic (or combinatorially equivalent) if

$$\cap\{X_i^{f(i)} : i \in I\} \neq \emptyset \Leftrightarrow \cap\{Y_i^{f(i)} : i \in I\} \neq \emptyset$$

for each subset I of J and for every function $f : I \to \{0, 1\}$.

Also, recall that two families $\{X_j : j \in J\}$ and $\{Y_j : j \in J\}$ of subsets of nonempty ground sets U and V, respectively, are weakly combinatorially isomorphic (or weakly combinatorially equivalent) if

$$\cap\{X_i^{f(i)} : i \in I\} \neq \emptyset \Leftrightarrow \cap\{Y_i^{f(i)} : i \in I\} \neq \emptyset$$

for each finite subset I of J and for every function $f : I \to \{0, 1\}$.

Give an example of two families of convex figures in the plane $\mathbf{R}^2$, which are weakly combinatorially isomorphic but are not combinatorially isomorphic.

3. Let U be any ground set and let X_1, X_2, X_3, X_4 be four subsets of U such that

$$X_1 \cap X_2 \cap X_3 \neq \emptyset,$$

$$X_1 \cap X_3 \cap X_4 \neq \emptyset,$$

$$X_1 \cap X_2 \cap X_4 \neq \emptyset,$$

$$X_2 \cap X_3 \cap X_4 \neq \emptyset,$$

$$X_1 \cap X_2 \cap X_3 \cap X_4 = \emptyset.$$

By using a straightforward argument (without any referring to Helly's celebrated theorem on intersections of convex sets), show that there exists no Euler–Venn diagram of the family $\{X_1, X_2, X_3, X_4\}$, consisting of convex subsets D_1, D_2, D_3, D_4 of the plane $\mathbf{R}^2$.

4. Recall that every finite subset of $\mathbf{R}^2$, whose cardinality is strictly greater than 2 and whose points are in general position, can be regarded as the set of all vertices of some simple polygon in $\mathbf{R}^2$ (for the multi-dimensional analog of this fact, see [1], [145], [190]).

Starting with the above-mentioned fact and using the method of induction, demonstrate (without referring to Lemma 4) that, for any natural number $n \geq 1$, there exists a family of n mutually independent simple polygons in $\mathbf{R}^2$.

In addition, evaluate (in terms of n) the algorithmic complexity of the construction of such a family.

5. Let X_r $(r = 1, 2, ..., n)$ be some subsets of a ground set U. Suppose that each set X_r either is represented in the form

$$X_r = \cup\{X_{i,r} : i = 1, 2, ..., k(r)\}$$

or is represented in the form

$$X_r = \cap\{X_{i,r} : i = 1, 2, ..., k(r)\}.$$

Show that the number of all atomic components of $\{X_1, X_2, ..., X_n\}$ does not exceed the number of all atomic components of

$$\{X_{i,r} : i = 1, 2, ..., k(r);\ r = 1, 2, ..., n\}.$$

6. Give detailed proofs of Lemmas 1 and 2.

7. Give a proof of Lemma 3.

8. Deduce from Theorem 3 that, for any natural number $m \geq 3$, there exists an infinite independent family of m-dimensional convex polyhedra in the Euclidean space $\mathbf{R}^m$.

9*. Try to establish the analogue of Theorem 4 for m-dimensional simple polyhedra in the Euclidean space $\mathbf{R}^m$.

10. Give a detailed proof of Theorem 5.

11. Let m be a natural number strictly greater than 2.

Demonstrate that there exists no uncountable independent family of m-dimensional simple polyhedra in the Euclidean space $\mathbf{R}^m$.

12*. Give a proof of Lemma 8 and, by using the method of transfinite recursion, establish the validity of Theorem 8.

13. A compact subset Q of the plane $\mathbf{R}^2$ is called a quasi-polygon if the interior of Q is nonempty and the boundary of Q is representable as a union of finitely or countably many non-degenerate line segments.

According to this definition, every polygon in $\mathbf{R}^2$ turns out to be a quasi-polygon (clearly, the converse assertion is not true in general).

By using Theorem 8, show that there exists an uncountable independent family of convex quasi-polygons in $\mathbf{R}^2$.

14. Let $\mathcal{A}$ denote the algebra of subsets of the interval $[0,1[$, generated by all those half-open (from the right) intervals which are contained in $[0,1[$. As has been shown in this chapter, the members of this algebra allow one to construct a geometric realization of any finite system of subsets of a ground set.

Actually, the algebra $\mathcal{A}$ is very poor and mostly plays an auxiliary role for constructing much more complicated classes of point sets. For instance, the Borel σ-algebra of $[0,1[$ is of paramount importance in descriptive set theory, mathematical analysis, general topology, and probability theory. It is usually denoted by the symbol $\mathcal{B} = \mathcal{B}([0,1[)$. In fact, the σ-algebra $\mathcal{B}$ is generated by the family of all open subsets of $[0,1[$, hence is also generated by $\mathcal{A}$.

Denote by $\mathcal{U}$ the algebra generated by the family of all open subsets of $[0,1[$.

Check that:

(i) $\mathcal{A} \subset \mathcal{U} \subset \mathcal{B}$;

(ii) $\mathcal{A} \neq \mathcal{U}$;

(iii) any set $X \in \mathcal{U}$ admits a representation in the form

$$X = \cup\{U_i \cap F_i : 1 \leq i \leq k\},$$

where k is some natural number (certainly, depending on X), all U_i are open sets in $[0,1[$ and all F_i are closed sets in $[0,1[$.

Deduce from (iii) that each member of $\mathcal{U}$ is of type G_δ, i.e., is representable as the intersection of countably many open subsets of $[0,1[$.

Conclude, by virtue of the Baire category theorem, that the set of all rational numbers in $[0,1[$ belongs to $\mathcal{B}$ but does not belong to $\mathcal{U}$.

CHAPTER 17

A geometric form of the Axiom of Choice

As the reader could see, in several preceding sections of our lecture course we essentially used certain forms of the Axiom of Choice (**AC**). Also, the reader was able to recognize that the usage of this axiom sometimes yields unexpected and extraordinary results in geometry of Euclidean spaces and even in the plane geometry (see, for instance, Chapters 6, 7, 9, and 15 of the present book).

Of course, **AC** has a much wider spectrum of its applications in other areas of contemporary mathematics such as topology, algebra, functional analysis, and so forth. It is reasonable to remind here that **AC** itself can be formulated in terms of objects defined within the frameworks of the above-mentioned mathematical disciplines.

For example, as was shown by J. L. Kelley in [165], Tychonov's celebrated theorem on the quasi-compactness of all topological products of quasi-compact spaces turns out to be a purely topological equivalent of **AC** (cf. Exercise 6 of this chapter).

The statement that every nonempty set can be endowed with a group structure is also equivalent to **AC** (see Exercise 10 of this chapter). We thus see that a purely algebraic equivalent of **AC** is possible in terms of groups.

Similarly, the fundamental theorem of the theory of vector spaces stating that every vector space possesses at least one basis turns out to be another algebraic equivalent of **AC**. This result was obtained by A. Blass [25], but there is a delicate moment here connected with the so-called Axiom of Foundation (**AF**) which is not relevant for traditional mathematical disciplines (see, e.g., [151] and [237] where many interesting aspects of **AF** are discussed).

There are also some statements of modern functional analysis which are closely related to **AC**. For instance, this can be said on the famous Hahn–Banach theorem concerning extensions of partial continuous linear functionals. Undoubtedly, the Hahn–Banach theorem (with its various geometric forms)

 DOI: 10.1201/9781003458708-17

is basic for the general theory of topological vector spaces and, in particular, for the theory of normed vector spaces.

One of the main and beautiful results in modern finite-dimensional convex analysis is Minkowski's theorem, according to which any compact convex subset X of the Euclidean space $\mathbf{R}^m$ coincides with the convex hull of the set of all extreme points of X (see Appendix 1).

This classical result admits a far going generalization in the form of the profound Krein–Milman theorem formulated as follows.

Krein–Milman Theorem. *Let E be a topological vector space over* $\mathbf{R}$ *such that its topological dual (conjugate) space E^* separates the points of E, and let X be a compact convex subset of E.*

Then the equality

$$X = \mathrm{cl}(\mathrm{conv}(\mathrm{extr}(X)))$$

holds true, i.e., X coincides with the closure of the convex hull of the set of all extreme points of X.

As an immediate corollary of this theorem, one obtains that if the given compact convex set $X \subset E$ is nonempty, then there exists at least one extreme point of X.

Another consequence of the Krein–Milman theorem, which will be of primer interest for our further purposes, is formulated as follows.

Theorem 1. *Let $(E, ||\cdot||)$ be an arbitrary normed vector space over* $\mathbf{R}$ *and let E^* denote the topological dual space endowed with its weak-* topology.*

Then the closed unit ball

$$B = \{f \in E^* : ||f|| \leq 1\}$$

has at least one extreme point.

Proof. Indeed, the ball B is compact in view of the definition of the weak-* topology and of Tychonov's product theorem (in this place **AC** is essentially used). Also, B is obviously convex in E^*. So, we can apply the Krein–Milman theorem to B and conclude that there exists at least one extreme point of B.

Theorem 1 has thus been proved. □

All standard proofs of the Krein–Milman theorem are heavily based on the Axiom of Choice or on one of its numerous logical equivalents (e.g., the Kuratowski–Zorn lemma). However, it was shown that this theorem is not effectively equivalent to **AC**. On the other hand and somewhat surprisingly, the assertion of Theorem 1 turns out to be effectively equivalent to **AC** (as was demonstrated in [13]), and our goal is to prove this equivalence below.

In our further considerations we need some notation and several simple auxiliary facts.

Let $\{T_i : i \in I\}$ be an arbitrary disjoint family of nonempty sets and let

$$T = \cup\{T_i : i \in I\}.$$

As usual, the symbol $\mathbf{R}^T$ stands for the family (in fact, vector space over $\mathbf{R}$) of all functions acting from T into $\mathbf{R}$.

We denote by E the set of all those functions $x \in \mathbf{R}^T$ which satisfy these two conditions:

(i) for any real number $\varepsilon > 0$, the set $\{t \in T : |x(t)| > \varepsilon\}$ is finite;

(ii) $\sum\{\sup_{t\in T_i} |x(t)| : i \in I\} < +\infty$.

Notice that if $x \in E$ and $\tau \in \mathbf{R}$, then $\tau x \in E$.

Also, if $x \in E$ and $y \in E$, then $x + y \in E$.

Both these facts are true in **ZF** set theory.

Therefore, E turns out to be a vector subspace of $\mathbf{R}^T$ (in the same theory).

Further, we introduce a norm $||\cdot||$ on E by the formula

$$||x|| = \sum\{\sup_{t\in T_i} |x(t)| : i \in I\} \qquad (x \in E).$$

One can readily check that $||\cdot||$ is a certain norm on E.

Let us try to describe the topological dual space E^*.

Consider the set F of all those functions $y \in \mathbf{R}^T$ for which

$$\sup\{\sum_{t\in T_i} |y(t)| : i \in I\} < +\infty.$$

Clearly, F is a vector subspace of $\mathbf{R}^T$ and we may equip F with a norm $||\cdot||$ defined by the formula

$$||y|| = \sup\{\sum_{t\in T_i} |y(t)| : i \in I\} \qquad (y \in F).$$

The following statement is valid in **ZF** theory.

Theorem 2. *The spaces E^* and F are canonically isomorphic. The canonical isomorphism between them is as follows:*

any $y \in F$ acts on E by the formula

$$y(x) = \sum\{y(t)x(t) : t \in T\} = \sum_{i\in I}(\sum_{t\in T_i} x(t)y(t)).$$

Conversely, every real-valued continuous linear functional f on E can be represented in the form of some $y \in F$ which is uniquely determined by f.

Proof. The argument is not difficult. We only have to be careful in details of our reasoning, always keeping in mind that the canonical isomorphism between E^* and F must be effective, i.e., not relying on any form of the Axiom of Choice.

First, let us check that each $y \in F$ may be considered as a real-valued continuous linear functional on the space E. For this purpose, first observe that the linearity of y is trivial, so only the continuity of y should be verified. Take any element $x \in E$. Obviously, for a fixed index $i \in I$, we may write

$$|\sum_{t\in T_i} x(t)y(t)| \leq \sum_{t\in T_i} |x(t)y(t)| \leq (\sup_{t\in T_i}|x(t)|)\sum_{t\in T_i}|y(t)|$$

$$\leq (\sup_{t\in T_i}|x(t)|)\sup\{\sum_{t\in T_j}|y(t)| : j \in I\} = ||y||(\sup_{t\in T_i}|x(t)|),$$

whence it immediately follows that

$$|y(x)| = |\sum_{i\in I}(\sum_{t\in T_i} x(t)y(t))| \leq ||y||(\sum_{i\in I}\sup_{t\in T_i}|x(t)|) = ||y||||x||,$$

which establishes the continuity of y.

Conversely, let f be an arbitrary real-valued linear continuous functional on E. Take any element $t_0 \in T$. Define an element $x_0 \in E$ by the formula: $x_0(t) = 1$ if $t = t_0$, and $x_0(t) = 0$ if $t \in T \setminus \{t_0\}$.

Afterwards, put $y(t_0) = f(x_0)$. Since $t_0 \in T$ was chosen arbitrarily, we can define analogously all values $y(t)$ for $t \in T$, and such a definition is effective, i.e., without appealing to the Axiom of Choice.

The reader can check in detail that the point $y \in \mathbf{R}^T$ introduced in this manner belongs to F and, for each point $x \in E$, one has

$$f(x) = y(x) = \sum\{y(t)x(t) : t \in T\} = \sum_{i\in I}(\sum_{t\in T_i} x(t)y(t)).$$

Observe that the procedure of checking the latter relation is free of using **AC**. □

Remark 1. For better understanding the formulation of Theorem 2, we suggest the reader to consider a very special case, where

$$I = \mathbf{N} = \cup\{\{n\} : n \in \mathbf{N}\},$$

$$E = \mathrm{l}_1 = \{(x_n)_{n\in\mathbf{N}} \in \mathbf{R}^{\mathbf{N}} : \sum\{|x_n| : n \in \mathbf{N}\} < +\infty\}.$$

In this case, the topological dual E^* coincides with the classical Banach space l_∞ consisting of all real-valued bounded sequences.

Now, we are ready to present the following remarkable result due to Bell and Fremlin (see [18]).

Theorem 3. *Within* **ZF** *set theory, these two statements are equivalent:*

(1) if $(E, ||\cdot||)$ is an arbitrary normed vector space (over the field $\mathbf{R}$) and E^* is the topological dual space equipped with its weak- topology, then the closed unit ball*

$$B = \{f \in E^* : ||f|| \leq 1\}$$

has at least one extreme point;

*(2) the Axiom of Choice (***AC***).*

Proof. As we already know, the implication (2) $\Rightarrow$ (1) holds true (cf. the proof of Theorem 1). We only need to establish the validity of the implication

$$(1) \Rightarrow (2).$$

So, suppose that (1) is satisfied and consider any family $\{T_i : i \in I\}$ of nonempty pairwise disjoint sets. Preserving the same notation as earlier, we will be dealing with the union

$$T = \cup\{T_i : i \in I\}$$

and with the corresponding spaces E, E^* and F described above. As has been mentioned, E^* and F are canonically isomorphic, hence they can be identified to each other. Denote by B the closed unit ball in F. This ball consists of all those functions $y \in F$ for which

$$||y|| = \sup\{\sum_{t \in T_i} |y(t)| : i \in I\} \leq 1.$$

In view of (1), there are extreme points of this ball and we denote by e one of such points. Obviously, we have

$$(\forall t \in T)(|e(t)| \leq 1).$$

Our goal is to show that, for each index $i \in I$, there exists a unique $t \in T_i$ satisfying the relation $e(t) \neq 0$. We will do it in two steps.

First, suppose on the contrary that, for some index $i_0 \in I$, we have

$$(\forall t)(t \in T_{i_0} \Rightarrow e(t) = 0).$$

In this case, fix $t_0 \in T_{i_0}$ and define two points $y \in B$ and $z \in B$ as follows:

$$y(t_0) = 1, \quad y(t) = e(t) \text{ for all } t \in T \setminus \{t_0\},$$

$$z(t_0) = -1, \quad z(t) = e(t) \text{ for all } t \in T \setminus \{t_0\}.$$

Clearly, we get

$$y \neq e, \quad z \neq e, \quad e = (1/2)(y + z),$$

which contradicts the extremeness property of e.

This contradiction shows that, for any index $i \in I$, there exists at least one $t \in T_i$ such that $e(t) \neq 0$.

Suppose now that, for some index $i_0 \in I$, there are two distinct members $r \in T_{i_0}$ and $s \in T_{i_0}$ satisfying the relations $e(r) \neq 0$ and $e(s) \neq 0$.

In this case, first observe that

$$|e(r)| + |e(s)| \leq 1,$$

and then define two points $y \in F$ and $z \in F$ as follows:

$$y(r) = e(r)(1 + |e(s)|),$$
$$y(s) = e(s)(1 - |e(r)|),$$
$$(\forall t)(t \in T \setminus \{r, s\} \Rightarrow y(t) = e(t)),$$
$$z(r) = e(r)(1 - |e(s)|),$$
$$z(s) = e(s)(1 + |e(r)|),$$
$$(\forall t)(t \in T \setminus \{r, s\} \Rightarrow z(t) = e(t)).$$

It can easily be verified that

$$y \neq e, \quad z \neq e, \quad e = (1/2)(y + z).$$

In addition, the relation

$$|y(r)| + |y(s)| = |z(r)| + |z(s)| = |e(r)| + |e(s)|$$

implies that

$$||y|| = ||z|| = ||e|| \leq 1,$$

so $y \in B$ and $z \in B$. We once again obtain a contradiction with the extremeness property of e.

Summarizing all the established above, we deduce that, for any $i \in I$, there exists exactly one $t \in T_i$ satisfying $e(t) \neq 0$.

In other words, we have a choice function for the family $\{T_i : i \in I\}$ of nonempty pairwise disjoint sets. Since this family was taken arbitrarily, we can conclude that (2), i.e., the Axiom of Choice, holds true.

This completes the proof of Theorem 3. □

Remark 2. We have already mentioned that the Krein–Milman theorem is not effectively equivalent to the Axiom of Choice. In this connection, let us recall that there are some weaker forms of **AC** which are formulated in terms of ultrafilters or in dual terms of prime ideals in Boolean algebras. In particular, it is known that these two assertions are effectively equivalent:

(a) Tychonov's theorem for compact (i.e., quasi-compact Hausdorff) spaces;

(b) Boolean prime ideal theorem stating that every Boolean algebra contains a prime ideal.

So, we see that the conjunction of (b) and the Krein–Milman theorem effectively implies **AC**.

Remark 3. Notice that the Hahn–Banach theorem can be effectively deduced from (b). Also, it was proved by J. Pawlikowski [281] that the Hahn–Banach theorem effectively implies the Banach–Tarski paradox. Consequently, in the theory **ZF** & **DC** the Hahn–Banach theorem implies the existence of a non-measurable subset of $\mathbf{R}^3$ with respect to the Lebesgue measure λ_3 (and the existence of a nonmeasurable subset of $\mathbf{R}$ with respect to the Lebesgue measure λ).

Although Theorem 3 is very impressive, it does not look as a purely geometric version of **AC**, because in its formulation we deal with certain infinite-dimensional real vector spaces. Generally, one might conjecture that it is impossible to give a geometric variant of **AC** in terms of finite-dimensional Euclidean spaces.

On the other hand, as shown by W. Sierpiński many years ago, for Cantor's Continuum Hypothesis (**CH**), the situation is slightly better in this respect. Recall that one of the standard forms of **CH** is the equality

$$2^{\omega} = \omega_1,$$

where ω denotes the least infinite cardinal number ($= \mathrm{card}(\mathbf{N})$) and ω_1 stands for the least uncountable cardinal number. Sierpiński found a statement of Euclidean geometry which is equivalent (however, within **ZFC** set theory) to the above equality. In order to present his result, let us introduce an auxiliary purely geometric notion.

Let v be a nonzero vector in the the space $\mathbf{R}^3$ and let Z be a subset of $\mathbf{R}^3$.

We shall say that Z is a finite set in direction v if every straight line in $\mathbf{R}^3$ parallel to v meets Z in finitely many points.

Sierpiński proved that the following statement is valid in **ZFC** theory.

Theorem 4. *These two assertions are equivalent:*

*(1) The Continuum Hypothesis (**CH**);*

(2) there exists a partition $\{Z_1, Z_2, Z_3\}$ of the space $\mathbf{R}^3$ such that

(a) Z_1 is finite in direction $v_1 = (1, 0, 0)$;

(b) Z_2 is finite in direction $v_2 = (0, 1, 0)$;

(c) Z_3 is finite in direction $v_3 = (0, 0, 1)$.

The proof of Theorem 4 is given in Sierpiński's monograph [316] where many other equivalents of **AC** and **CH** are thoroughly considered (in this context, see also [137] and [150]).

EXERCISES

1. Let $\{T_i : i \in I\}$ be an arbitrary disjoint family of nonempty sets and let

$$T = \cup\{T_i : i \in I\}.$$

Denote again by E the set of all those functions $x \in \mathbf{R}^T$ which satisfy the following two conditions:

(i) for any real number $\varepsilon > 0$, the set $\{t \in T : |x(t)| > \varepsilon\}$ is finite;

(ii) $\sum\{\sup_{t\in T_i}|x(t)| : i \in I\} < +\infty$.

Give a detailed proof of the fact that E is a vector subspace of $\mathbf{R}^T$.

For this purpose, use the inclusion relations

$$\{t \in T : |x(t) + y(t)| > \varepsilon\} \subset \{t \in T : |x(t)| + |y(t)| > \varepsilon\} \subset$$

$$\{t \in T : |x(t)| > \varepsilon/2\} \cup \{t \in T : |y(t)| > \varepsilon/2\},$$

where ε is an arbitrary strictly positive real number.

2. Working within **ZF** set theory, give a more detailed argument in the proof of Theorem 2, concerning the identity $f = y$, where f is an arbitrary real-valued linear continuous functional on E and $y \in F$ is defined via f.

3. Work in **ZF** theory and demonstrate that:

(a) there exists no set W satisfying the relation $(\forall x)(x \in W \Leftrightarrow x \notin x)$;

(b) there exists no set U satisfying the relation $(\forall x)(x \in U)$.

For (a), suppose on the contrary that W does exist and verify that both relations $W \in W$ and $W \notin W$ are false, which yields a contradiction.

For (b), use Cantor's inequality $\mathrm{card}(X) < \mathrm{card}(\mathcal{P}(X))$, where X is an arbitrary set and $\mathcal{P}(X)$ denotes the power set of X.

Remark 4. Assertion (a) of Exercise 3 is known as the negation of Russell's paradox and assertion (b) of the same exercise is known as the negation of Cantor's paradox.

4. Let $\{X_i : i \in I\}$ be an arbitrary family of sets.

Show, within **ZF** set theory, that

$$(\exists x)(x \notin \cup\{X_i : i \in I\}).$$

For this purpose, apply (b) of Exercise 3.

5. Let $\{(E_i, \mathcal{T}_i) : i \in I\}$ be a family of topological spaces and let, for each index $i \in I$, a set $F_i \subset E_i$ be closed in E_i.

Work in **ZF** theory and prove that the set $\prod\{F_i : i \in I\}$ is closed in the topological product space $\prod\{E_i : i \in I\}$.

6. Suppose that the following statement holds true in **ZF** theory:

The topological product space of any family $\{(E_i, \mathcal{T}_i) : i \in I\}$ of topological spaces such that $(\forall i \in I)(\mathrm{card}(\mathcal{T}_i) \le 3)$ is quasi-compact.

Prove that in this case the Axiom of Choice is valid in **ZF** theory.

For this purpose, consider an arbitrary family $\{X_i : i \in I\}$ of nonempty sets. In view of Exercise 4, there exists $x \notin \cup\{X_i : i \in I\}$. For each index $i \in I$, define

$$E_i = X_i \cup \{x\}, \qquad \mathcal{T}_i = \{\emptyset, E_i, \{x\}\},$$

and consider the quasi-compact topological product space

$$(E, \mathcal{T}) = \prod\{(E_i, \mathcal{T}_i) : i \in I\}.$$

Further, denote by $\mathcal{F}$ the family of all those subsets of E which are representable in the form $\prod\{Y_i : i \in I\}$, where either $Y_i = E_i$ or $Y_i = X_i$, and the set

$$\{i \in I : Y_i \neq E_i\}$$

is finite.

Using Exercise 5, check that $\mathcal{F}$ is a centered family of nonempty closed subsets of E, so $\cap\mathcal{F} \neq \emptyset$.

Deduce that any element of $\cap\mathcal{F}$ belongs to the set $\prod\{X_i : i \in I\}$, which implies **AC**.

Conclude from the proved above that these two statement are equivalent within **ZF** set theory:

(a) the Axiom of Choice;

(b) the topological product of an arbitrary family of quasi-compact spaces is quasi-compact.

7. Work in **ZF** theory and verify that there exists no set of all ordinal numbers.

For this purpose, take into account the fact that, for any family of ordinal numbers, there is an ordinal number strictly greater than all members of the family.

Remark 5. The fact mentioned in Exercise 7 is known as Burali-Forti paradox.

8. Work in **ZF** theory and prove that, for any set X, there exists a least ordinal number $h(X)$ which cannot be injectively mapped into X (this $h(X)$ is usually called the Hartogs ordinal number for X).

Argue as follows. Consider the family $\mathcal{W}$ of all those elements of the power set $\mathcal{P}(X \times X)$, which are the graphs of well-orderings of some subsets of X. Then, for each element $w \in W$, define $f(w)$ as the unique ordinal isomorphic to the well-ordered set $\mathrm{pr}_1(w) \subset X$. In view of Exercise 7, there exist ordinal numbers not belonging to the set $\{f(w) : w \in W\}$. Check that the least of such ordinals is as required.

9*. Preserving the notation of Exercise 8, using the above-mentioned property of $h(X)$, and applying the principle of transfinite induction, demonstrate within **ZFC** theory that X can be made well-ordered (Zermelo's theorem).

To do this, first introduce some choice function

$$\phi : \mathcal{P}(X) \setminus \{\emptyset\} \to X,$$

i.e., $\phi(Y) \in Y$ for any nonempty set $Y \subset X$. Then pick an element $z \notin X$ and define by transfinite recursion a family $\{x_\xi : \xi < h(X)\}$ so that:

$x_\xi = \phi(X \setminus \{x_\zeta : \zeta < \xi\})$ if $X \setminus \{x_\zeta : \zeta < \xi\} \neq \emptyset$,

$x_\xi = z$ if $X \setminus \{x_\zeta : \zeta < \xi\} = \emptyset$.

Finally, deduce the existence of a least ordinal $\xi_0 < h(X)$ such that $x_{\xi_0} = z$ and conclude that

$$X = \{x_\zeta : \zeta < \xi_0\}.$$

Therefore, X admits a well-ordering isomorphic to the ordinal ξ_0.

10*. Work in **ZF** theory and show that these two assertions are equivalent:

(a) the Axiom of Choice (**AC**);

(b) every nonempty set can be endowed with the group structure.

For this purpose, argue as follows.

To establish the implication (a) ⇒ (b), it suffices to observe that in **ZFC** theory for any infinite cardinal number $\mathbf{a}$, there exists a vector space U over the field $\mathbf{Q}$ of all rational numbers, such that $\mathrm{card}(U) = \mathbf{a}$.

To establish the converse implication (b) ⇒ (a), suppose (b) and take any set X with a set Y satisfying these two conditions:

(i) $X \cap Y = \emptyset$ and Y is well-ordered by some relation $\leq$;

(ii) there is no injection from Y into X.

The existence of Y is provable in **ZF** theory and, actually, follows from the result of Exercise 8.

According to (b), the set $X \cup Y$ can be endowed with some group operation $\cdot$. For each element $x \in X$, verify that there are two elements $y_1 \in Y$ and $y_2 \in Y$ such that $x \cdot y_1 = y_2$. Starting with this fact and keeping in mind that the product set $Y \times Y$ is well-ordered with respect to the lexicographic ordering produced by $\leq$, define effectively an injective mapping from X into $Y \times Y$ and conclude that X is also a well-orderable set.

11. A real vector space E is called a partially pre-ordered vector space, with respect to a binary relation $\leq$ on E, if the following four conditions hold:

(1) $x \leq x$ for all $x \in E$ (the reflexivity of $\leq$);

(2) $x \leq y$ and $y \leq z$ imply $x \leq z$ (the transitivity of $\leq$);

(3) $x \leq y$ and $z \in E$ imply $x + z \leq y + z$ (the invariance of $\leq$ under all translations of E);

(4) $x \leq y$ and $t \in [0, +\infty[$ imply $tx \leq ty$ (the invariance of $\leq$ under all positive homotheties of E with center $0 \in E$).

Deduce from these conditions that:

(a) $x \leq x_1$ and $y \leq y_1$ imply $x + y \leq x_1 + y_1$;

(b) $0 \leq x$ and $0 \leq y$ imply $0 \leq x + y$.

Conclude that the set $C = \{x \in E : 0 \leq x\}$ turns out to be a convex cone in E with vertex 0.

Conversely, if C is a convex cone in E with vertex 0, then the binary relation $\leq$ defined by the formula

$$x \leq y \Leftrightarrow y - x \in C$$

is a partial pre-ordering of E determined by the cone C.

In other words, to speak of a real partially pre-ordered vector space means to speak of a real vector space in which some convex cone is indicated.

If a pre-ordering $\leq$ of E is such that $x \leq y$ and $y \leq x$ imply $x = y$, then $\leq$ is a partial ordering of E, and we come to a real partially ordered vector space $(E, \leq)$.

Check that a partial pre-ordering determined by a convex cone $C \subset E$ is a partial ordering of E if and only if the equality

$$C \cap (-C) = \{0\}$$

holds true.

12*. Prove the following fundamental result of M. Riesz on extensions of positive linear functionals:

Let E be a real vector space, let U be a vector subspace of E, let C be a convex cone in E with vertex 0, and suppose that:

(1) $C + U = E$;

(2) $f : U \to \mathbf{R}$ is a linear functional positive on $C \cap U$, i.e., for all $x \in C \cap U$ one has $f(x) \geq 0$.

Then there exists a linear functional $f^* : E \to \mathbf{R}$ extending f and such that $f^*|C$ is positive on C, i.e., for all $x \in C$ one has $f^*(x) \geq 0$.

Argue as follows. Denote by $\preceq$ the partial pre-ordering of E determined by C. First, observe that condition (1) is equivalent to the following:

For any vector $x \in E$, there exists a vector $u_1 \in U$ such that $u_1 \preceq x$ and there exists a vector $u_2 \in U$ such that $x \preceq u_2$.

Actually, the latter condition says that the set U is simultaneously coinitial and cofinal in the pre-ordered set $(E, \preceq)$.

By using the standard method of transfinite induction (or the well-known Kuratowski–Zorn lemma) it suffices to demonstrate that if $x \in E \setminus U$, then there exists a linear functional $f^* : V \to \mathbf{R}$ such that:

(a) V is a vector subspace of E generated by $\{x\} \cup U$;

(b) the restriction $f^*|(C \cap V)$ is positive.

For proving the existence of f^* with properties (a) and (b), consider the following two sets:

$$A = \{y \in U : y \preceq x\}, \qquad B = \{z \in U : x \preceq z\}.$$

Both A and B are nonempty. If $y \in A$ and $z \in B$, then $z - y \in C \cap U$ and $f(y) \leq f(z)$ in view of (2). Therefore,

$$\sup\{f(y) : y \in A\} \leq \inf\{f(z) : z \in B\}.$$

Choose arbitrarily τ from the segment

$$[\sup\{f(y) : y \in A\}, \inf\{f(z) : z \in B\}]$$

and define a linear functional $f^* : V \to \mathbf{R}$ by putting

$$f^*(tx + u) = t \cdot \tau + f(u) \qquad (t \in \mathbf{R},\ u \in U).$$

Verify that f^* is as required. For this purpose, take any vector $tx+u \in C$, where $u \in U$.

Suppose first that $t > 0$. Then $-u/t \preceq x$, whence it follows that

$$f^*(-u/t) = f(-u/t) \leq \tau = f^*(x), \qquad f^*(tx + u) \geq 0.$$

Analogously, supposing that $t < 0$, write $x \preceq -u/t$, whence it follows that

$$f^*(x) = \tau \leq f(-u/t) = f^*(-u/t), \qquad f^*(tx+u) \geq 0.$$

So $f^*|(C \cap V)$ is positive, and this completes the proof of the existence of the required extension f^*.

13. Show that condition (1) in Exercise 12 is essential for the validity of the Riesz theorem.

To see this circumstance, consider $E = \mathbf{R}^2 = \mathbf{R} \times \mathbf{R}$ and let U and C be defined as follows:

$$U = \{(x,y) \in \mathbf{R}^2 : y = 0\} = \mathbf{R} \times \{0\},$$

$$C = \{(x,y) \in \mathbf{R}^2 : y \geq 0, \ (\forall t < 0)((x,y) \neq (t,0))\}.$$

Consider also the linear functional $f : U \to \mathbf{R}$ defined by

$$f((x,0)) = x \qquad ((x,0) \in U).$$

Verify that f is positive on the convex cone $C \cap U$, but the same f does not admit a linear extension $f^* : \mathbf{R}^2 \to \mathbf{R}$ positive on C.

14. Infer from the Riesz extension theorem the following result of Krein.

Let E be a real vector space, C be a convex cone in E with vertex 0, and let $x \in E \setminus (-C)$ be such that $\mathbf{R}x + C = E$.

Then there exists a linear functional $f^* : E \to \mathbf{R}$ positive on C and satisfying the equality $f^*(x) = 1$.

For this purpose, consider the vector subspace $U = \mathbf{R}x$ of E. Let $f : U \to \mathbf{R}$ be a linear functional such that $f(tx) = t$ for $t \in \mathbf{R}$. Check that the restriction $f|(C \cap U)$ is positive. Suppose otherwise, i.e., $f(tx) < 0$ for some $tx \in C$. By virtue of the definition of f, one must have $t < 0$. Consequently, $-1/t > 0$ and

$$-x = (-1/t)(tx) \in C, \quad x \in -C,$$

which contradicts the assumption that $x \notin -C$. Now, applying the Riesz extension theorem to U, f, and C, obtain the required result.

15*. Let E be again a real vector space. A functional $p : E \to \mathbf{R}$ is called sublinear on E if the following two conditions hold:

(a) $p(tx) = tp(x)$ for any real number $t \geq 0$ and for any $x \in E$;

(b) $p(x+y) \leq p(x) + p(y)$ for all $x \in E$ and for all $y \in E$.

These conditions immediately imply $p(0) = 0$ and

$$0 = p(0) = p(x + (-x)) \le p(x) + p(-x), \quad -p(-x) \le p(x), \quad -p(x) \le p(-x).$$

It also follows from the above conditions that every sublinear functional p is a convex functional, i.e., one has

$$p(tx + (1-t)y) \le tp(x) + (1-t)p(y) \qquad (0 \le t \le 1, \ x \in E, \ y \in E).$$

Deduce from the Riesz extension theorem the following standard version of the Hahn–Banach theorem:

Let E be a real vector space with a sublinear functional $p : E \to \mathbf{R}$, let U be a vector subspace of E, and let $f : U \to \mathbf{R}$ be a linear functional such that $f(u) \le p(u)$ for all $u \in U$.

Then there exists a linear functional $f^* : E \to \mathbf{R}$ which extends f and satisfies $f(x) \le p(x)$ for all $x \in E$.

For this purpose, in the product vector space $\mathbf{R} \times E$ consider the two sets

$$C = \{(t, x) : t \in \mathbf{R}, \ x \in E, \ p(x) \le t\}, \qquad V = \mathbf{R} \times U.$$

Verify that C is a convex cone in $\mathbf{R} \times E$ with vertex $(0, 0)$ and V is a vector subspace of $\mathbf{R} \times E$. Moreover, the equality

$$C + V = \mathbf{R} \times E$$

holds true. Further, define a linear functional $g : V \to \mathbf{R}$ by putting

$$g((t, u)) = t - f(u) \qquad (t \in \mathbf{R}, \ u \in U).$$

The relation $f(u) \le p(u)$ for any $u \in U$ implies that the restriction of g to $C \cap V$ is positive. So, in this situation one can apply the Riesz extension theorem. Consequently, there exists a linear functional

$$g^* : \mathbf{R} \times E \to \mathbf{R}$$

which extends g and is positive on the cone C. Now, consider the linear functional $f^* : E \to \mathbf{R}$ defined by the formula

$$f^*(x) = -g^*((0, x)) \qquad (x \in E).$$

Check that f^* is as required. Clearly, f^* is an extension of f. Further, take any vector $x \in E$. By the definition of C, one has

$$(p(x), x) \in C, \qquad g^*((p(x), x)) \ge 0$$

or, keeping in mind the linearity of g^*, the relation

$$g^*((p(x), x)) = g^*((0, x)) + g^*((p(x), 0)) = g^*((0, x)) + g((p(x), 0)) =$$
$$-f^*(x) + p(x) \ge 0$$

holds true, which is equivalent to $f^*(x) \le p(x)$.

This ends the proof of the Hahn–Banach theorem.

Remark 6. In practice, the most typical situation is when a given sublinear functional $p : E \to \mathbf{R}$ is a seminorm on E. This means that, for all $t \in \mathbf{R}$ and for all $x \in E$, the equality

$$p(tx) = |t|p(x)$$

is valid. In its turn, the latter implies that $p(-x) = p(x)$ and, therefore,

$$0 = p(x + (-x)) \leq p(-x) + p(x) = 2p(x), \quad p(x) \geq 0.$$

It is easy to see that in the case of a seminorm p on E, the inequality $f(x) \leq p(x)$ for a real-valued linear functional f on E and for all $x \in E$ is equivalent to the inequality $|f(x)| \leq p(x)$.

APPENDIX 1

Convex sets in real vector spaces

This Appendix is devoted to some general facts and theorems concerning convex sets in vector spaces over $\mathbf{R}$. The results about convex sets presented here are helpful in many branches of pure and applied mathematics. In particular, they are substantially exploited in this book. Of course, in this rather concise Appendix it is impossible to give an extensive and more or less complete survey of the theory of convex sets and their fundamental properties. For more information around this topic, we refer the reader to numerous works and manuals, e.g., such as [4], [20], [24], [38], [45], [52], [68], [77], [107], [120], [122], [123], [126], [232], [249], [298], [306], [308], [356], [360].

As was already said, the vector spaces considered below are assumed to be over the standard field $(\mathbf{R}, +, \cdot)$ of all real numbers. This assumption is natural, because we are primarily interested in various useful properties of convex sets lying in such spaces. Briefly speaking, we intend to discuss some aspects of the convex structure which is canonically produced by the standard vector structure over $(\mathbf{R}, +, \cdot)$.

First of all, we wish to recall the basic definition of a convex set. It needless to say that this notion plays an extremely important role in many areas of contemporary mathematics.

If E is a real vector space and x and y are any two points of E, then the set

$$[x, y] = \{tx + (1 - t)y \ : \ 0 \leq t \leq 1\}$$

is usually called the line segment with the endpoints x and y.

If $x = y$, then this segment is degenerate and coincides with the singleton $\{x\}$.

Naturally, the point

$$z = (x + y)/2$$

is called the mid-point of $[x, y]$.

DOI: 10.1201/9781003458708-A1

Let X be a subset of E. We say that X is a convex set if

$$(\forall x)(\forall y)((x \in X \ \& \ y \in X) \Rightarrow [x, y] \subset X)$$

or, in other words, for any two points x and y from X, the corresponding line segment $[x, y]$ is entirely contained in X.

The following sets provide simple examples of convex subsets of E:

(a) empty set ($= \emptyset$) and the entire space E;

(b) every line segment in E (in particular, any singleton in E);

(c) all affine hyperplanes in E, i.e., the sets of the form

$$\Gamma(f, t) = \{x \in E : f(x) = t\},$$

where $t \in \mathbf{R}$ and $f : E \to \mathbf{R}$ is a linear functional not identically equal to zero;

(d) all sets representable in one of the forms

$$C(f, t) = \{x \in E : f(x) \leq t\},$$

$$C'(f, t) = \{x \in E : f(x) \geq t\},$$

$$O(f, t) = \{x \in E : f(x) < t\},$$

$$O'(f, t) = \{x \in E : f(x) > t\},$$

where $t \in \mathbf{R}$ and $f : E \to \mathbf{R}$ is again a linear functional not identically equal to zero.

$C(f, t)$ and $C'(f, t)$ are called the two closed half-spaces determined by the affine hyperplane $\Gamma(f, t)$, while $O(f, t)$ and $O'(f, t)$ are called the two open half-spaces determined by $\Gamma(f, t)$.

The next two properties of convex sets follow directly from the definition.

(*) If $\{X_i : i \in I\}$ is an arbitrary family of convex subsets of E, then their intersection $\cap\{X_i : i \in I\}$ is a convex set, too.

(**) If $\{X_i : i \in I\}$ is a family of convex subsets of E such that, for any two indices $i \in I$ and $j \in I$, there exists an index $k \in I$ satisfying the relation $X_i \cup X_j \subset X_k$, then the union $\cup\{X_i : i \in I\}$ is a convex set, too.

Let E be a vector space and let X be a subset of E. The convex hull of X, denoted by $\mathrm{conv}(X)$, is defined by the formula

$$\mathrm{conv}(X) = \cap\{Y \ : \ X \subset Y \subset E, \ Y \text{ is a convex set in } E\}.$$

In other words, $\mathrm{conv}(X)$ is the smallest (with respect to inclusion) convex set containing X.

Naturally, a set $X \subset E$ is called convexly independent if the relation

$$(\forall x \in X)(x \notin \mathrm{conv}(X \setminus \{x\})$$

holds true.

Example 1. Let $\mathbf{R}^m$ denote the m-dimensional Euclidean space. A set $P \subset \mathbf{R}^m$ is called a convex m-dimensional polyhedron in $\mathbf{R}^m$ if P is the convex hull of a finite family of points of $\mathbf{R}^m$ which do not lie in an affine hyperplane of $\mathbf{R}^m$. Another (equivalent) definition can be formulated as follows. A subset P of $\mathbf{R}^m$ is a convex m-dimensional polyhedron in $\mathbf{R}^m$ if P is bounded, has nonempty interior and can be represented as the intersection of a finite family of closed half-spaces of $\mathbf{R}^m$. It can easily be observed that the first definition of a polyhedron is formulated only in terms of affine geometry. The second definition appeals also to the topological structure of $\mathbf{R}^m$. Nevertheless, as mentioned above, these two definitions are logically equivalent.

If we deal with an arbitrary (possibly, infinite-dimensional) vector space E, then it is more convenient to use the first definition of a convex polyhedron, i.e., to define a convex polyhedron P in E as the convex hull of a nonempty finite family of points lying in E. In this case, we say that P is m-dimensional and write $\dim(P) = m$ if m is the dimension of the smallest (with respect to inclusion) affine linear manifold in E containing P.

Sometimes, the empty set $\emptyset$ is also regarded as a polyhedron whose dimension is equal to -1.

Further, in the standard manner the notion of a face of a convex polyhedron P can be introduced (cf. Exercise 14). All faces turn out to be convex polyhedra whose dimensions are less than or equal to $\dim(P)$. The faces of dimension 0 are called vertices of P, the faces of dimension 1 are called edges of P, the faces of dimension $\dim(P) - 1$ are called facets of P.

The family of all edges of P is usually called the 1-skeleton of P.

An important special case of a convex polyhedron is a simplex defined as follows.

Example 2. Let E be a vector space and let $\{x_0, x_1, ..., x_k\}$ be a finite subset of E whose cardinality equals $k+1$. Suppose, in addition, that the dimension of the smallest affine linear manifold containing $\{x_0, x_1, ..., x_k\}$ is equal to k (this obviously means that the vectors $x_1 - x_0,\ x_2 - x_0,\ \ldots,\ x_k - x_0$ are linearly independent). Then the set

$$S = \mathrm{conv}(\{x_0, x_1, ..., x_k\})$$

is called the simplex generated by $x_0, x_1, ..., x_k$. The points $x_0, x_1, ..., x_k$ turn out to be the vertices of this simplex S and k is its dimension. The empty set is regarded as a simplex, too, and, as was indicated above, its dimension equals -1. Furthermore, it is easy to describe all faces of S. Namely, for each subset X of $\{x_0, x_1, ..., x_k\}$, the set $\mathrm{conv}(X)$ is a face of S and, conversely, every face of S has the form conv(X), where X is some subset of $\{x_0, x_1, ..., x_k\}$. In particular, the empty set is a face of S and the entire simplex S is a face of itself, too. It is convenient to introduce the standard notation for the k-dimensional simplex $S = \text{conv}(\{x_0, x_1, ..., x_k\})$. In our lecture course we prefer the notation

$$S = [x_0, x_1, ..., x_k],$$

because it is compatible with the notation of the line segment $[x, y]$ with end-points x and y. Obviously, if $x = y$, then $[x, y] = [x] = \{x\}$.

Example 3. Let M be a subset of the Euclidean space $\mathbf{R}^m$. We say that M is a polyhedron in $\mathbf{R}^m$ (not necessarily convex) if M can be represented in the form

$$M = \cup\{M_i : i \in I\},$$

where $\{M_i : i \in I\}$ is a finite family of convex polyhedra lying in the same space (the dimensions of these polyhedra may differ from each other). It is not hard to show that a set T is a polyhedron in $\mathbf{R}^m$ if and only if T admits a representation in the form $T = \cup\{T_j : j \in J\}$, where $\{T_j : j \in J\}$ is a finite family of simplices in $\mathbf{R}^m$. We leave a non-difficult proof of this fact to the reader.

It is reasonable to underline once more that the concept of a convex set in a real vector space is purely algebraic, so does not rely on any topological concepts. Nevertheless, for substantial development of convex geometry and convex analysis, topological methods and constructions play a very significant role. Moreover, the interplay between the vector and topological structures turns out to be utterly useful for obtaining many nontrivial results and facts in the general theory of convex sets.

It is clear that if one wants to apply topological methods to convex sets, then some consistency conditions must take place for the vector and topological structures (in other words, these structures have to be somehow compatible). This purpose can be reached with the aid of the notion of a topological vector space.

We wish to recall here the corresponding definition (see, e.g., [45] or [306]).

Let E be a ground set equipped with a vector structure over $\mathbf{R}$ and with a topological structure.

This E is said to be a topological vector space if the corresponding algebraic operations

$$(x, y) \to x + y \quad (x \in E, \ y \in E),$$

$$(t, x) \to tx \quad (t \in \mathbf{R}, \ x \in E)$$

are continuous on $E \times E$ and on $\mathbf{R} \times E$ respectively.

Let us emphasize that, as a rule, topological vector spaces under consideration below are assumed to be Hausdorff.

Example 4. Let E be an arbitrary finite-dimensional (Hausdorff) topological vector space and let $m = \dim(E)$ denote its dimension. Then it can be demonstrated that E is isomorphic to the canonical space $\mathbf{R}^m$. In other words, the family of all Euclidean spaces $\mathbf{R}^m$ ($m \in \mathbf{N}$) realizes the class of all finite-dimensional (Hausdorff) topological vector spaces. The reader can try to establish this important fact by using induction on $m = \dim(E)$. An essentially stronger result (for vector spaces over more general fields) can be found in [45]. For infinite-dimensional vector spaces, the situation is radically different, i.e., if E is an infinite-dimensional vector space, then there are many Hausdorff topologies compatible with the vector structure of E (see again [45], [306]).

The following geometric notions are extremely important in the theory of topological vector spaces and in some applied mathematical disciplines.

Let E be a topological vector space, let X and Y be two subsets of E, and let Γ be a closed affine hyperplane in E. We say that Γ separates X and Y if these sets lie in the distinct closed half-spaces determined by Γ. Furthermore, we say that Γ strongly separates X and Y if these sets lie in the distinct open half-spaces determined by Γ (see (d)).

The statement given below can be regarded as a geometric form of the fundamental Hahn–Banach theorem on extensions of linear functionals (see, e.g., [45]).

Theorem 1. *Let E be a topological vector space and let A and B be two nonempty convex subsets of E satisfying the following conditions:*

(1) $A \cap B = \emptyset$;

(2) at least one of these sets is open.

Then there exists a closed affine hyperplane in E separating A and B. In addition, if both these sets are open, then there exists a closed affine hyperplane in E strongly separating them.

As an immediate consequence of this theorem, we get the following statement:

If A is an open convex subset of E and x is a boundary point for A, then there exists a closed affine hyperplane Γ in E passing through x and having no common points with A.

Actually, Γ turns out to be a supporting hyperplane for the closure of A (cf. Theorem 4 below).

For the sake of completeness, we also present the standard analytic form of the Hahn–Banach theorem (see Exercise 15 from Chapter 17).

Theorem 2. *Let E be a vector space and let $p : E \to \mathbf{R}$ be a functional on E satisfying the following relations:*

$$p(x + y) \leq p(x) + p(y) \qquad (x \in E,\ y \in E),$$

$$p(tx) = tp(x) \qquad (x \in E,\ t \geq 0).$$

Suppose, in addition, that a linear functional $g_0 : F \to \mathbf{R}$ is given on some vector subspace F of E, such that $g_0(x) \leq p(x)$ for all points $x \in F$.

Then there exists at least one linear functional $g : E \to \mathbf{R}$ extending g_0 such that $g(x) \leq p(x)$ for all points $x \in E$.

Theorem 2 is usually applied in those cases when p is some seminorm given on a vector space E (cf. [45], [306]). In this connection, let us recall that a seminorm on E is any mapping

$$p : E \to [0, +\infty[$$

satisfying the relations

$$p(x + y) \leq p(x) + p(y) \qquad (x \in E,\ y \in E),$$

$$p(tx) = |t|p(x) \qquad (t \in \mathbf{R},\ x \in E).$$

If, in addition, the equality $p(x) = 0$ is equivalent to $x = 0$, then p is called a norm on E (the standard notation for a norm is $||\cdot||$).

The following definition describes an important class of topological vector spaces.

We say that a topological vector space E is locally convex if each point of E has a fundamental system (i.e., local base) of convex neighborhoods.

Evidently, in this definition it suffices to require only the existence of a local base of convex neighborhoods for the neutral element 0 of E.

Another characterization of all locally convex topological vector spaces is also well known:

A topological vector space E is locally convex if and only if its topology can be defined by a family of seminorms (see, e.g., [45]).

If the topology of E is given by a norm, then E is called a normed vector space. Finally, according to the classical definition, any complete normed vector space is called a Banach space.

Notice that the importance of locally convex topological vector spaces is implied by the fact that there are sufficiently many linear continuous functionals on such spaces. To be more precise, consider any topological vector space E and denote by E^* the family of all continuous linear functionals on E. Obviously, E^* forms a vector space with respect to the natural algebraic operations. This vector space is usually called the conjugate (or adjoint, or dual) space for E.

Let G be a subset of E^*. We say that G separates points of E if, for any two distinct points $x \in E$ and $y \in E$, there exists at least one functional $g \in G$ such that $g(x) \neq g(y)$.

The next simple (but rather useful) statement is true.

Theorem 3. *If E is a locally convex topological vector space, then E^* separates points of E.*

This result is one of the immediate consequences of the geometric form of the Hahn-Banach theorem. In connection with Theorem 3, let us remark that the converse assertion does not hold in general. In other words, there are examples of a topological vector space F such that F^* separates the points of F, but F is not locally convex.

Let E be a topological vector space and let X be a subset of E.

We say that a closed affine hyperplane Γ in E is a supporting hyperplane for (of) X if the conjunction of these two relations is fulfilled:

(i) $X \cap \Gamma \neq \emptyset$;

(ii) all points of X lie in one closed half-space determined by Γ.

Following [45], we say that a set $Y \subset E$ is a convex body (in E) if Y is closed, convex, and has nonempty interior.

Example 5. Let $\mathbf{R}^m$ denote, as usual, the m-dimensional Euclidean space and let X be a closed convex subset of this space. Then X is a convex body in $\mathbf{R}^m$ if and only if the smallest affine linear manifold containing X coincides with $\mathbf{R}^m$. In particular, any m-dimensional simplex is a convex body in $\mathbf{R}^m$. More generally, any m-dimensional convex polyhedron in $\mathbf{R}^m$ is a convex body in $\mathbf{R}^m$.

For convex bodies in topological vector spaces, we have the following two important statements which are frequently used in various geometric arguments and considerations.

Theorem 4. *Let E be a topological vector space and let X be a convex body in E.*

Then, for any boundary point $x \in X$, there exists at least one supporting hyperplane of X passing through x.

Theorem 5. *Let E be a topological vector space and let X be a convex body in E.*

Then the equality $X = \cap\{C_i : i \in I\}$ is valid, where $\{C_i : i \in I\}$ denotes the family of all those closed half-spaces in E which contain X and are determined by the supporting hyperplanes for X.

Actually, the two statements presented above can readily be deduced from the geometric form of the Hahn–Banach theorem.

Example 6. Evidently, there are many compact convex bodies in the Euclidean space $\mathbf{R}^m$. For instance, all m-dimensional convex polyhedra and all closed m-dimensional balls in this space are such bodies. Further, a convex body X in $\mathbf{R}^m$ is compact if and only if X is bounded. On the other hand, let E be a topological vector space in which there exists at least one compact body. Then it can easily be observed that E is locally compact and, consequently, E is finite-dimensional, i.e., isomorphic to $\mathbf{R}^m$ for some natural number m. Thus, there are no compact bodies in infinite-dimensional topological vector spaces.

In connection with various properties of supporting hyperplanes for compact sets, let us indicate one simple statement.

Theorem 6. *Let E be a topological vector space and let X be a nonempty compact subset of E.*

Then, for each closed affine hyperplane Γ in E, there exists at least one closed affine hyperplane Γ_1 in E parallel to Γ and supporting for the set X.

Proof. The argument is very simple. Since Γ is closed, it is determined by a nonzero linear continuous functional $g : E \to \mathbf{R}$, i.e., for some $r \in \mathbf{R}$, the hyperplane Γ is described by the formula

$$\Gamma = \{x \in E \ : \ g(x) = r\}.$$

Now, let us put

$$r_1 = \sup_{x \in X} g(x),$$

$$r_2 = \inf_{x \in X} g(x).$$

Since g is continuous on the compact set X, the values r_1 and r_2 belong to $g(X)$. Consequently, the hyperplanes Γ_1 and Γ_2 determined, respectively, by the equations

$$\Gamma_1 = \{x \in E : g(x) = r_1\},$$

$$\Gamma_2 = \{x \in E : g(x) = r_2\},$$

are supporting for X and, clearly, they are parallel to the original hyperplane Γ. Of course, the case when the latter two hyperplanes coincide cannot be ignored. This case is realized if and only if the set X lies in some closed affine hyperplane of E parallel to Γ. □

Theorem 6 implies, in particular, that if X is a compact convex body in a finite-dimensional topological vector space E, then there exist exactly two supporting hyperplanes for X, each of which is parallel to a given (closed) affine hyperplane Γ in E.

Let T be a compact convex body in the space $\mathbf{R}^m$, where $m \geq 1$, and suppose that the origin 0 of $\mathbf{R}^m$ belongs to the interior $\mathrm{int}(T)$ of T. Fix a unit vector e in $\mathbf{R}^m$ and consider the half-line $l(e)$ in the same space, whose direction coincides with e and whose endpoint is 0. Then, according to Theorem 6, there exists a unique affine hyperplane $\Gamma(e)$ in $\mathbf{R}^m$ such that:

(1) $\Gamma(e)$ is a supporting hyperplane for T;

(2) $\Gamma(e)$ is orthogonal to e;

(3) $\Gamma(e)$ intersects $l(e)$.

Denote by $h(e)$ the distance between 0 and $\Gamma(e)$. Clearly, in this manner we obtain a certain function $h = h_T$ acting from the unit sphere $\mathbf{S}_{m-1}$ of $\mathbf{R}^m$

into the open interval $]0, +\infty[$. The above-mentioned function is usually called the Minkowski support function for (of) T.

Let us notice that h completely determines the body T. In other words, if T_1 and T_2 are any two compact convex bodies in $\mathbf{R}^m$ such that $0 \in \text{int}(T_1)$ and $0 \in \text{int}(T_2)$, then the following assertions are equivalent:

($'$) the Minkowski support function for T_1 is identical with the Minkowski support function for T_2;

($''$) $T_1 = T_2$.

The reader can easily deduce the equivalence of ($'$) and ($''$) from the separation principle for convex sets (it suffices to apply the weakest version of this principle, stating that if a point does not belong to a closed convex set, then there exists a closed affine hyperplane strongly separating the point and the set).

The next notion plays a key role in modern convex analysis.

Let E be a vector space (over $\mathbf{R}$) and let Z be a convex subset of E.

We say that a point $z \in Z$ is extreme for Z if there exists no non-degenerate line segment in Z whose mid-point is z.

The set of all extreme points for Z will be denoted by the symbol $\text{extr}(Z)$.

Note that a set $X \subset E$ is convexly independent if and only if the equality

$$X = \text{extr}(\text{conv}(X))$$

holds true.

Example 7. Let $Z = [x_0, x_1, ..., x_k]$ be any k-dimensional simplex in a vector space E. Then it is not hard to see that $\text{extr}(Z) = \{x_0, x_1, ..., x_k\}$. The analogous result is valid for any convex polyhedron P in the space E. Actually, quite often the vertices of P are defined as its extreme points. Such a definition is completely compatible with geometric intuition.

Example 8. Let B be any closed ball in the space $\mathbf{R}^m$, where $m \geq 1$, and let S be the boundary of B. Then it is evident that $\text{extr}(B) = S$.

If now E is a finite-dimensional normed vector space and T is any bounded convex body in E, then there always are extreme points of T (for a much stronger result, see Theorem 8 below). It is interesting to notice that, in general, the analogous statement fails to be true for an infinite-dimensional Banach space F. Namely, it may happen that there is no extreme point of the closed unit ball in F (we suggest the reader verify this fact).

Let E be a topological vector space and let X be a subset of E.

By definition, the closed convex hull of X is the intersection of all those closed convex sets in E which contain X (i.e., the closed convex hull of X is the smallest closed convex set containing X).

Let us mention that the closure of any convex subset of E is convex, too. By starting with this simple observation, it can easily be checked that the closed convex hull of a set $X \subset E$ coincides with $\text{cl}(\text{conv}(X))$.

One of the most important results in convex analysis is the following theorem due to Krein and Milman (see, for instance, [45] or [306]).

Theorem 7. *Let E be a locally convex topological vector space and let X be a compact convex subset of E.*

Then X coincides with the closed convex hull of the set of all extreme points of X, i.e., the equality

$$X = \mathrm{cl}(\mathrm{conv}(\mathrm{extr}(X)))$$

holds true.

This result implies, in particular, that if X is an arbitrary nonempty compact convex subset of E, then there exists at least one extreme point of X (cf. Example 8 above). Moreover, in each supporting hyperplane of X there is at least one extreme point of X. Let us stress that the Krein–Milman theorem is valid for all locally convex topological vector spaces but fails to hold for some Hausdorff topological vector spaces.

On the other hand, for finite-dimensional topological vector spaces, this theorem can be substantially strengthened. Namely, we have the following classical result of Minkowski.

Theorem 8. *Let E be a finite-dimensional topological vector space and let X be a compact convex subset of E.*

Then X coincides with the convex hull of the set of all extreme points of X, i.e., the equality

$$X = \mathrm{conv}(\mathrm{extr}(X))$$

is fulfilled.

The proof of Theorem 8 can be carried out by induction on $\dim(E)$ and is not difficult. We leave the corresponding details to the reader (take into account the fact that if Γ is a supporting hyperplane of X, then any extreme point of $\Gamma \cap X$ is simultaneously an extreme point of X). Notice also that, for proving the Krein-Milman theorem, essentially non-elementary set-theoretical methods are necessary. As a rule, they are based on the Axiom of Choice (**AC**) or on some of its logical equivalents, e.g. the Kuratowski–Zorn lemma, which cannot be avoided here (in this context, see [18]).

The next statement is relative to the Krein–Milman theorem and turns out to be helpful in certain situations.

Theorem 9. *Let E be a locally convex topological vector space and let Y be a compact subset of E. Suppose, in addition, that the closed convex hull of Y is also compact in E.*

Then each extreme point of $\mathrm{cl}(\mathrm{conv}(Y))$ *belongs to Y; in other words, the inclusion*

$$\mathrm{extr}(\mathrm{cl}(\mathrm{conv}(Y))) \subset Y$$

is satisfied.

For the proof of Theorem 9, see e.g. [45] or [306].

Example 9. Let T be an arbitrary closed convex set in the Euclidean plane $\mathbf{R}^2$. It is easy to show that the set of all extreme points of T is closed in $\mathbf{R}^2$. Consequently, if T is compact, then the set extr(T) is also compact. This geometric fact fails to be true for the space $\mathbf{R}^m$, where $m > 2$. Namely, there exist many compact convex bodies T in $\mathbf{R}^m$, where $m > 2$, for which the sets extr(T) are not closed in $\mathbf{R}^m$. We suggest the reader construct such a body, e.g., in the space $\mathbf{R}^3$.

Let us return, for a while, to arbitrary vector spaces (over $\mathbf{R}$) and to convex sets lying in them. Let E be a vector space and let X be a subset of E. In many cases, one needs a more descriptive characterization of the convex hull of X. Some valuable information on the structure of conv(X) yields the formula

$$\mathrm{conv}(X) = \cup\{\mathrm{conv}(X_i) : i \in I\},$$

where $\{X_i : i \in I\}$ denotes the family of all finite subsets of X (the usual notation for this family is $[X]^{<\omega}$, where ω stands for the least infinite cardinal number). The above equality is directly implied by property (**) formulated at the beginning of the Appendix. In fact, this equality allows one to describe the structure of conv(X) as follows:

The set conv(X) consists of all those elements $x \in E$ which can be represented in the form

$$x = t_1x_1 + t_2x_2 + ... + t_kx_k,$$

where k is a natural number (in general, depending on x), $x_1, x_2, ..., x_k$ are some points from X and $t_1, t_2, ..., t_k$ are some real numbers such that

$$0 \leq t_1, \ \ 0 \leq t_2, \ \ . \ . \ . \ , \ 0 \leq t_k, \quad t_1 + t_2 + \ . \ . \ . \ + t_k = 1.$$

In other words, conv(X) consists of all possible convex combinations of points from X.

If a vector space E is finite-dimensional, then the above-mentioned formula for conv(X) can be essentially simplified. More precisely, we have the following classical statement due to Carathéodory (see, e.g., [4], [20], [36], [38], [68], [298]).

Theorem 10. *Let m be a natural number, let E be a vector space with* dim(E) $= m$, *and let X be an arbitrary subset of E.*

Then the equality

$$\mathrm{conv}(X) = \cup\{\mathrm{conv}(X_j) : j \in J\}$$

holds true, where $\{X_j : j \in J\} = [X]^{\leq m+1}$ denotes the family of all those subsets of X whose cardinalities are less than or equal to $m + 1$.

In some special cases of X, the statement of Theorem 10 can be improved. For instance, if X is a connected set in the space $\mathbf{R}^m$ $(m \geq 1)$, then

$$\mathrm{conv}(X) = \cup\{\mathrm{conv}(X_j) : j \in J\},$$

where $\{X_j : j \in J\} = [X]^{\leq m}$ denotes the family of all those subsets of X whose cardinalities do not exceed m (see Exercise 15).

Also, Carathéodory's theorem can be reformulated in a purely geometric form. Namely, we have the next statement.

Theorem 11. *Let E be a topological vector space with $dim(E) = m$ and let P be an m-dimensional convex polyhedron in E.*

Then there exists a finite family $\{T_i : i \in I\}$ of m-dimensional simplices in E such that:

(1) $P = \cup\{T_i : i \in I\}$;

(2) the interiors of these simplices are pairwise disjoint;

(3) for each index $i \in I$, the set of all vertices of T_i is contained in the set of all vertices of P.

This statement can easily be proved by induction on m (actually, the family $\{T_i : i \in I\}$ may be assumed to be a triangulation of P). We recommend the reader carry out all necessary details of the corresponding argument. Notice that, from the geometric point of view, Theorem 11 is almost trivial. It is also obvious that Theorem 10 is a direct consequence of Theorem 11.

Let us give one useful application of Carathéodory's theorem.

Theorem 12. *If E is a finite-dimensional topological vector space and X is a compact subset of E, then the set* conv(X) *is also compact in E.*

Proof. We may suppose without loss of generality that E coincides with the space $\mathbf{R}^m$, where $m \in \mathbf{N}$. Denote by T the compact convex subset of $\mathbf{R}^{m+1}$ consisting of all those points $(t_1, t_2, ..., t_{m+1}) \in \mathbf{R}^{m+1}$ which satisfy the relations

$$0 \leq t_1, \ 0 \leq t_2, \ \ldots, \ 0 \leq t_{m+1}, \quad t_1 + t_2 + ... + t_{m+1} = 1.$$

It can easily be observed that T is a regular m-dimensional simplex in $\mathbf{R}^{m+1}$ (however, we do not need this fact below). Further, the product $T \times X^{m+1}$ is a compact set in the space

$$\mathbf{R}^{m+1} \times \mathbf{R}^{m(m+1)} = \mathbf{R}^{(m+1)^2}.$$

Let us introduce a mapping

$$\phi : T \times X^{m+1} \to \mathbf{R}^m$$

defined by the formula

$$\phi(t_1, t_2, ..., t_{m+1}, x_1, x_2, ..., x_{m+1}) = t_1x_1 + t_2x_2 + ... + t_{m+1}x_{m+1}.$$

Clearly, ϕ is a continuous mapping. According to Carathéodory's theorem, the range of ϕ coincides with the convex hull of X. Thus, conv(X) is the range of a continuous mapping given on a compact set, so conv(X) is compact as well. □

The next statement can be established in a similar manner.

Theorem 13. *Let E be a topological vector space and let $\{X_i : i \in I\}$ be a finite family of compact convex subsets of E.*

Then the convex hull of the set $\cup\{X_i : i \in I\}$ is compact, too.

A particular case of Theorem 13 is the assertion that the convex hull of any finite family of points in $\mathbf{R}^m$ is compact, i.e., every convex polyhedron in $\mathbf{R}^m$ turns out to be compact. Of course, this is a very simple statement concerning properties of convex polyhedra in Euclidean spaces. There are a lot of profound results about such polyhedra (see, for instance, [4], [20], [29], [123], [126], [130], [249], [260], [278], [301], [302], [364]). Among them the following exceedingly beautiful statement due to Minkowski should be mentioned especially.

Theorem 14. *Let $\{s_i : 1 \leq i \leq k\}$ be a finite family of strictly positive real numbers and let $\{e_i : 1 \leq i \leq k\}$ be an injective family of unit vectors in the space $\mathbf{R}^m$ $(m \geq 2)$, which satisfy the relation*

$$\sum\{s_i e_i : 1 \leq i \leq k\} = 0$$

and do not lie in an affine hyperplane of $\mathbf{R}^m$.

Then there exists a m-dimensional convex polyhedron P in $\mathbf{R}^m$ with the facets F_1, F_2, ..., F_k such that, for each natural index $i \in [1, k]$, the number s_i coincides with the $(m-1)$-dimensional volume of F_i and e_i is an exterior normal vector of F_i.

For a proof of Minkowski's theorem, see e.g. [4].

EXERCISES

1. Let E be an arbitrary vector space (over $\mathbf{R}$).

By using the existence of an algebraic (linear) basis in E, show that the vector structure of E can be enriched up to the structure of a normed vector space, i.e., there exists at least one norm $||\cdot||$ on E.

Moreover, demonstrate that such a norm can be chosen so that it is induced by some inner product $\langle\cdot\rangle$ in E, in other words, E becomes a pre-Hilbert space.

On the other hand, give an example of a vector space E whose structure cannot be enriched to the structure of a complete normed vector (i.e., Banach) space.

Verify that the topological vector space $\mathbf{R}^{\mathbf{N}}$ is Polish (i.e., complete, metrizable, separable) but does not admit a norm compatible with the topology of $\mathbf{R}^{\mathbf{N}}$.

For this purpose, use the following geometric property of an arbitrary normed vector space $(E, ||\cdot||)$: if B_1 and B_2 are any two balls in E, then B_1 is contained in some homothetic image of B_2.

2. Let E be a vector space (over $\mathbf{R}$) and let G denote the family of all linear functionals on E.

Check that there exists a locally convex topology $\mathcal{T}$ compatible with the vector structure of E and having the property that all functionals from G are continuous with respect to $\mathcal{T}$.

Remark 1. In fact, the required topology $\mathcal{T}$ coincides with the weak topology $\sigma(E, G)$ (see, e.g., [45] or [306]).

3*. Let $(E, ||\cdot||)$ be a normed vector space (over $\mathbf{R}$).

Prove that the following two assertions are equivalent:

(a) E is finite-dimensional;

(b) every linear functional defined on E is continuous.

For this purpose, argue as follows.

The implication (a) $\Rightarrow$ (b) is easily verified.

To establish the converse implication (b) $\Rightarrow$ (a), take into account the fact that any linear functional defined on a vector subspace of E can be extended (by using the Kuratowski–Zorn lemma) to a linear functional defined on the whole of E.

In addition, verify that the existence of a discontinuous linear functional on a real infinite-dimensional separable Hilbert space H essentially relies on an uncountable form of the Axiom of Choice. More concretely, start with the fact that **ZF** & **DC** theory is consistent with the assumption that all subsets of $\mathbf{R}$ have the Baire property, then construct in the same theory a Borel isomorphism between $\mathbf{R}$ and H which preserves the Baire category of sets, and finally demonstrate that any linear functional on H is continuous under the above-mentioned assumption.

4. For an arbitrary vector space E over $\mathbf{R}$, show that the following two statements are equivalent:

(a) E is finite-dimensional;

(b) there exists a unique Hausdorff topology compatible with the vector structure of E.

5*. Let E be a vector space (over $\mathbf{R}$) and let z be a point of E.

A nonempty subset C of E is called a cone in E with vertex z if

$$(\forall x \in C)(\forall t \geq 0)(z + t(x - z) \in C).$$

Notice that, according to this definition, one has $z \in C$.

A cone C is called proper if $C \neq E$ and is called non-degenerate if $C \neq \{z\}$.

If C is a cone with vertex x and $y \in E$, then the set $C + (y - x)$ is a cone with vertex y.

The set of all vertices of a given cone C is a certain affine submanifold of E.

If $\{C_j : j \in J\}$ is a family of cones with vertex z, then both sets

$$\cup\{C_j : j \in J\}, \qquad \cap\{C_j : j \in J\}$$

are cones with vertex z.

Naturally, a cone C is convex if it is a convex subset of E.

Give a characterization of all maximal (with respect to inclusion) proper convex cones in E.

For this purpose, examine the closed half-spaces in E determined by all possible affine hyperplanes in E.

Let E be a locally convex topological vector space, let C be a closed convex cone in E with vertex 0, and let l be a closed ray in E with endpoint 0. Suppose that $C \cap l = \{0\}$.

Show that

(i) there exists a closed hyperplane Γ in E such that $0 \in \Gamma$ and one of the open half-spaces determined by Γ contains $l \setminus \{0\}$ and has no common points with C.

For this purpose, use Theorem 1 on separation of two disjoint convex sets.

Further, let E be a real pre-Hilbert space and let C be a non-degenerate proper closed convex cone in E with vertex 0. Denote by C' the family of all those closed rays in E which have endpoint 0 and form with all rays from C non-acute angles.

Verify that:

(ii) C' is a non-degenerate proper closed convex cone in E;

(iii) $(C')' = C$.

For establishing (iii), use (i).

6. Let X and Y be any two nonempty disjoint closed convex subsets of a real finite-dimensional topological vector space E.

Prove that there exists an affine hyperplane in E separating X and Y.

Moreover, check that, in general, such a hyperplane does not strongly separate these sets (on the other hand, cf. Exercise 9 below).

Infer from the above result that if A and B are any two nonempty convex subsets of E without common points, then there exists an affine hyperplane in E separating A and B.

7. Let E be an infinite-dimensional separable Hilbert space over $\mathbf{R}$.

Verify that there are two nonempty disjoint closed convex sets X and Y in E such that the set

$$X - Y = \{x - y : x \in E,\ y \in E\}$$

is everywhere dense in E.

Conclude that there exists no closed affine hyperplane in E separating X and Y.

8. Give an example of an infinite-dimensional Banach space E such that the closed unit ball in E possesses no extreme point.

For this purpose, consider the Banach space of all those real-valued sequences which converge to zero.

9. Let E be a locally convex topological vector space, let X be a nonempty closed convex subset of E, and let Y be a nonempty compact convex subset of E.

Demonstrate that if X and Y are disjoint, then there exists a closed affine hyperplane in E strongly separating X and Y.

10*. The standard proof of the Krein–Milman theorem can be found, e.g., in the well-known monograph [45].

Specify this proof and check that the same result remains true for all compact convex sets in any topological vector space E having the property that E^* separates the points of E.

11*. Let X be a compact convex subset of the space $\mathbf{R}^m$ and let x be a point of X. This x is called a point of strong convexity of X if there exists a supporting hyperplane Γ for X such that $\Gamma \cap X = \{x\}$.

Observe that each point of strong convexity of X is simultaneously an extreme point of X (the converse assertion does not hold in general).

Prove S. Straszewicz's theorem stating that:

(a) X coincides with the closed convex hull of the set of all points of strong convexity of X;

(b) the set of all points of strong convexity of X is everywhere dense in the set $\mathrm{extr}(X)$.

For proving (a), assume without loss of generality that $\dim(X) = m$ and first establish that the closed convex hull of the set of all points of strong convexity of X contains X. This inclusion is implied by the following fact:

If Γ is an affine hyperplane in $\mathbf{R}^m$ such that an open half-space determined by Γ has nonempty intersection with X, then this half-space contains some point of strong convexity of X.

For proving (b), apply (a) and Carathéodory's theorem.

12*. Let $(E, ||\cdot||)$ be an arbitrary infinite-dimensional normed vector space. Demonstrate that there exists a partition $\{X, Y\}$ of E such that:

(a) each of the sets X and Y is convex in E;

(b) both X and Y are everywhere dense in E.

More precisely, by using the principle of transfinite induction, give some transfinite construction of the above-mentioned sets X and Y.

Finally, observe that if $(E, ||\cdot||)$ is a finite-dimensional normed vector space, then a partition $\{X, Y\}$ of E with the properties (a) and (b) does not exist.

13. Let E be a vector space and let $\{z_i : i \in I\}$ be a nonempty finite family of pairwise distinct points in E.

Recall that the geometric barycenter z of $\{z_i : i \in I\}$ is defined by the formula

$$z = (1/\text{card}(I)) \sum\{z_i : i \in I\}.$$

Let now $n \geq 2$ be a natural number and let $X = \{x_1, x_2, ..., x_n\}$ be an injective finite family of points in E. Fix a nonzero natural number $k < n$ and consider an arbitrary partition $\{Y, Z\}$ of X such that $\text{card}(Y) = k$. Let y denote the geometric barycenter of Y and let z denote the geometric barycenter of Z.

Show that the line segment $[y, z]$ contains the geometric barycenter x of X.

14. Let P be an m-dimensional convex polyhedron in the space $\mathbf{R}^m$.

A set of the form $\Gamma \cap P$, where Γ is an affine hyperplane in $\mathbf{R}^m$ not intersecting the interior of P, is called a proper face of P.

Demonstrate that:

(a) each proper face of P is a convex polyhedron whose vertices are some vertices of P and whose dimension is strictly less than m (consequently, the total number of faces of P is finite);

(b) the intersection of any two faces of P is again a face of P;

(c) the boundary of P is representable as the union of all $(m-1)$-dimensional faces (i.e., facets) of P.

Remark 2. The above definition of a face makes sense for every compact body X in $\mathbf{R}^m$ which, in general, is not a polyhedron. In this case, even the total number of facets of X may be infinite (but always is at most countable).

15*. Let X be a set in the space $\mathbf{R}^m$, where $m \geq 1$, and let x be a point of $\text{conv}(X)$ such that x does not belong to the convex hull of any subset of X with fewer than $m+1$ elements.

Prove that there are at least $m+1$ connected components of X (in particular, X is not connected).

For this purpose, first check that there exist some points $x_0, x_1, ..., x_m$ in X which are in general position and such that x belongs to the interior of the simplex $S = [x_0, x_1, ..., x_m]$. For each natural index $i \in [0, m]$, consider the polyhedral angle Q_i generated by the rays which have common endpoint x and pass through the points $x_0, x_1, \ldots, x_{i-1}, x_{i+1}, \ldots, x_m$ respectively. Further, denote by Q_i' the image of Q_i under the central symmetry of $\mathbf{R}^m$ with center x, and verify that the following two relations hold true:

(a) x_i belongs to the interior of Q_i' ;

(b) the boundary of Q_i' does not have common points with X.

Conclude from (a) and (b) that the points $x_0, x_1, ..., x_m$ lie, respectively, in pairwise distinct connected components of X.

16*. Let X be a set in the space $\mathbf{R}^m$ ($m \geq 1$) and let x be an arbitrary point of the interior of $\text{conv}(X)$.

Prove the Steinitz theorem stating that there exists a subset Y of X such that:

(a) $\text{card}(Y) \leq 2m$;

(b) x belongs to the interior of $\text{conv}(Y)$.

For this purpose, first find an m-dimensional convex polyhedron P all vertices of which belong to X and the interior of which contains x. Let $\{T_j : j \in J\}$ be a triangulation of the boundary of P into $(m-1)$-dimensional simplices, without adding new vertices.

Show that there exist a straight line l in $\mathbf{R}^m$ passing through x and two distinct simplices T' and T'' from $\{T_j : j \in J\}$ such that l has a common point with the relative interior of T' and has a common point with the relative interior of T''.

Finally, check that the union of all vertices of T' and T'' can be taken as the required subset Y of X.

17*. Let X be an arbitrary set in the space $\mathbf{R}^m$ with $\text{card}(X) \geq m+2$.

Prove Radon's theorem stating that:

(a) there exists at least one partition $\{X_1, X_2\}$ of X such that

$$\text{conv}(X_1) \cap \text{conv}(X_2) \neq \emptyset;$$

(b) a partition of X with the above-mentioned property is unique if and only if $\text{card}(X) = m+2$ and all points of X are in general position in $\mathbf{R}^m$.

For establishing both (a) and (b), use induction on m.

18*. Let $\{X_i : i \in I\}$ be a finite family of convex subsets of $\mathbf{R}^m$ such that, for each set $J \subset I$ with $\text{card}(J) \leq m+1$, the intersection of the partial family $\{X_j : j \in J\}$ is nonempty.

Prove Helly's theorem stating that the intersection of the entire given family $\{X_i : i \in I\}$ is also nonempty.

For this purpose, use induction on $\text{card}(I)$ and the result of Exercise 17.

In addition, deduce from Helly's theorem that if a family $\{Y_k : k \in K\}$ of compact convex subsets of $\mathbf{R}^m$ has the property that, for each set $J \subset K$ with $\text{card}(J) \leq m+1$, the intersection of the partial family $\{Y_j : j \in J\}$ is nonempty, then the intersection of the entire given family $\{Y_k : k \in K\}$ is also nonempty.

19*. Let $S = [x_0, x_1, ..., x_m]$ be an arbitrary m-dimensional simplex in the space $\mathbf{R}^m$ and let $\{T_j : j \in J\}$ be a finite triangulation of S into m-dimensional simplices. Denote by V the set of all vertices of this triangulation. For any vertex $v \in V$, consider the face $[x_{i_1}, x_{i_2}, ..., x_{i_k}]$ of S having minimal dimension and containing v. Assign a value $f(v)$ to v so that $f(v) \in \{i_1, i_2, ..., i_k\}$.

Let f^* denote the canonical extension of f defined on the family of all faces of the simplices T_j $(j \in J)$. In other words, for any face $T = [v_0, v_1, ..., v_n]$ of the triangulation $\{T_j : j \in J\}$, one has

$$f^*(T) = \{f(v_0), f(v_1), ..., f(v_n)\}.$$

A simplex $T_j = [v_0, v_1, ..., v_m]$ of this triangulation is called admissible (or rainbow) if

$$f^*(T_j) = \{f(v_0), f(v_1), ..., f(v_m)\} = \{0, 1, ..., m\}.$$

Prove Sperner's combinatorial lemma stating that the number of admissible simplices is always odd (in particular, it differs from zero).

For this purpose, in each simplex T_j choose an interior point w_j and also pick a point w not belonging to S. On the set $W = \{w\} \cup \{w_j : j \in J\}$ define the structure of a graph. Namely, two distinct points w_j and $w_{j'}$ connect by an edge if and only if the corresponding simplices T_j and $T_{j'}$ have a common facet T such that

$$f^*(T) = \{0, 1, ..., m-1\}.$$

Also, a point w_j connect by an edge with the point w if and only if the corresponding simplex T_j has a facet T' contained in $[x_0, x_1, ..., x_{m-1}]$ and such that

$$f^*(T') = \{0, 1, ..., m-1\}.$$

Finally, apply to the obtained graph Exercise 1 from Appendix 5, and use induction on m for establishing the required result.

Remark 3. Sperner's lemma plays an important role in topology and its applications. In particular, by starting with this lemma, an elegant proof of the celebrated Brouwer theorem on fixed points can be given. At present, many nontrivial purely geometric applications of Brouwer's theorem are known (see, e.g., [4], [20], [123], [242], [261]).

20*. Let $S = [x_0, x_1, ..., x_m]$ be an m-dimensional simplex in the space $\mathbf{R}^m$ and let, for each natural index $i \in [0, m]$, the symbol S_i denote the facet of S opposite of the vertex x_i. Assume also that $\{F_0, F_1, ..., F_m\}$ is a family of closed subsets of S having the following two properties:

(a) $S = F_0 \cup F_1 \cup ... \cup F_m$ and $x_i \in F_i$ for any vertex x_i of S;

(b) $F_i \cap S_i = \emptyset$ for any facet S_i of S.

Demonstrate that $F_0 \cap F_1 \cap ... \cap F_m \neq \emptyset$.

For this purpose, take a natural number $k > 0$ and construct a finite triangulation $\{T_{k,j} : j \in J(k)\}$ of S into m-dimensional simplices such that the diameters of all simplices of this triangulation do not exceed $1/k$. Further, for every vertex v of the above-mentioned triangulation, put $f(v) = i$ where i is the least natural index from $[0, m]$ such that $v \in F_i$. By using Sperner's lemma, find a simplex $T_{k,j(k)}$ for which the relation

$$f^*(T_{k,j(k)}) = \{0, 1, ..., m\}$$

is satisfied. Then consider the sequence of simplices $\{T_{k,j(k)} : k < \omega\}$ and, keeping in mind the compactness of S, check that this sequence has some limit point $x \in S$. Finally, show that the point x belongs to all sets from the family $\{F_0, F_1, ..., F_m\}$.

21*. Again, let $S = [x_0, x_1, ..., x_m]$ be an m-dimensional simplex in the space $\mathbf{R}^m$ and let, for each natural index $i \in [0, m]$, the symbol S_i denote the facet of S opposite of the vertex x_i. Assume that $\{P_0, P_1, ..., P_m\}$ is a family of closed subsets of S having the following two properties:

(a) $S = P_0 \cup P_1 \cup ... \cup P_m$ and $x_i \notin P_i$ for every vertex x_i of S;

(b) $S_i \subset P_i$ for any facet S_i of S.

Demonstrate that $P_0 \cap P_1 \cap ... \cap P_m \neq \emptyset$.

For this purpose, suppose to the contrary that $P_0 \cap P_1 \cap ... \cap P_m = \emptyset$. Then infer that, for sufficiently small real number $\varepsilon > 0$, the relation

$$U_\varepsilon(P_0) \cap U_\varepsilon(P_1) \cap ... \cap U_\varepsilon(P_m) = \emptyset$$

also must be valid, where the symbol $U_\varepsilon(P_i)$ denotes the open ε-neighborhood of the set P_i $(i = 0, 1, ..., m)$. Further, introduce the sets

$$F_i = S \setminus U_\varepsilon(P_i) \qquad (i = 0, 1, ..., m)$$

and verify that, for sufficiently small $\varepsilon > 0$, the family $\{F_0, F_1, ..., F_m\}$ of closed subsets of S has the properties (a) and (b) of Exercise 20. Therefore, according to Exercise 20, one gets $F_0 \cap F_1 \cap ... \cap F_m \neq \emptyset$, which contradicts the equality

$$S = P_0 \cup P_1 \cup ... \cup P_m.$$

The contradiction obtained yields the required result.

22*. Let Y_0, Y_1, ..., Y_{m+1} be convex subsets of the space $\mathbf{R}^m$ such that any $(m+1)$-element subfamily of $\{Y_0, Y_1, ..., Y_{m+1}\}$ has nonempty intersection.

Show that $Y_0 \cap Y_1 \cap ... \cap Y_{m+1} \neq \emptyset$ (cf. Exercise 18).

Argue as follows. For each natural index $i \in [0, m+1]$, pick a point

$$y_i \in Y_0 \cap ... \cap Y_{i-1} \cap Y_{i+1} \cap ... \cap Y_{m+1}$$

and introduce the convex polyhedron $Q = \text{conv}(\{y_0, y_1, ..., y_{m+1}\})$ in $\mathbf{R}^m$.

Check (e.g., using Carathéodory's theorem or directly by induction on m) that the equality

$$Q = Q_0 \cup Q_1 \cup ... \cup Q_{m+1}$$

always holds true, where

$$Q_i = \text{conv}(\{y_0, ..., y_{i-1}, y_{i+1}, ..., y_{m+1}\}) \qquad (i = 0, 1, ..., m+1).$$

Let $S = [x_0, x_1, ..., x_{m+1}]$ be an arbitrary $(m+1)$-dimensional simplex in the space $\mathbf{R}^{m+1}$, let $\phi : \mathbf{R}^{m+1} \to \mathbf{R}^m$ denote the affine mapping such that

$$\phi(x_0) = y_0, \quad \phi(x_1) = y_1, \quad . \, . \, . \, , \quad \phi(x_{m+1}) = y_{m+1},$$

and let

$$P_i = S \cap \phi^{-1}(Q_i) \qquad (i = 0, 1, ..., m+1).$$

Applying to S and to the family $\{P_0, P_1, ..., P_{m+1}\}$ the result of Exercise 21, find a point $x \in S$ belonging to all sets P_i $(i = 0, 1, ..., m+1)$ and, keeping in mind the relations

$$y_i \in Y_j \qquad (\{i, j\} \subset \{0, 1, ..., m+1\}, \ \ i \neq j),$$

conclude that $y = \phi(x) \in Y_0 \cap Y_1 \cap ... \cap Y_{m+1}$, which yields the required result.

Remark 4. The argument presented above is due to M. Krasnoselskii.

23. By starting with the result of the previous exercise, give a different proof of Helly's theorem (see Exercise 18).

Namely, show that if $\{Y_j : j \in J\}$ is a finite family of convex subsets of $\mathbf{R}^m$ having the property that

$$Y_{j_1} \cap Y_{j_2} \cap ... \cap Y_{j_k} \neq \emptyset$$

for any natural number $k \leq m+1$ and for any indices j_1, j_2, ..., j_k from J, then

$$\cap\{Y_j : j \in J\} \neq \emptyset.$$

For this purpose, use induction on card(J).

24. A subset X of the space $\mathbf{R}^m$ is called a P-set if X can be represented as the union of a sequence of convex polyhedra.

Verify that:

(a) if $m = 1$, then every closed convex subset of $\mathbf{R}^m$ is a P-set;

(b) if $m = 2$, then a closed convex set $X \subset \mathbf{R}^m$ is a P-set if and only if extr(X) is at most countable;

(c) in $\mathbf{R}^3$ there exists a closed convex set Y such that card(extr(Y)) $= 1$ and Y is not a P-set;

(d) if $m \geq 2$, then a compact convex set $Z \subset \mathbf{R}^m$ is a P-set if and only if extr(Z) is at most countable.

25. Let $\mathbf{c}$ denote the cardinality of the continuum and let $m \geq 2$ be a natural number.

Demonstrate that the family of all convex sets in $\mathbf{R}^m$ is of cardinality $2^{\mathbf{c}}$ while the family of all Borel convex sets in $\mathbf{R}^m$ is of cardinality $\mathbf{c}$.

26*. Let P be a convex n-gon in the plane $\mathbf{R}^2$. Consider the family of all triangulations of P without extra vertices and denote by t_n the total number of such triangulations.

Find an analytic expression for the value t_n.

For this purpose, argue as follows. At the first step, establish the validity of the recursive formula

$$t_{n+1} = t_n + t_3 t_{n-1} + t_4 t_{n-2} + ... + t_{n-2} t_4 + t_{n-1} t_3 + t_n.$$

At the second step, check that

$$(2n-2)t_{2n+1} = (2n+1)(t_3 t_{2n} + t_4 t_{2n-1} + ... + t_n t_{n+3} + t_{n+1} t_{n+2}).$$

At the third step verify that

$$(2n-3)t_{2n} = n(t_3 t_{2n-1} + t_4 t_{2n-2} + ... + t_{2n-2} t_4 + t_{2n-1} t_3).$$

Then infer from these three recursive formulas that

$$(2n+1)t_{2n+2} = (8n-2)t_{2n+1},$$

$$nt_{2n+1} = (4n-3)t_{2n}.$$

Finally, taking into account the above facts, show that

$$t_n = (2n-4)!/((n-1)!(n-2)!).$$

Actually, for any integer $n \geq 1$, one has $t_{n+2} = C_n$, where C_n is Catalan's number (see, e.g., [48]).

27. Let $m \geq 2$ be a natural number and let $B_1, \ B_2, \ . \ . \ . \ , \ B_{m-1}$ be any closed convex sets in the space $\mathbf{R}^m$ such that the origin 0 of $\mathbf{R}^m$ does not belong to the union of these sets.

Demonstrate that there exists a straight line in $\mathbf{R}^m$ passing through 0 and having no common points with the union of all B_i $(i = 1, 2, ..., m-1)$.

For this purpose, use induction on m.

28. Let $m \geq 2$ be a natural number and let $r > 0$ be a real number.

Prove that there are pairwise disjoint m-dimensional closed balls

$$B_1, \ B_2, \ . \ . \ . \ , \ B_m$$

in the space $\mathbf{R}^m$ such that:

(a) the origin 0 of $\mathbf{R}^m$ does not belong to $B_1 \cup B_2 \cup ... \cup B_m$;

(b) every straight line in $\mathbf{R}^m$ passing through 0 intersects at least one of the balls $B_1, \ B_2, \ . \ . \ . \ , \ B_m$;

(c) the distance between the centers of any two of these balls is strictly greater than r.

29. Let P be an m-dimensional convex polyhedron in the space $\mathbf{R}^m$, where $m \geq 2$, and let x be a point in the interior of P.

Show that there exists a facet F of P such that the perpendicular dropped from x to the affine hyperplane carrying F intersects the relative interior of F.

Give an example of an m-dimensional convex polyhedron $Q \subset \mathbf{R}^m$ with arbitrarily many facets and of a point $x \in \text{int}(Q)$ having the property that there is only one facet F of Q such that the perpendicular dropped from x to the affine hyperplane carrying F intersects the relative interior of F.

For this purpose, first consider the case $m = 2$ and then use induction on m.

30. Let P be an m-dimensional convex polyhedron in the space $\mathbf{R}^m$. Suppose that, for any distinct facets $F_1, F_2, \ldots, F_{m+1}$ of P, there exists a point $x \in P$ such that all perpendiculars dropped from x to the affine hyperplanes in $\mathbf{R}^m$ carrying $F_1, F_2, \ldots, F_{m+1}$, respectively, intersect the relative interiors of these facets.

Demonstrate that there exists a point $z \in P$ such that the perpendicular dropped from z to the affine hyperplane carrying any facet F of P intersects the relative interior of F.

For this purpose, apply Helly's theorem (see Exercises 18 and 23) to an appropriate finite family of convex subsets of P.

31. Let P be a regular tetrahedron (respectively, a cube) in the space $\mathbf{R}^3$.

Is it possible to make a hole within P so that a tetrahedron (respectively, cube) congruent to P could freely move through this hole?

Consider the analogous question for a regular m-simplex and for an m-cube in the Euclidean space $\mathbf{R}^m$, where $m > 3$.

32. Let P be a convex polyhedron in a finite-dimensional (Hausdorff) topological vector space E and let X be a subset of P.

Show that the following two assertions are equivalent:

(a) $X = \text{extr}(P)$;

(b) X is the smallest (with respect to inclusion) subset of P having the property that, for any linear functional g on E, one has

$$\sup(g(P)) \in g(X), \qquad \inf(g(P)) \in g(X).$$

Also, for $m \geq 2$, give an example of a compact m-dimensional convex set $P \subset \mathbf{R}^m$ such that there is a set X satisfying (b) and properly contained in $\text{extr}(P)$.

33*. Let n be a nonzero natural number and let

$$A = (a_{ij})_{1 \le i \le n, 1 \le j \le n}$$

be an $(n \times n)$-matrix, all terms of which belong to $\mathbf{R}$.

Recall that this matrix is doubly stochastic if:

(1) $a_{ij} \ge 0$ for all natural indices i and j from $[1, n]$;

(2) $\sum_{1 \le j \le n} a_{ij} = 1$ for each natural number $i \in [1, n]$;

(3) $\sum_{1 \le i \le n} a_{ij} = 1$ for each natural number $j \in [1, n]$.

Obviously, any $(n \times n)$-matrix over $\mathbf{R}$ may be regarded as a certain point in the Euclidean space

$$E = \mathbf{R}^n \times \mathbf{R}^n \times \; ... \; \times \mathbf{R}^n = \mathbf{R}^{n^2}.$$

Check that the set $S(n)$ of all doubly stochastic $(n \times n)$-matrices is convex in E.

Also, check that the same set is closed and bounded in E. Consequently, $S(n)$ turns out to be a compact convex subset of E.

A doubly stochastic $(n \times n)$-matrix A is called a permutation $(n \times n)$-matrix if:

(a) for each natural number $i \in [1, n]$, the terms a_{ij} $(1 \le j \le n)$ of A are equal to zero, except only one of them (which equals 1);

(b) for each natural number $j \in [1, n]$, the terms a_{ij} $(1 \le i \le n)$ of A are equal to zero, except only one of them (which equals 1).

Demonstrate that any matrix $A \in S(n)$ is representable as a convex combination of permutation $(n \times n)$-matrices and infer that $S(n)$ turns out to be a convex polyhedron in E (Birkhoff's theorem).

For proving this result, use Hall's combinatorial theorem (see Exercise 1 of Appendix 3) and induction on the number of all those terms a_{ij} of A which are strictly positive.

Moreover, by applying Carathéodory's theorem, show that every $A \in S(n)$ can be represented as a convex combination of $n^2 - 2n + 2$ permutation $(n \times n)$-matrices.

34. A set Z in a topological vector space E is called locally conical at a point $z \in Z$ if there exist a neighborhood $U(z)$ and a cone C in E with vertex z such that

$$U(z) \cap Z = U(z) \cap C.$$

Observe that, according to this definition, Z is locally conical at each of its isolated points.

Verify that:

(a) if $\{Z_i : i \in I\}$ is a finite family of sets in E, all of which are locally conical at a point z, then the union of this family and the intersection of this family are both locally conical at the same z;

(b) a compact subset Z of a finite-dimensional topological vector space E is a polyhedron if and only if Z is locally conical at all points $z \in Z$.

For establishing (b), use induction on $\dim(E)$.

Let now $X \subset \mathbf{R}^m$ be a closed convex set whose boundary contains the origin 0 of $\mathbf{R}^m$ and let t be a strictly positive real number distinct from 1. Suppose that there exists a neighborhood $U(0)$ of 0 for which the equality

$$U(0) \cap X = U(0) \cap tX$$

holds true.

Prove that in this case the given set X is locally conical at 0.

35*. Show that if a two-dimensional compact convex set $X \subset \mathbf{R}^2$ is locally conical at all of its extreme points, then X is a convex polygon.

Demonstrate that there exists a compact convex body $Y \subset \mathbf{R}^3$ satisfying the following conditions:

(a) Y is locally conical at all of its extreme points;

(b) Y is not a convex polyhedron in $\mathbf{R}^3$.

Let Z be a nonempty convex compact set in the Euclidean space $\mathbf{R}^m$. Suppose that there exists a real number $\varepsilon > 0$ possessing the following property:

If $z \in \text{bd}(Z)$ is any point of strict convexity of Z, then Z is conical at z in the ε-neighborhood of z.

Prove that Z is a convex polyhedron in $\mathbf{R}^m$.

For this purpose, apply Straszewicz's theorem on the points of strict convexity of compact convex sets in $\mathbf{R}^m$ (see Exercise 11 from this Appendix).

36*. Let S be a nonempty compact convex subset of the space $\mathbf{R}^m$. Suppose that S has the following property:

For every point $x \in \mathbf{R}^m$, the set $(x + S) \cap S$ either is empty or is a singleton or is a homothetic copy of S (with strictly positive homothety coefficient).

Demonstrate that S is a simplex whose dimension is less than or equal to m (this is an interesting characterization of simplices in $\mathbf{R}^m$ obtained by Rogers and Shephard).

For this purpose, applying the results of Exercises 34 and 35, first show that S is locally conical at all of its strict convexity points, then infer that S is a convex polyhedron, and finally use induction on m.

37*. Let $\{S_k : k < \omega\}$ be a sequence of subsets of the space $\mathbf{R}^m$ such that:

(a) if $k < r < \omega$, then $S_r \subset S_k$, i.e., this sequence is decreasing by the inclusion relation;

(b) each set S_k $(k < \omega)$ is a simplex whose dimension is less than or equal to m.

Prove that the set $\cap\{S_k : k < \omega\}$ is a simplex in $\mathbf{R}^m$ (Kolmogorov's theorem).

For this purpose, first verify that the Rogers–Shephard characterization of simplices, presented in Exercise 36, is invariant under the operation of taking intersections of decreasing sequences of nonempty compact convex sets, and then derive from this circumstance the required result.

38*. Let $X \subset \mathbf{R}^m$ be a nonempty compact convex set and let t be a real number from the open interval $]0, 1[$. Suppose that, for every $x \in \mathbf{R}^m$, the set $(x + tX) \cap X$ is centrally symmetric.

Prove that the given set X is a parallelepiped (whose dimension is less than or equal to m).

For this purpose, by using again Exercises 34 and 35, first demonstrate that X is a convex polyhedron and then argue by induction on m.

Remark 5. The result of Exercise 38 was obtained in the paper [170]. Further investigations in this direction are presented in [322].

39*. Let $\{P_k : k < \omega\}$ be a sequence of subsets of $\mathbf{R}^m$ such that:

(a) if $k < r < \omega$, then $P_r \subset P_k$, i.e., this sequence is decreasing by the inclusion relation;

(b) each set P_k $(k < \omega)$ is a parallelepiped whose dimension is less than or equal to m.

Prove that the set $\cap\{P_k : k < \omega\}$ is a parallelepiped (Gruber's theorem).

For this purpose, first check that the characterization of a parallelepiped, presented in Exercise 38, is invariant under the operation of taking intersections of decreasing sequences of nonempty compact convex sets, and then deduce from this circumstance the required result.

40. Identify the Euclidean plane $\mathbf{R}^2$ with the field $\mathbf{C}$ of all complex numbers and suppose that a finite family of points

$$\{z_1, z_2, \ldots, z_n\} \subset \mathbf{C} \setminus \{0\}$$

is given such that all these points lie in one half-plane determined by some straight line $l \subset \mathbf{R}^2$ passing through 0.

Show that there exists a straight line $l' \subset \mathbf{R}^2$ also passing through 0 and such that all points $1/z_1, 1/z_2, \ldots, 1/z_n$ lie in one half-plane determined by this l'.

41. Let $z_1, z_2, \ldots, z_n$ be all the roots of a non-constant polynomial f over the field $\mathbf{C}$, and let f' denote, as usual, the derivative of f.

Demonstrate that the formula

$$f'(z)/f(z) = 1/(z - z_1) + 1/(z - z_2) + ... + 1/(z - z_n)$$

holds true for all those numbers $z \in \mathbf{C}$ which differ from $z_1, z_2, \ldots, z_n$.

In order to establish the validity of the-above-mentioned formula, keep in mind the equality

$$f(z) = (z - z_1)(z - z_2)...(z - z_n)$$

and differentiate both sides of it.

42*. Let a polynomial

$$f(z) = z^n + a_1 z^{n-1} + ... + a_{n-1}z + a_n$$

be given over the field $\mathbf{C}$ and let f' denote again the derivative of f.

By using the results of Exercises 40 and 41, prove that the set $Z(f')$ of all roots of f' is contained in the convex hull of the set $Z(f)$ of all roots of f (this is a classical theorem of Gauss).

Deduce from the Gauss theorem that if all roots of f belong to $\mathbf{R}$ (i.e., are real numbers), then all roots of f' also belong to $\mathbf{R}$.

43*. In the Euclidean space $\mathbf{R}^m$ consider a finite system of linear inequalities

$$\langle e_i, x \rangle \ \leq \ b_i \qquad (i = 0, 1, 2, ..., n),$$

where $\{e_i : i \in \{0, 1, 2, ..., n\}\}$ is a family of pairwise distinct nonzero vectors in $\mathbf{R}^m$ and $\{b_i : i \in \{0, 1, 2, ..., n\}\}$ is a family of real numbers.

The linear inequality $\langle e_0, x \rangle \ \leq \ b_0$ is called a consequence of the system of inequalities

$$\langle e_i, x \rangle \ \leq \ b_i \qquad (i = 1, 2, ..., n)$$

if this system is consistent (i.e., admits at least one solution) and, for every solution x_0 of it, one has $\langle e_0, x_0 \rangle \ \leq b_0$.

Prove the Minkowski–Farkas theorem stating that the following two assertions are equivalent:

(a) the inequality $\langle e_0, x \rangle \ \leq \ b_0$ is a consequence of the above system;

(b) there exist real numbers $t_1 \geq 0$, $t_2 \geq 0$, . . . , $t_n \geq 0$ such that

$$t_1 e_1 + t_2 e_2 + ... + t_n e_n = e_0,$$

$$t_1 b_1 + t_2 b_2 + ... + t_n b_n \leq b_0.$$

Moreover, show that the numbers t_1, t_2, . . . , t_n can be chosen so that at most m of them differ from zero.

For this purpose, first give a geometric interpretation of the equivalence of (a) and (b), and then use Helly's theorem, Exercise 5, and induction on m.

44*. Preserving the notation of the previous exercise, obtain A. D. Alexandrov's theorem which states that the following two assertions are equivalent:

(a) in $\mathbf{R}^m$ a system of linear inequalities of the form

$$\langle e_i, x \rangle \ \leq \ b_i \qquad (i = 1, 2, ..., n)$$

is consistent;

(b) for any non-negative real numbers t_1, t_2, . . . , t_n, the relation

$$t_1 e_1 + t_2 e_2 + ... + t_n e_n = 0$$

implies the inequality $t_1 b_1 + t_2 b_2 + ... + t_n b_n \geq 0$.

In order to show the required equivalence, argue as follows.

The implication (a) $\Rightarrow$ (b) is almost trivial.

Further, assume for a moment (b) and the negation of (a). Without loss of generality, suppose that the system of inequalities

$$\langle e_i, x \rangle \ \leq \ b_i \qquad (i = 1, 2, ..., n-1)$$

is consistent. Then demonstrate that there exists a real $\varepsilon > 0$ such that the inequality

$$\langle x, -e_n \rangle \leq -b_n - \varepsilon$$

is a consequence of the above system. Finally, use the Minkowski–Farkas theorem presented in Exercise 43, obtain a contradiction, and conclude that the implication (b) $\Rightarrow$ (a) must hold.

45. Let P and Q be two convex polygons in the plane $\mathbf{R}^2$, separated (in general, not strongly separated) by some straight line in $\mathbf{R}^2$.

Prove that the disjunction of the following two assertions is valid:

(a) there exists a straight line in $\mathbf{R}^2$ carrying one of the sides of P and separating P and Q;

(b) there exists a straight line in $\mathbf{R}^2$ carrying one of the sides of Q and separating P and Q;

Verify that, in general, the analogous result fails to be true for two separated convex polyhedra U and V in the space $\mathbf{R}^3$. More precisely, it may happen that there is no affine hyperplane carrying one of the facets of U (or one of the facets of V) and separating U and V.

APPENDIX 2

Real-valued convex functions

This Appendix may be treated as a natural continuation of Appendix 1. Here we briefly discuss some general facts and theorems concerning real-valued convex functions defined on convex subsets of vector spaces over **R**.

Analogously to various important properties of convex sets, the results about convex functions presented here turn out to be useful in many branches of pure and applied mathematics. Let us repeat once more that in one short section of the book it is impossible to give an extensive and sufficiently complete survey of the theory of convex functions and their analytical properties. For more information around this topic, we again refer the reader to numerous works and manuals, e.g., such as [4], [24], [45], [49], [107], [123], [130], [249], [298].

Let E be a vector space, let Z be a convex subset of E, and let f be a mapping acting from Z into **R**.

Recall that f is lower convex (in short, convex) if, for any two points x and y from Z and for each real $t \in]0,1[$, one has

$$f(tx + (1-t)y) \leq tf(x) + (1-t)f(y).$$

Furthermore, f is called strongly lower convex (in short, strongly convex) if the above-mentioned inequality is strong for any two distinct points x and y from Z and for each real $t \in]0,1[$, i.e.,

$$f(tx + (1-t)y) < tf(x) + (1-t)f(y).$$

In the similar manner, the notion of an upper convex (of a strongly upper convex) function can be introduced. Namely, for an upper convex function f, one has the reverse inequality

$$f(tx + (1-t)y) \geq tf(x) + (1-t)f(y)$$

DOI: 10.1201/9781003458708-A2

and, for a strongly upper convex function f, one has the reverse strong inequality

$$f(tx + (1-t)y) > tf(x) + (1-t)f(y).$$

The upper convex functions (respectively, strongly upper convex functions) are usually called concave functions (respectively, strongly concave functions).

Note that every linear functional on E is simultaneously a convex function and a concave function on E.

Observe that, for a convex set $X \subset E$, a function $f : X \to \mathbf{R}$ is convex if and only if the set

$$\{(x,t) \in X \times \mathbf{R} \ : \ f(x) \leq t\}$$

is a convex subset of the product vector space $E \times \mathbf{R}$.

A very nice property of convex functions given on a finite-dimensional Euclidean space is formulated in the following statement.

Theorem 1. *Any convex function $f : U \to \mathbf{R}$, where U is an open convex subset of a finite-dimensional Hausdorff topological vector space E, is continuous.*

The proof can be carried out by induction on m (the corresponding details are left to the reader). Both conditions that U is open in E and that E is finite-dimensional are essential in the formulation of Theorem 1.

Notice that differential properties of convex functions on the same space E (which, in fact, is isomorphic to $\mathbf{R}^m$, where $m = \dim(E)$) are much more delicate. The discussion of such properties may be found, e.g., in [298].

Remark 1. In the case $m = 1$, it is well known that monotone functions on the real line $\mathbf{R}$ are closely connected with real-valued convex and concave functions on $\mathbf{R}$ (cf. Exercise 10).

Recall that a subset Z of a real vector space E is mid-point convex if, for any two points x and y from Z, their mid-point $(x+y)/2$ also belongs to Z.

A real-valued function f defined on a mid-point convex set Z is called mid-point convex if the inequality

$$f((x+y)/2) \leq (f(x) + f(y))/2$$

holds true for any two points x and y from Z.

In the similar manner, real-valued mid-point concave functions are introduced.

Clearly, if a function f is convex, then f is mid-convex.

Also, f is convex (respectively, mid-point convex) if and only if $-f$ is concave (respectively, mid-point concave).

If a function $f : E \to \mathbf{R}$ is convex and concave simultaneously, then f is an affine function.

In general, the mid-convexity of f does not imply the convexity of f, but this implication is valid under some additional assumptions on f (see Exercise 1).

Theorem 2. *The following three assertions are true:*

(1) *if a function f is convex and a real number t is non-negative, then tf is convex;*

(2) *if f and g are convex functions and their domains coincide, then their sum $f + g$ is convex;*

(3) *if a function f is convex and $g : \mathbf{R} \to \mathbf{R}$ is convex and increasing, then the composition $g \circ f$ is convex.*

An easy proof of this theorem is left to the reader.

Remark 2. It may happen that a function f is strictly convex but its square f^2 is strictly concave. So the family of convex functions is not closed under the usual multiplication operation (cf. Exercise 16).

Theorem 3. *If a function $f : Z \to \mathbf{R}$ is convex, then the following inequality (due J. Jensen) takes place:*

$$f(t_1x_1 + t_2x_2 + ... + t_nx_n) \leq t_1f(x_1) + t_2f(x_2) + ... + t_nf(x_n),$$

where n is a nonzero natural number, $t_1, t_2, ..., t_n$ are non-negative real numbers such that $\sum\{t_i : 1 \leq i \leq n\} = 1$, and $x_1, x_2, \ldots, x_n$ are any points of Z.

Proof. We argue by induction on n. For $n = 1$ and $n = 2$, there is nothing to prove. Suppose that $n \geq 2$ and Jensen's inequality holds true for n. Take arbitrarily non-negative real numbers $t_1, t_2, \ldots, t_n, t_{n+1}$ such that

$$\sum\{t_i : 1 \leq t_i \leq n + 1\} = 1$$

and choose arbitrarily points $x_1, x_2, \ldots, x_n, x_{n+1}$ from Z. Since $n \geq 2$, there exists $t_i < 1$. We may assume, without loss of generality, that $t_{n+1} < 1$. Consider the point

$$x = (t_1x_1 + t_2x_2 + ... + t_nx_n)/(1 - t_{n+1})$$

which belongs to Z in view of the convexity of Z. Obviously, we can write

$$f(t_{n+1}x_{n+1} + (1 - t_{n+1})x) \leq t_{n+1}f(x_{n+1}) + (1 - t_{n+1})f(x).$$

By virtue of the inductive assumption for n, we get

$$f(x) \leq (t_1f(x_1) + t_2f(x_2) + ... + t_nf(x_n))/(1 - t_{n+1}),$$

whence it readily follows that

$$f(t_1x_1 + t_2x_2 + ... + t_{n+1}x_{n+1}) \leq t_1f(x_1) + t_2f(x_2) + ... + t_{n+1}f(x_{n+1}).$$

Jensen's inequality has thus been proved by the induction method. □

Remark 3. Clearly, one has the analogous inequality for real-valued concave functions on Z. Namely, if $f : Z \to \mathbf{R}$ is concave, then

$$f(t_1x_1 + t_2x_2 + ... + t_nx_n) \geq t_1f(x_1) + t_2f(x_2) + ... + t_nf(x_n),$$

where n is a nonzero natural number, $t_1, t_2, ..., t_n$ are non-negative real numbers such that $\sum\{t_i : 1 \leq i \leq n\} = 1$, and $x_1, x_2, \ldots, x_n$ are any points of Z.

Remark 4. Jensen's inequality for convex functions and its dual for concave functions are extremely useful for obtaining many other deep inequalities in mathematical analysis. For instance, the classical inequality between the geometrical and arithmetical means of positive real numbers $x_1, x_2, \ldots, x_n$, i.e., the relation

$$(x_1x_2...x_n)^{1/n} \leq (x_1 + x_2 + ... + x_n)/n$$

is a straightforward consequence of the fact that the function

$$x \to \ln(x) \qquad (x > 0)$$

is concave on the interval $]0, +\infty[$ (for a more profound use of Jensen's inequality, see e.g. Exercise 20).

Theorem 4. *If g is a convex function and differs from a constant, then g cannot attain its supremum at an interior point of the domain of g.*

Proof. Suppose on the contrary that $\sup(g) = g(z)$, where z is an interior point of $\mathrm{dom}(g)$. Let x be any point of $\mathrm{dom}(g)$ distinct from z and let $y \in \mathrm{dom}(f)$ be such that z is an interior point of the line segment $[x, y]$. Then, for some $t \in]0, 1[$, we have $z = tx + (1 - t)y$. In view of the convexity of g, we may write

$$g(z) \leq tg(x) + (1 - t)g(y) \leq g(z),$$

whence it immediately follows that $g(x) = g(y) = g(z)$, i.e., g is a constant function.

The obtained contradiction ends the proof. □

Consider now an arbitrary convex function $f : [a, b] \to \mathbf{R}$, where $[a, b]$ is a non-degenerate segment in $\mathbf{R}$.

Let x, y, and z be three interior points of $[a, b]$ such that $x < y < z$. Evidently, y admits a representation in the form

$$y = ((y - x)/(z - x))z + ((z - y)/(z - x))x.$$

Using Jensen's inequality, we get

$$f(y) \leq ((y - x)/(z - x))f(z) + ((z - y)/(z - x))f(x).$$

It is not difficult to check that the above relation implies

$$(f(y) - f(x))/(y - x) \leq (f(z) - f(x))/(z - x) \leq (f(z) - f(y))/(z - y).$$

The latter inequalities enable one to infer that:

(i) if y tends to x, then the value $(f(y) - f(x))/(y - x)$ decreases and is bounded from below; therefore, there exists a right derivative $f'_r(x)$;

(ii) $f'_r(x) \leq (f(z) - f(x))/(z - x) \leq f'_r(z)$, so the function

$$x \to f'_r(x) \qquad (a < x < b)$$

is increasing.

The analogous two assertions hold true for the left derivative f'_l of f.

Since there are $f'_r(x)$ and $f'_l(x)$ at each interior point x of $[a, b]$, one can conclude that f is continuous at all points of the open interval $]a, b[$ (cf. Theorem 1).

It should also be mentioned here that the equality $f'_r(x) = f'_l(x)$ is valid for all, except countably many, points of $[a, b]$. Therefore, f is differentiable on a co-countable subset of $[a, b]$ (in this connection, see Exercises 3 and 12).

Remark 5. In this lecture course we do not intend to consider many other important analytical properties of convex functions. They are thoroughly discussed in various monographs and text-books devoted to convex analysis and its applications (see, e.g., [298]).

EXERCISES

1*. Let U be an open convex set in the space $\mathbf{R}^m$, where $m \geq 1$, and let $f : U \to \mathbf{R}$ be a function satisfying the mid-point convexity condition, i.e.,

$$f((x + y)/2) \leq (f(x) + f(y))/2 \qquad (x \in U, \ y \in U).$$

Prove Sierpiński's theorem stating that if f is measurable with respect to the m-dimensional Lebesgue measure λ_m on $\mathbf{R}^m$, then f is convex and, consequently, continuous on U.

For this purpose, use the generalized Steinhaus property of λ_m, which means that if X and Y are any two sets in $\mathbf{R}^m$ such that

$$X \in \mathrm{dom}(\lambda_m), \quad Y \in \mathrm{dom}(\lambda_m), \quad \lambda_m(X) > 0, \quad \lambda_m(Y) > 0,$$

then the Minkowski sum $X + Y = \{x + y : x \in X, \ y \in Y\}$ of these sets has nonempty interior (the analogous property is valid for the Haar measure; in this connection, see e.g. [141]).

2*. Formulate and prove the analogue of Exercise 1 in terms of the Baire property of a function $f : U \to \mathbf{R}$ satisfying the mid-point convexity condition.

Remark 6. Extensive information on the Baire property (which may be treated as a certain topological version of measurability) is presented in [240].

3. Recall that, according to the classical theorem of real analysis, any monotone function acting from **R** into **R** is continuous at all points of **R** except countably many of them.

Starting with this theorem, demonstrate that if $f : \mathbf{R} \to \mathbf{R}$ is an arbitrary convex function, then the set of all those points $x \in \mathbf{R}$ for which the derivative $f'(x)$ does exist is co-countable (i.e., is the complement of a countable subset of **R**).

Further, according to another fundamental result of real analysis, which is due to Lebesgue, any monotone function ϕ acting from **R** into **R** is differentiable almost everywhere, i.e. the set of all those points x from **R**, at which the derivative $\phi'(x)$ does exist, is the complement of a Lebesgue measure zero subset of **R**.

Starting with this result, prove that if $g : \mathbf{R} \to \mathbf{R}$ is an arbitrary convex function, then the second derivative $g''(x)$ exists for almost all (in the sense of the Lebesgue measure λ) points $x \in \mathbf{R}$. In other words, g is twice differentiable λ-almost everywhere in **R**.

Remark 7. It should be noticed that the latter result admits an extension to all convex functions defined on the space $\mathbf{R}^m$ (or, more generally, defined on a convex subset of $\mathbf{R}^m$ with nonempty interior). Namely, it turns out that any such function possesses the total second differential at almost all points of its domain.

4. Let f be a convex function acting from $\mathbf{R}^m$ into **R**, where $m \geq 2$.

By applying the method of induction on m and using Fubini's classical theorem, give a direct proof of the following statement:

The first total differential of f exists at λ_m-almost all points of $\mathbf{R}^m$, where λ_m denotes the standard m-dimensional Lebesgue measure on $\mathbf{R}^m$.

Formulate and prove the analogous result in terms of the Baire category.

5*. Let $\{f_n : n < \omega\}$ be a sequence of real-valued convex functions on **R** and suppose that

$$f(x) = \lim_{n\to\infty} f_n(x) \qquad (x \in \mathbf{R}).$$

Verify that $f : \mathbf{R} \to \mathbf{R}$ is a convex function.

In addition, assuming that f and f_n $(n < \omega)$ are differentiable at all points of **R**, demonstrate that

$$f'(x) = \lim_{n\to\infty} f'_n(x) \qquad (x \in \mathbf{R}).$$

Give a geometric interpretation of these two facts; formulate and prove the analogous results for convex functions of m real variables, where a natural m is greater than 1.

6. Let $h : X \to \mathbf{R}$ be a mid-point convex function on a convex X and let $n > 0$ be a natural number.

Demonstrate that the inequality

$$h((x_1 + x_2 + ... + x_n)/n) \leq (h(x_1) + h(x_2) + ... + h(x_n))/n$$

holds true for any points $x_1, x_2, \ldots, x_n$ from X.

For this purpose, use Cauchy type induction on n. In other words, argue by the following scheme:

(a) if the inequality is valid for n, then it is also valid for $2n$;

(b) if the inequality is valid for $n + 1$, then it is also valid for n.

Deduce from the above that if h is additionally continuous, then h is a convex function (cf. Exercise 1).

7. Let Z be a convex subset of a real vector space E and let f be a real-valued function on Z.

Show that the following two assertions are equivalent:

(a) f is a convex function;

(b) the set $\{(z, t) \in Z \times \mathbf{R} : f(z) \leq t\}$ is convex in the product space $E \times \mathbf{R}$.

Infer from the above equivalence that if $\{f_i : i \in I\}$ is a family of convex (respectively, concave) functions on Z and $f = \sup\{f_i : i \in I\}$ (respectively, $f = \inf\{f_i : i \in I\}$), then f is a convex (respectively, concave) function on Z.

8. Prove that there exists a mid-point convex function $f : \mathbf{R} \to \mathbf{R}$ which is not convex.

For this purpose, consider $\mathbf{R}$ as a vector space over $\mathbf{Q}$ and, using a basis of this space (i.e., a Hamel basis), define an additive function $f : \mathbf{R} \to \mathbf{Q}$ which trivially is mid-point convex, but is not convex.

9. Give an example of a convex function $f : [a, b] \to \mathbf{R}$ such that f is discontinuous at the endpoints of $[a, b]$.

10. Let $g : [r, +\infty[\to \mathbf{R}$ be a function, where r is a real constant.

Show that these two assertions are equivalent:

(a) g is a convex function and has a right derivative at r;

(b) there exists an increasing function $\phi : [r, +\infty[\to \mathbf{R}$ such that

$$g(x) = g(r) + \int_r^x \phi(t)dt \qquad (x \in [r, +\infty[).$$

Here the integration is meant in the usual Riemann sense.

Conclude once again that if (a) holds true, then g is differentiable on a co-countable subset of $[r, +\infty[$.

For this purpose, keep in mind the following facts:

(i) the function ϕ of (b) is continuous at all points of a co-countable subset of $[r, +\infty[$;

(ii) the derivative $g'(x)$ exists and is equal to $\phi(x)$ whenever x belongs to $[r, +\infty[$ and is a continuity point of ϕ.

11. Give an example of a continuous function $h : [0, 1] \to \mathbf{R}$ such that:

(a) h is a convex function;

(b) h is not representable in the form

$$h(x) = h(0) + \int_0^x \psi(t)dt \qquad (x \in [0, 1]),$$

where $\psi : [0, 1] \to \mathbf{R}$ is an increasing function.

12. The widely known theorem of elementary geometry states that the sum of all external angles of a convex polygon in $\mathbf{R}^2$ is equal to 2π.

Starting with this theorem, show once again that any convex function f given on a segment $[a, b]$ is differentiable on a co-countable subset of $[a, b]$ (another way to obtain the same result is indicated in Exercise 3).

13. Consider the vector space $C([0, 1])$ of all real-valued continuous functions f defined on the unit segment $[0, 1]$. Equip this space with the topology of uniform convergence by using the standard sup-norm

$$||f|| = \sup\{|f(x)| : x \in [0, 1]\}.$$

As is well known, $C([0, 1])$ becomes a separable Banach space with respect to $|| \cdot ||$.

Check that the set $Conv([0, 1])$ consisting of all continuous convex functions on $[0, 1]$ is a closed subset of $C([0, 1])$, so $Conv([0, 1])$ turns out to be a complete separable metric space (i.e., Polish space).

14*. Introduce the following notation:

F_0 = the family of all those functions from $C([0, 1])$ which are differentiable at least at one point of $[0, 1]$;

F_1 = the family of all those functions from $Conv([0, 1])$ which are differentiable at each interior point of $[0, 1]$.

Show that:

(a) F_0 is a first category subset of the space $C([0,1])$ (the theorem of Banach and Mazurkiewicz);

(b) F_1 is a co-meager subset of $Conv([0,1])$, i.e, the set $Conv([0,1]) \setminus F_1$ is of first category in $Conv([0,1])$.

Argue step by step. At the beginning, for every natural number n and for every rational number $q > 0$, define

$$K_{n,q} = \{(f,x) \in C([0,1]) \times [0,1] :$$
$$(\forall t)(0 < |t| < q \Rightarrow (|f(x+t) - f(x)|)/|t| \leq n)\}.$$

Verify that the set $K_{n,q}$ is closed in the product space $C([0,1]) \times [0,1]$. Then, using the compactness of $[0,1]$, prove that the set $\mathrm{pr}_1(K_{n,q})$ is closed and nowhere dense in $C([0,1])$.

Finally, check the validity of the inclusion

$$F_0 \subset \cup\{\mathrm{pr}_1(K_{n,q}) : n \in \mathbf{N},\ q \in \mathbf{Q},\ q > 0\},$$

from which (a) immediately follows.

Concerning (b), argue in a similar manner. First, for any natural number $n > 0$, denote

$$a_n = 1/3^n, \quad b_n = 1 - 1/3^n.$$

If $t \in [a_n, b_n]$ and $f \in Conv([0,1])$, consider two straight lines $L_1(t,f)$ and $L_2(t,f)$ in $\mathbf{R}^2$ defined by their canonical equations for $(x,y) \in \mathbf{R}^2$. The equation corresponding to $L_1(t,f)$ is

$$y - f(t) = f'_r(t)(x - t)$$

and the equation corresponding to $L_2(t,f)$ is

$$y - f(t) = f'_l(t)(x - t).$$

Also, denote by $\alpha(t,f)$ the non-obtuse angle between $L_1(t,f)$ and $L_2(t,f)$.

Now, take a rational number $q > 0$ and define

$$P_{q,n} = \{(f,t) : Conv([0,1]) \times [a_n, b_n] : \alpha(t,f) \geq q\}.$$

Show that the set $P_{q,n}$ is closed in the product space $Conv([0,1]) \times [0,1]$. Then, using again the compactness of $[0,1]$, establish the closedness and nowhere density of $\mathrm{pr}_1(P_{q,n})$ in the space $Conv([0,1])$. Finally, take into account the relation

$$F_1 = Conv([0,1]) \setminus \cup\{\mathrm{pr}_1(P_{q,n}) : q \in \mathbf{Q},\ q > 0,\ n \in \mathbf{N},\ n > 0\},$$

which trivially implies the validity of (b).

Remark 8. Let us denote:

F_2 = the family of all those functions from $Conv([0,1])$ which are twice differentiable everywhere on $[0,1]$.

In connection with (b) of Exercise 14, it should be noticed that according to Gruber's theorem, the family F_2 is of first category in the space $Conv([0,1])$ (for more details about various strange phenomena which occur in convex and discrete geometry, see [365]).

15. Let $f : [a,b] \to \mathbf{R}$ be a convex function.

Prove that the following two assertions are equivalent:

(i) there exists a convex function $f^* : \mathbf{R} \to \mathbf{R}$ extending f;

(ii) there exists a right derivative of f at the point a and there exists a left derivative of f at the point b.

16. Verify that the function

$$x \to -x^{1/3} \qquad (x > 0)$$

is strictly convex and that the function

$$x \to x^{2/3} \qquad (x > 0)$$

is strictly concave.

Conclude that the family of all convex functions on $]0,+\infty[$ is not closed under the standard multiplication operation.

17. Let $f : [a,b] \to \mathbf{R}$ be a convex function.

Demonstrate that if this f is differentiable at all interior points of $[a,b]$, then the derivative f' is continuous on $]a,b[$.

Formulate and prove the analogous result for a convex function of m real variables, where a natural m is greater than 1.

Remark 9. It is widely known that a function $g : \mathbf{R} \to \mathbf{R}$ can be differentiable at all points of $\mathbf{R}$, but the derivative g' of g can be discontinuous at many points of $\mathbf{R}$. In this context, Exercise 17 shows that all convex functions on $\mathbf{R}$ have a certain nice property of their derivatives.

18. Let f and g be two functions acting from $\mathbf{R}$ into $\mathbf{R}$. Suppose that, for all points $x \in \mathbf{R}$ and $y \in \mathbf{R}$ and for each $q \in [0,1]$, the inequality

$$f(qx + (1-q)y) \leq qg(x) + (1-q)g(y)$$

is satisfied. Consider the set

$$\Gamma^*(g) = \{(x,t) \in \mathbf{R}^2 : g(x) \leq t\}$$

and let $\text{conv}(\Gamma^*(g))$ denote, as usual, the convex hull of $\Gamma^*(g)$.

Check that $\text{conv}(\Gamma^*(g))$ is identical with the union of the family of all those triangles whose vertices belong to the set $\Gamma^*(g)$.

Show also that the boundary of $\text{conv}(\Gamma^*(g))$ can be regarded as the graph of some convex function $\phi : \mathbf{R} \to \mathbf{R}$ for which the relation

$$f(x) \leq \phi(x) \leq g(x) \qquad (x \in \mathbf{R})$$

is fulfilled (in other words, the convex function ϕ separates the given functions f and g).

19*. Let f be a function acting from $\mathbf{R}$ into $\mathbf{R}$ and let ε be a strictly positive real number.

This f is called an ε-convex function if, for any points $x \in \mathbf{R}$ and $y \in \mathbf{R}$ and for each $q \in [0, 1]$, the inequality

$$f(qx + (1 - q)y) \leq qf(x) + (1 - q)f(y) + \varepsilon$$

holds true.

By applying the result of the previous exercise, demonstrate that if f is an ε-convex function, then there exists a convex function $\phi : \mathbf{R} \to \mathbf{R}$ such that

$$f(x) \leq \phi(x) \leq f(x) + \varepsilon \qquad (x \in \mathbf{R}).$$

Further, define a function $\psi : \mathbf{R} \to \mathbf{R}$ by the formula

$$\psi(x) = \phi(x) - \varepsilon/2 \qquad (x \in \mathbf{R})$$

and show that $|\psi(x) - f(x)| \leq \varepsilon/2$ for all $x \in \mathbf{R}$.

In other words, for every ε-convex function f, there exists a convex function ψ such that the norm $||\psi - f||$ is less than or equal to $\varepsilon/2$ (this nice result is due to Hyers and Ulam).

Remark 10. Let f be a function acting from $\mathbf{R}$ into $\mathbf{R}$ and suppose that, for a real $\varepsilon > 0$, we have the relation

$$f((x + y)/2) \leq (f(x) + f(y))/2 + \varepsilon \qquad (x \in \mathbf{R},\ y \in \mathbf{R}).$$

In general, we cannot assert that there exists a convex function $\psi : \mathbf{R} \to \mathbf{R}$ satisfying the inequalities $\psi(x) - \varepsilon \leq f(x) \leq \psi(x) + \varepsilon$ for each $x \in \mathbf{R}$. Indeed, if f is additive and discontinuous, then f is a Lebesgue nonmeasurable function and, obviously, it cannot be uniformly approximated by convex functions (which trivially are measurable in the Lebesgue sense).

20*. Let $n \geq 1$ be a natural number and let $a_1, a_2, \ldots, a_n$ be any strictly positive real numbers. Consider a function f defined by

$$f(x) = ((a_1^x + a_2^x + ...a_n^x)/n)^{1/x} \qquad (x > 0).$$

Prove that this f is increasing on $]0, +\infty[$.

Argue as follows. It suffices to demonstrate that $f'(x) \geq 0$ whenever $x > 0$. For this purpose, write

$$f(x)^x = (a_1^x + a_2^x + ...a_n^x)/n \qquad (x > 0)$$

and, taking the derivatives of both sides of the above formula, obtain

$$f(x)^x(\ln(f(x)) + xf'(x)/f(x)) = (a_1^x\ln(a_1) + a_2^x\ln(a_2) + ...a_n^x\ln(a_n))/n.$$

Then, using the convexity of the function

$$g(x) = x\ln(x) \qquad (x > 0)$$

and the corresponding Jensen inequality, verify that

$$f(x)^x\ln(f(x)) \leq (a_1^x\ln(a_1) + a_2^x\ln(a_2) + ...a_n^x\ln(a_n))/n$$

for all $x > 0$, which implies at once that $f'(x) \geq 0$ whenever $x > 0$.

In particular, the quadratic mean of the numbers $a_1, a_2, \ldots, a_n$ is greater than or equal to the arithmetic mean of the same numbers, and the equality of these two means holds true if and only if $a_1 = a_2 = ... = a_n$ (the latter fact was essentially used in the proof of Lemma 3 of Chapter 4).

Demonstrate that $\lim_{x \to 0+} f(x) = (a_1 a_2 ... a_n)^{1/n}$, i.e., this limit coincides with the geometric mean of the numbers $a_1, a_2, \ldots, a_n$.

21. Let g be a linear functional given on a real topological vector space E.

Show that the following three assertions are equivalent:

(a) g is continuous on E;

(b) there exists a nonempty open set $V \subset E$ such that the restriction $g|V$ is bounded from above;

(c) the set $g^{-1}(0)$ is closed in E.

22. Give a detailed proof of Theorem 1 by applying induction on m.

Argue as follows. First, verify that if a function $f : U \to \mathbf{R}$ is convex and U is an open convex subset of $\mathbf{R}^m$, then f is locally bounded from above on U (for this purpose, use Jensen's inequality). Afterwards, starting with the local boundedness from above of f, infer the continuity of f.

23. Check that the analog of Theorem 1 does not hold even in the case of a real infinite-dimensional separable Hilbert space H.

For this purpose, take into account Exercise 3 of Appendix 1, according to which there exist many everywhere discontinuous linear functionals on H, and observe that all of them are convex functions on H.

24*. Let E be a (Hausdorff) topological vector space over $\mathbf{R}$, let U be a nonempty open convex subset of E, and let $f : U \to \mathbf{R}$ be a mid-point convex function.

Prove that the following conditions are equivalent:

(a) the restriction of f to some nonempty open subset of U is bounded from above;

(b) f is locally bounded on U;

(c) f is continuous at some point of U;

(d) f is continuous on U.

For this purpose, keep in mind the result of Exercise 6.

Conclude that any of these four conditions implies the convexity of f on U.

APPENDIX 3

The Principle of Inclusion and Exclusion

Many interesting facts from finite combinatorics are known which can be successfully applied in different practical situations and to concrete objects of everyday life.

For example, suppose that k is a natural number and there are k many professional areas (i.e., specialities) and a group X of persons each of which wants to work in accordance with one of these specialities.

Let us enumerate all given professions by 1, 2, ..., k and, for any index $i \in \{1, 2, ..., k\}$, denote by X_i the set of all those persons from the group X which have adequate skills in the profession with number i.

Then Hall's celebrated theorem states that the following two assertions are equivalent:

(a) there exist pairwise distinct members $x_1, x_2, \ldots, x_k$ of X such that every x_i corresponds to the profession i, i.e., $x_i \in X_i$;

(b) for any set $I \subset \{1, 2, ..., k\}$, the inequality

$$\text{card}(I) \leq \text{card}(\cup\{X_i : i \in I\})$$

holds true.

We thus get a useful criterion (of combinatorial nature) for solving the problem of vacancies. Sometimes, this problem is also called the matching problem (see, for instance, [48], [53], [118], [134]).

Another extremely helpful combinatorial result is the so-called Inclusion and Exclusion Principle (see again the same references). In the present Appendix we want to discuss this principle and to point out some of its applications. Actually, our final goal is to give a proof of the Euler–Poincaré formula for multi-dimensional convex polyhedra, by starting with the above-mentioned purely combinatorial principle.

DOI: 10.1201/9781003458708-A3

First, we need several simple concepts from general lattice theory and the relevant notation for intersections of certain finite families of sets.

Recall that a partially ordered set $(E, \preceq)$ is a lattice if, for any two elements x and y from E, there exist $\sup(x, y)$ and $\inf(x, y)$.

Accordingly, a family $(\mathcal{L}, \subset)$ of sets is called a lattice of sets if, for any two members X and Y of $\mathcal{L}$, one has $X \cup Y \in \mathcal{L}$ and $X \cap Y \in \mathcal{L}$.

This definition readily implies (by induction on k) that

$$X_1 \cup X_2 \cup ... \cup X_k \in \mathcal{L}, \quad X_1 \cap X_2 \cap ... \cap X_k \in \mathcal{L}$$

whenever $k \in \mathbf{N} \setminus \{0\}$ and $X_1 \in \mathcal{L}$, $X_2 \in \mathcal{L}$, . . . , $X_k \in \mathcal{L}$.

We will be dealing below only with those lattices of sets $\mathcal{L}$ which satisfy the relation $\emptyset \in \mathcal{L}$. This circumstance does not essentially restrict our further consideration, because we always may add the empty set to any lattice of sets.

Let $\mathcal{L}$ be a lattice of sets and let $h : \mathcal{L} \to \mathbf{R}$ be a function.

We shall say that h is a modular function if $h(\emptyset) = 0$ and

$$h(X \cup Y) + h(X \cap Y) = h(X) + h(Y)$$

for any two sets $X \in \mathcal{L}$ and $Y \in \mathcal{L}$.

Notice that:

(*) the family of all modular functions on $\mathcal{L}$ forms a vector space over the field $(\mathbf{R}, +, \cdot)$;

(**) if T is a subset of $\mathbf{R}$ having a limit point $t_0 \in \mathbf{R}$ and $\{h_t : t \in T\}$ is a family of modular functions on $\mathcal{L}$ such that there exists $h = \lim_{t \in T, t \to t_0} h_t$, then h is also a modular function on $\mathcal{L}$.

As usual, for a set E denote by $\mathcal{P}(E)$ the power set of E (i.e., the family of all subsets of E). The pair $(\mathcal{P}(E), \subset)$ gives a standard example of a lattice of sets. If E is finite, then the mapping h defined by

$$h(X) = \mathrm{card}(X) \qquad (X \in \mathcal{P}(E))$$

is a canonical modular function on $\mathcal{P}(E)$.

More generally, we may consider an arbitrary set E and the family

$$\mathcal{F}in(E) = [E]^{<\omega}$$

of all finite subsets of E. In this case, the pair $(\mathcal{F}in(E), \subset)$ is also a lattice of sets and the mapping h defined analogously by

$$h(X) = \mathrm{card}(X) \qquad (X \in \mathcal{F}in(E))$$

is a canonical modular function on $\mathcal{F}in(E)$.

As a straightforward consequence of the definition of modular functions, we obtain the following statement.

Theorem 1. *Every modular function $h : \mathcal{L} \to \mathbf{R}$ is additive, i.e.,*

$$h(X \cup Y) = h(X) + h(Y)$$

for any two disjoint sets $X \in \mathcal{L}$ and $Y \in \mathcal{L}$.

Moreover, h is finitely additive, i.e., if $\{X_1, X_2, ..., X_k\}$ is an arbitrary finite family of pairwise disjoint sets from $\mathcal{L}$, then the equality

$$h(X_1 \cup X_2 \cup ... \cup X_k) = h(X_1) + h(X_2) + ... + h(X_k)$$

takes place.

Proof. Trivial induction on k. □

In finite combinatorics we have the fundamental Principle of Inclusion and Exclusion that extends Theorem 1 to the general case when the given sets

$$X_1 \in \mathcal{L}, \quad X_2 \in \mathcal{L}, \quad . . . , \quad X_k \in \mathcal{L}$$

are not assumed to be pairwise disjoint.

Let us introduce some notation which will be convenient for formulating and proving this principle.

Let j be an arbitrary natural number from the set $\{1, 2, ..., k\}$ and let

$$s = \{i_1, \ i_2, \ . . . , \ i_j\}$$

be a sequence of pairwise distinct indices from $\{1, 2, ..., k\}$. In other words, s is a j-element subset of $\{1, 2, ..., k\}$. The set-theorists usually express this circumstance in the form $s \in [\{1, 2, ..., k\}]^j$.

For the sake of brevity, we also denote $s \in \mathcal{S}_j$, where $\mathcal{S}_j = [\{1, 2, ..., k\}]^j$ is the family of all j-element subsets of $\{1, 2, ..., k\}$.

Further, suppose that we are given a k-sequence of sets $X_1, \ X_2, \ . . . , \ X_k$. For any set

$$s = \{i_1, \ i_2, \ . . . , \ i_j\} \in \mathcal{S}_j,$$

where $j \in \{1, 2, ..., k\}$, consider the corresponding sets $X_{i_1}, \ X_{i_2}, \ . . . , \ X_{i_j}$ from the family $\{X_1, X_2, ..., X_k\}$ and put $X_s = X_{i_1} \cap X_{i_2} \cap ... \cap X_{i_j}$.

Under the introduced notation, the Principle of Inclusion and Exclusion takes the form presented below.

Theorem 2. *Let $(\mathcal{L}, \subset)$ be a lattice of sets and let $h : \mathcal{L} \to \mathbf{R}$ be a modular function on $\mathcal{L}$. Further, let $X_1, \ X_2, \ ..., \ X_k$ be some members from $\mathcal{L}$, where a natural $k \geq 1$.*

Then the following formula holds true:

$$h(X_1 \cup X_2 \cup ... \cup X_k) = \sum\{h(X_s) : s \in \mathcal{S}_1\} - \sum\{h(X_s) : s \in \mathcal{S}_2\}$$

$$+ \ . . . \ + (-1)^{k+1} \sum\{h(X_s) : s \in \mathcal{S}_k\}.$$

Proof. First of all, notice that the last sum on the right-hand side of this formula coincides with $h(X_1 \cap X_2 \cap ... \cap X_k)$.

We proceed by induction on k. For $k = 1$ the statement is trivial and for $k = 2$ the same statement is reduced to the definition of a modular function.

Suppose now that the assertion has already been proved for a natural number $k \geq 2$ and let us show its validity for $k+1$. Take any $(k+1)$-sequence

$$X_1,\ X_2,\ ,\ \ldots,\ X_k,\ X_{k+1}$$

of members from $\mathcal{L}$ and denote

$$Y_1 = X_1 \cap X_{k+1},\ Y_2 = X_2 \cap X_{k+1},\ \ldots,\ Y_k = X_k \cap X_{k+1},$$

$$Y = X_1 \cup X_2 \cup ... \cup X_k.$$

Obviously, we have $h(X_1 \cup X_2 \cup ... \cup X_k \cup X_{k+1}) = h(Y \cup X_{k+1})$.

Now, by virtue of the definition of modular functions, we may write

$$h(Y \cup X_{k+1}) = h(Y) + h(X_{k+1}) - h(Y \cap X_{k+1}).$$

Keeping in mind that

$$Y \cap X_{k+1} = Y_1 \cup Y_2 \cup ... \cup Y_k$$

and applying the inductive assumption to $h(Y) = h(X_1 \cup X_2 \cup ... \cup X_k)$ and to $h(Y_1 \cup Y_2 \cup ... \cup Y_k)$, we come (after easy calculations) to the required formula for $h(X_1 \cup X_2 \cup ... \cup X_k \cup X_{k+1})$.

This completes the proof of Theorem 2. □

Remark 1. One can conclude from the formulation of Theorem 2 that the value of a modular function h on the union of a finite family of sets, all of which belong to a lattice $\mathcal{L}$, is computable via the values of h on the intersections of all nonempty subfamilies of the given family.

Let us present two simple (and fairly standard) examples vividly illustrating the usefulness of the Principle of Inclusion and Exclusion.

Example 1. For a natural number k, take the set $K = \{1, 2, ..., k\}$ and consider the family $G(k)$ of all bijections of K onto itself. $G(k)$ becomes a group with respect to the ordinary composition operation $\circ$ of mappings. This group is usually called the group of all permutations of K, or the symmetric group of K (accordingly, the symbol $\mathrm{Sym}(K)$ is frequently used instead of $G(k)$). Let us denote

$$T(k) = \{f \in G(k) : (\forall i \in \{1, 2, ..., k\})(f(i) \neq i)\}.$$

In other words, $T(k)$ consists of those permutations of K which move all elements of K or, equivalently, of those permutations of K which do not have fixed points.

The Principle of Inclusion and Exclusion allows us to calculate the number of elements of $T(k)$. Indeed, it is not difficult to show with the aid of this principle that

$$\text{card}(T(k)) = k!/0! - k!/1! + k!/2! - ... + (-1)^k k!/k! =$$

$$k!(1/0! - 1/1! + 1/2! - ... + (-1)^k/k!).$$

Notice, by the way, that the expression $1/0! - 1/1! + 1/2! - ... + (-1)^k/k!$ is a very good approximation of the fraction $1/\mathrm{e}$, where $\mathrm{e} = 2,718281...$ denotes, as usual, Napier's constant. Since we have

$$\lim_{k\to\infty}(1/0! - 1/1! + 1/2! - ... + (-1)^k/k!) = 1/\mathrm{e},$$

the asymptotic behavior of $\text{card}(T(k))$ is the same as the asymptotic behavior of $k!/\mathrm{e}$ (when k tends to infinity).

Example 2. Let k be a natural number and consider again the set

$$K = \{1, 2, ..., k\}.$$

Some members of K are co-prime with k and some are not. This property of co-primeness decomposes K into two disjoint parts: the first part consists of all those numbers which are co-prime with k and the second part consists of all those numbers which are not co-prime with k. We are interested to know how many members of K belong to the first part. Following Euler, denote the cardinality of the first part by $\phi(k)$. So the posed question is reduced to the problem of finding a suitable formula for $\phi(k)$. The Principle of Inclusion and Exclusion readily solves this problem. Namely, k can be represented in the canonical form

$$k = p_1^{a_1} p_2^{a_2} ... p_n^{a_n},$$

where p_1, p_2,..., p_n are some pairwise distinct prime numbers and a_1, a_2,..., a_n are some strictly positive natural numbers. An easy argument based on the Principle of Inclusion and Exclusion yields the well-known number-theoretic formula

$$\phi(k) = k(1 - 1/p_1)(1 - 1/p_2)...(1 - 1/p_n).$$

If k itself is a prime number, i.e., $k = p$, where p is prime, then we obviously have $\phi(k) = p - 1$ which is also confirmed by the above much more general formula.

Many other purely combinatorial applications of the Principle of Inclusion and Exclusion can be found in various text-books of finite combinatorics (see, e.g., [48], [53], [118]).

We now turn our attention to a somewhat extraordinary geometric application of this principle.

Let $\mathbf{R}^m$ be, as usual, the m-dimensional Euclidean space and let $\mathcal{Q}_m$ denote the class of all relatively open convex polyhedra in $\mathbf{R}^m$ (it is reasonable to recall in this place that a relatively open convex polyhedron is defined as the relative interior of some convex polyhedron in $\mathbf{R}^m$ whose dimension does not exceed m).

In the sequel, we will be dealing with the lattice of sets generated by $\mathcal{Q}_m$. We will denote this lattice by the symbol $\mathcal{P}_m$.

It can readily be checked that all convex polyhedra in $\mathbf{R}^m$ (with dimensions less than or equal to m) belong to $\mathcal{P}_m$.

The following example is concerned with the most simple case when $m = 1$ and completely explains the situation in this case.

Example 3. Take the real line $\mathbf{R}^1 = \mathbf{R}$ and consider the class $\mathcal{P}_1$ of all those subsets of $\mathbf{R}$ which are representable as the unions of finite families consisting of nonempty open bounded intervals and singletons. It is easy to see that $\mathcal{P}_1$ is closed under the standard set-theoretical operations $\cup$ and $\cap$, i.e., $\mathcal{P}_1$ is a lattice of subsets of $\mathbf{R}$.

Now, we wish to define a certain function h $(= h_1)$ on this lattice. We do it as follows.

First of all, we put $h(\emptyset) = 0$.

If $x \in \mathbf{R}$, then we put $h(\{x\}) = 1$.

If X is a nonempty open bounded interval in $\mathbf{R}$, then we put $h(X) = -1$.

Observe now that if Z is any member of $\mathcal{P}_1$, then it admits a representation in the form of a finite union of nonempty open bounded intervals and singletons, all of which are pairwise disjoint. We define the value $h(Z)$ by additivity. Of course, it should be verified that this definition is correct, i.e., it does not depend on a representation of Z in the above-mentioned form. We leave this detail to the reader. Notice only that, according to the definition of h, if Y is a half-open bounded interval in $\mathbf{R}$, then $h(Y) = 0$ and if Y is a closed bounded interval, then $h(Y) = 1$.

In fact, $\mathcal{P}_1$ coincides with the ring generated by the family of all segments in $\mathbf{R}$ (including the degenerate ones). In particular, $\mathcal{P}_1$ turns out to be closed under the operation of set-theoretical difference.

Starting with the function $h = h_1$, we are going to construct a modular function h_m on the class $\mathcal{P}_m$ which is an m-dimensional analogue of $\mathcal{P}_1$. More precisely, we remember that $\mathcal{P}_m$ is the lattice generated by the family of all relatively open convex polyhedra in the space $\mathbf{R}^m$, and we want to introduce an appropriate modular function h_m on this lattice.

First, we need the following auxiliary proposition.

Lemma 1. *Let the space* $\mathbf{R}^{m+1}$ *be represented in the canonical form of a direct product* $\mathbf{R}^{m+1} = \mathbf{R}^m \times \mathbf{R}$.

Then the relation $\mathcal{P}_m = \{Z \cap \mathbf{R}^m : Z \in \mathcal{P}_{m+1}\}$ *takes place.*

The proof of this lemma is almost trivial and hence is omitted here.

The next auxiliary statement is crucial for obtaining the main result of this Appendix.

Lemma 2. *For any natural number $m \geq 1$, there exists a modular function h_m defined on the class $\mathcal{P}_m$ and such that*

$$h_m(P) = (-1)^j$$

for any (nonempty) j-dimensional relatively open convex polyhedron $P \subset \mathbf{R}^m$, and

$$h_m(Q) = 1$$

for any (nonempty closed) convex polyhedron $Q \subset \mathbf{R}^m$ whose dimension does not exceed m.

Proof. We argue by induction on m. First of all, Example 3 shows us that there exists a function h_1 on the class $\mathcal{P}_1$.

Suppose that the existence of a modular function h_m has already been proved for the lattice $\mathcal{P}_m$.

Let us try to establish the existence of h_{m+1} (for $\mathcal{P}_{m+1}$) with the required properties.

Again, represent the space $\mathbf{R}^{m+1}$ in the form of a direct product

$$\mathbf{R}^{m+1} = \mathbf{R}^m \times \mathbf{R}.$$

Then, by virtue of Lemma 1, we have $\mathcal{P}_m = \{Z \cap \mathbf{R}^m : Z \in \mathcal{P}_{m+1}\}$.

Take any nonempty set $X \in \mathcal{P}_{m+1}$. Then, according to the definition of $\mathcal{P}_{m+1}$, we may write

$$X = X_1 \cup X_2 \cup .... \cup X_k,$$

where $k \geq 1$ and every X_i $(i = 1, 2, ..., k)$ is a relatively open convex polyhedron in $\mathbf{R}^{m+1}$.

For any $t \in \mathbf{R}$, consider in the space $\mathbf{R}^{m+1}$ the affine hyperplane

$$G(t) = \mathbf{R}^m \times \{t\}.$$

Since we have the equality

$$G(t) \cap X = (G(t) \cap X_1) \cup (G(t) \cap X_2) \cup ... \cup (G(t) \cap X_k),$$

we may apply the function h_m to $G(t) \cap X$. Denoting

$$Y_1(t) = G(t) \cap X_1, \quad Y_2(t) = G(t) \cap X_2, \quad . . . , \quad Y_k(t) = G(t) \cap X_k$$

and taking into account the modularity of h_m, we may write

$$h_m(Y_1(t) \cup Y_2(t) \cup ... \cup Y_k(t)) =$$

$$\sum\{h_m(Y_s(t)) : s \in \mathcal{S}_1\} - \sum\{h_m(Y_s(t)) : s \in \mathcal{S}_2\} +$$

$$... + (-1)^{k+1} \sum\{h_m(Y_s(t)) : s \in \mathcal{S}_k\},$$

where the above notation is in accordance with the notation of Theorem 2.

Now, for any $t \in \mathbf{R}$, we put

$$g(Y_s(t)) = \lim_{\tau \to t+} h_m(Y_s(\tau)).$$

It is not difficult to show the existence of $\lim_{\tau \to t+} h_m(Y_s(\tau))$ for each $t \in \mathbf{R}$, so the function g is well-defined.

Further, we introduce a function h_{m+1} by the formula

$$h_{m+1}(X) = h_{m+1}(X_1 \cup X_2 \cup ... \cup X_k) =$$

$$\sum_{t \in \mathbf{R}} (\sum \{h_m(Y_s(t)) - g(Y_s(t)) : s \in \mathcal{S}_1\} -$$

$$\sum \{h_m(Y_s(t)) - g(Y_s(t)) : s \in \mathcal{S}_2\} + ...$$

$$+ (-1)^{k+1} \sum \{h_m(Y_s(t)) - g(Y_s(t)) : s \in \mathcal{S}_k\}).$$

Notice that the correctness of this definition is based on the following two simple facts.

Fact 1. There are only finitely many values of $t \in \mathbf{R}$ for which at least one of the above-mentioned k algebraic sums is nonzero.

This circumstance can easily be seen. Indeed, every set $Y_s(t)$ is a relatively open convex polyhedron in the affine hyperplane $G(t)$, namely,

$$Y_s(t) = G(t) \cap X_s.$$

The difference

$$h_m(Y_s(t)) - g(Y_s(t)) \qquad (t \in \mathbf{R})$$

may be nonzero only for those hyperplanes $G(t)$ which contain at least one vertex of X_s. But the total number of these vertices is obviously finite. Therefore, there are only finitely many such affine hyperplanes.

Fact 2. The value $h_{m+1}(X)$ does not depend on the choice of a representation

$$X = X_1 \cup X_2 \cup ... \cup X_k.$$

This means that if, for $X \in \mathcal{P}_{m+1}$, we have any two representations

$$X = X_1 \cup X_2 \cup ... \cup X_k = X'_1 \cup X'_2 \cup ... \cup X'_r,$$

then we necessarily get the equality

$$h_{m+1}(X_1 \cup X_2 \cup ... \cup X_k) = h_{m+1}(X'_1 \cup X'_2 \cup ... \cup X'_r).$$

The validity of this assertion can readily be demonstrated by induction on m. Indeed, the two above-mentioned representations of X yield, for each $t \in \mathbf{R}$, the corresponding two representations of the section $G(t) \cap X$, and we are able to apply the inductive assumption to this section.

Now, the equality $h_{m+1}(P) = (-1)^j$ holds true for every j-dimensional relatively open convex polyhedron P in $\mathbf{R}^{m+1}$, where $0 \leq j \leq m+1$.

This fact can easily be justified by induction on j, and we leave it to the reader.

In a similar manner, the case of any (nonempty closed) convex polyhedron $Q \subset \mathbf{R}^{m+1}$ can be considered. Indeed, the equality $h_{m+1}(Q) = 1$ is readily obtained by induction on $\dim(Q)$.

Finally, it directly follows from the definition of h_{m+1} that it is a modular function on the lattice $\mathcal{P}_{m+1}$ (cf. the properties (*) and (**) of modular functions).

This completes the proof of Lemma 2. □

After establishing Lemmas 1 and 2, we are able to prove the following fundamental statement due to Poincaré and Schläfli.

Theorem 3. *Let P be an m-dimensional convex polyhedron in the space $\mathbf{R}^m$ and let, for each index $j \in \{-1, 0, 1, 2, ..., m-1, m\}$, the symbol f_j denote the number of all j-dimensional faces of P.*

Then the formula

$$\sum\{(-1)^j f_j : j \in \{-1, 0, 1, 2, ..., m-1, m\}\} = 0$$

holds true.

Proof. Obviously, the given polyhedron P can be represented as the union of all its relatively open faces. Observe that these faces are pairwise disjoint (since they are relatively open). Recall also that the unique (-1)-dimensional face of P is the empty set $\emptyset$ and the unique m-dimensional closed face of P is P itself.

Consider the modular function h_m described in Lemma 2.

According to Theorem 1, our modular function h_m is additive.

According to Lemma 2, we have

$$h_m(P) = 1$$

and $h_m(F) = (-1)^j$ for any j-dimensional relatively open face F of P, where $0 \leq j \leq m$.

Thus, we may write

$$1 = h_m(P) = f_0 - f_1 + ... + (-1)^{m-1} f_{m-1} + (-1)^m f_m,$$

whence the required formula follows in view of the trivial equality $f_{-1} = 1$ (here we stipulate $(-1)^{-1} = -1$ and $(-1)^0 = 1$). Notice also that $f_m = 1$.

Theorem 3 has thus been proved. □

The obtained important relation

$$-f_{-1} + f_0 - f_1 + ... + (-1)^{m-1} f_{m-1} + (-1)^m f_m = 0$$

is usually called the Euler–Poincaré formula for any m-dimensional convex polyhedron P in the Euclidean space $\mathbf{R}^m$.

Remark 2. The above approach for obtaining the Euler–Poincaré formula is due to Hadwiger (see his work [129] and the extensive monograph [130]). The modular function h_m can be extended to a modular function defined on a lattice of subsets of $\mathbf{R}^m$ which is much wider than the lattice generated by all relatively open convex polyhedra in the space $\mathbf{R}^m$. Here we do not intend to discuss this interesting topic and refer the reader to [129] and [130] (see also [312]).

Remark 3. A somewhat paradoxical circumstance should be underlined in connection with the just described method of proving the Euler–Poincaré formula. This formula is concerned with finitely many geometric figures and with certain combinatorial relationships between them. But the method suggested by Hadwiger appeals to some limit process for introducing the function g, which is a key tool for obtaining the desired result. At first sight, it seems rather strange that for deriving a combinatorial formula, one needs the help of limit processes. However, this fact is not surprising from the historical point of view. In this context, it is reasonable to recall that several centuries ago Euler created the method of generating functions (based on power series), with the aid of which he solved many purely combinatorial and number-theoretic problems (see, for instance, [48], [53], [134]).

Remark 4. Actually, neither Poincaré nor Schläfli was able to give a completely rigorous proof of the Euler–Poincaré formula. In both cases their arguments had logical gaps. Probably, the first elementary proof of this formula was presented by Hadwiger [129] in 1955.

EXERCISES

1*. Let $\{X_i : i \in I\}$ be an arbitrary finite family of sets.

Prove Hall's combinatorial theorem stating that the following two assertions are equivalent:

(a) there exists a family $\{x_i : i \in I\}$ of pairwise distinct elements such that $x_i \in X_i$ for each index $i \in I$;

(b) for any subset J of I, the inequality

$$\mathrm{card}(J) \leq \mathrm{card}(\cup\{X_i : i \in J\})$$

holds true.

For obtaining the equivalence (a) $\Leftrightarrow$ (b), first observe that the implication (a) $\Rightarrow$ (b) is trivial. In order to establish the converse implication (b) $\Rightarrow$ (a), argue by induction on $\mathrm{card}(I)$. For this purpose, consider the two possible cases:

(i) for each nonempty proper subset J of I, the strong inequality

$$\mathrm{card}(J) < \mathrm{card}(\cup\{X_i : i \in J\})$$

is satisfied;

(ii) there exists a nonempty proper subset J of I such that

$$\mathrm{card}(J) = \mathrm{card}(\cup\{X_i : i \in J\}).$$

In the first case, fix an index $j \in I$, choose an element $x_j \in X_j$ and consider the family of sets $\{X_i \setminus \{x_j\} : i \in I \setminus \{j\}\}$. Check that the inductive assumption is applicable to this family.

In the second case, fix a nonempty proper subset J of I such that

$$\mathrm{card}(J) = \mathrm{card}(\cup\{X_i : i \in J\}).$$

Then denote $Y = \cup\{X_i : i \in J\}$ and verify that the inductive assumption is applicable to both families of sets $\{X_i : i \in J\}$ and $\{X_i \setminus Y : i \in I \setminus J\}$.

Remark 5. Notice that Hall's theorem remains valid for some infinite families of sets. Namely, if I is an arbitrary set of indices and $\{X_i : i \in I\}$ is a family of finite sets, then we have again the equivalence of (a) and (b). This fact can be established by using Tychonov's theorem on topological products of compact spaces. Also, let us mention that Hall's theorem is closely connected with the Dilworth theorem on a decomposition of a partially ordered set into its subchains (see, e.g., [134], [186], [275]).

2. Let $\mathcal{L}$ be a lattice of sets and let h be a modular function on $\mathcal{L}$.

Let $k \geq 1$ be a natural number and let $X_1, X_2, \ldots, X_k$ be some members of $\mathcal{L}$.

Prove the following formula:

$$h(X_1 \cap X_2 \cap ... \cap X_k) = \sum\{h(X'_s) : s \in \mathcal{S}_1\} - \sum\{h(X'_s) : s \in \mathcal{S}_2\}$$

$$+ ... + (-1)^{k+1} \sum\{h(X'_s) : s \in \mathcal{S}_k\},$$

where for every $s = \{i_1, i_2, ..., i_j\} \in \mathcal{S}_j$ the notation

$$X'_s = X_{i_1} \cup X_{i_2} \cup ... \cup X_{i_j}$$

is used.

Remark 6. Actually, the above-mentioned formula and the Principle of Inclusion and Exclusion are two dual statements. So, each of them can be deduced from the other.

3. Complete the details in the proof of Theorem 2.

4. Give a detailed proof of the fact presented in Example 1.

5. Give a detailed proof of the fact presented in Example 2.

6. Let $\mathcal{P}_1$ denote the lattice described in Example 3.

(a) Show that $\mathcal{P}_1$ coincides with the ring of sets generated by all half-open bounded intervals in $\mathbf{R}$.

(b) Show also, by using a direct argument, that $\mathcal{P}_1$ is a ring of sets generated by the family of all segments on $\mathbf{R}$.

7. Check the validity of Lemma 1.

In addition, demonstrate that the function h_m described in the proof of Lemma 2 is invariant under the group of all affine transformations of $\mathbf{R}^m$.

8. Consider the function h_2 on the class $\mathcal{P}_2$ (see Lemma 2). Actually, this function is defined on the lattice generated by all convex polygons in $\mathbf{R}^2$ (including the degenerate ones).

Let $n \geq 3$ be a natural number and let P be a simple polygon in $\mathbf{R}^2$ with precisely n vertices.

Verify that $P \in \mathrm{dom}(h_2)$ and $h_2(P) = 1$.

For this purpose, keep in mind Exercise 2 from Chapter 14 and use induction on n.

Further, let k be a natural number and let $\{P_1, P_2, ..., P_k\}$ be a disjoint family of simple polygons in $\mathbf{R}^2$ all of which are contained in $\mathrm{int}(P)$.

Verify that the equality

$$h_2(P \setminus \mathrm{int}(P_1 \cup P_2 \cup ... \cup P_k)) = 1 - k$$

is valid.

For this purpose, use induction on k.

9. Consider the unit cube $[0,1]^m$ in the space $\mathbf{R}^m$, where $m \geq 2$. The vertices and edges of this cube form a certain graph denoted by (V, E).

Prove the existence of a simple cycle in (V, E) whose length is 2^m, and give a set-theoretical interpretation of this fact (argue by induction on m).

10*. Let E be a ground set and let G be a group of transformations of E.

For any element $e \in E$, introduce the notation

$$St(e) = \{g \in G : g(e) = e\}.$$

The set $St(e)$ is a subgroup of G which is usually called the stabilizer of e.

Further, for any transformation $g \in G$, denote

$$Inv(g) = \{e \in E : g(e) = e\}.$$

The set $Inv(g)$ is usually called the set of all fixed (invariant) points of g.

Finally, for any element $e \in E$, define

$$G(e) = \{g(e) : g \in G\}.$$

Recall that the set $G(e)$ is the G-orbit of e.

It is well known that the family $\{G(e) : e \in E\}$ forms a partition of the given ground set E. The cardinality of this partition will be denoted below by the symbol $\mathrm{orb}(G)$.

Show that, for each element $e \in E$, the following equality is fulfilled:

$$\mathrm{card}(G(e)) \cdot \mathrm{card}(St(e)) = \mathrm{card}(G).$$

Consequently, in the case of a finite set E the equality

$$\mathrm{card}(E) = \sum\{\mathrm{card}(G)/\mathrm{card}(St(x)) : x \in X\}$$

holds true, where X is an arbitrary selector of the family of all G-orbits.

Starting with the above facts, demonstrate that if E is finite, then the following Burnside formula is valid:

$$\mathrm{card}(G) \cdot \mathrm{orb}(G) = \sum\{\mathrm{card}(Inv(g)) : g \in G\}$$

For this purpose, consider the set of pairs

$$\{(g, e) : g \in G, \ e \in E, \ g(e) = e\}$$

and calculate in two different ways (separately with respect to $g \in G$ and with respect to $e \in E$) the cardinality of this set.

11*. Consider the two-dimensional unit sphere $\mathbf{S}_2 \subset \mathbf{R}^3$ equipped with the group of all its (proper) rotations about the origin of $\mathbf{R}^3$.

Let G be any nontrivial finite subgroup of this group.

Show that the disjunction of the following five assertions is satisfied:

(a) G is a cyclic group of order $n \geq 2$, i.e., the elements of G are rotations $g_0, g_1, \ldots, g_{n-1}$ about one straight line in $\mathbf{R}^3$ passing through the origin, and the corresponding rotation angles are

$$0, \ 2\pi/n, \ \ldots, \ 2\pi(n-1)/n;$$

(b) G coincides with the dihedral group of order $2n$; this group consists of all those rotations of $\mathbf{S}_2$, which map onto itself some regular n-gon inscribed in a large circle of $\mathbf{S}_2$ (in particular, the dihedral group contains all π-rotations of the same n-gon about its axes of symmetry);

(c) G coincides with the proper symmetry group of a regular tetrahedron inscribed in $\mathbf{S}_2$;

(d) G coincides with the proper symmetry group of a cube (or octahedron) inscribed in $\mathbf{S}_2$;

(e) G coincides with the proper symmetry group of a dodecahedron (or icosahedron) inscribed in $\mathbf{S}_2$.

For proving this result, first observe that any non-identity rotation has exactly two fixed points on $\mathbf{S}_2$. Denote by E the set of all those points in $\mathbf{S}_2$ which are fixed under some nontrivial rotation from G. The set E is finite and the Burnside formula from Exercise 10 yields in this case

$$\mathrm{card}(G) \cdot \mathrm{orb}(G) = \mathrm{card}(E) + 2(\mathrm{card}(G) - 1).$$

Remembering that

$$\mathrm{card}(E) = \sum\{\mathrm{card}(G)/\mathrm{card}(St(x)) : x \in X\}$$

for any selector X of the family of all G-orbits, deduce the equality

$$2 - 2/\mathrm{card}(G) = \sum\{1 - 1/\mathrm{card}(St(x)) : x \in X\},$$

whence it follows that the number of summands on the right-hand side of this equality does not exceed 3 or, equivalently, the number of G-orbits in E does not exceed 3. It remains to consider all possibilities for the numbers

$$\mathrm{card}(St(x_1)) \leq \mathrm{card}(St(x_2)) \leq \mathrm{card}(St(x_3)),$$

where $X = \{x_1, x_2\}$ or $X = \{x_1, x_2, x_3\}$.

12*. Let E be a metric space and let μ be a Borel probability measure on this space.

Recall that a sequence $\{\mu_n : n < \omega\}$ of Borel probability measures on E weakly converges to μ if

$$\mathrm{limsup}_{n\to\infty}\mu_n(X) \leq \mu(X)$$

for every closed set $X \subset E$ or, equivalently,

$$\mathrm{liminf}_{n\to\infty}\mu_n(Y) \geq \mu(Y)$$

for every open set $Y \subset E$.

Suppose that ν is a Borel probability measure on E and $\{\nu_n : n < \omega\}$ is some sequence of Borel probability measures on E.

Let $\mathcal{F}$ be a family of Borel subsets of E satisfying the following three conditions:

(a) the intersection of any finite subfamily of $\mathcal{F}$ also belongs to $\mathcal{F}$;

(b) each open set in E can be represented as the union of a countable subfamily of $\mathcal{F}$;

(c) $\lim_{n\to\infty}\nu_n(X) = \nu(X)$ for every set $X \in \mathcal{F}$.

Prove that the given sequence $\{\nu_n : n < \omega\}$ weakly converges to ν.

For this purpose, argue as follows. Let $\{Z_1, Z_2, ..., Z_k\}$ be an arbitrary finite family of sets from $\mathcal{F}$. Denoting $Z = Z_1 \cup Z_2 \cup ... \cup Z_k$, keeping in mind (a), and using the Inclusion and Exclusion Principle, verify that

$$\lim_{n\to\infty}\nu_n(Z) = \nu(Z).$$

Further, let U be an arbitrary open set in E. According to (b), one may write

$$U = \cup\{X_k : k < \omega\}$$

for some countable family $\{X_k : k < \omega\} \subset \mathcal{F}$. Let ε be a strictly positive real number. There exists a natural index k_0 such that

$$\nu(X_0 \cup X_1 \cup ... \cup X_{k_0}) > \nu(U) - \varepsilon.$$

According to the fact shown above, the relation

$$\nu(U) - \varepsilon < \lim_{n\to\infty}\nu_n(X_0 \cup X_1 \cup ... \cup X_{k_0})$$

holds true. This relation implies the inequality

$$\nu(U) - \varepsilon \leq \mathrm{liminf}_{n\to\infty}\nu_n(U).$$

Finally, since $\varepsilon > 0$ can be taken arbitrarily small, one gets

$$\nu(U) \leq \mathrm{liminf}_{n\to\infty}\nu_n(U),$$

which means that the sequence of probability measures $\{\nu_n : n < \omega\}$ weakly converges to ν.

13. Let E be a separable metric space, ν be a Borel probability measure on E, and let $\{\nu_n : n < \omega\}$ be a sequence of Borel probability measures on E.

Suppose that $\mathcal{F}$ is a family of Borel subsets of E satisfying the following three conditions:

(a) the intersection of any finite subfamily of $\mathcal{F}$ also belongs to $\mathcal{F}$;

(b) for each point $x \in E$ and for any neighborhood $U(x)$ of x, there exists a set $Y \in \mathcal{F}$ such that $Y \subset U(x)$ and x belongs to the interior of Y;

(c) $\lim_{n\to\infty}\nu_n(X) = \nu(X)$ for every set $X \in \mathcal{F}$.

Demonstrate that the sequence $\{\nu_n : n < \omega\}$ weakly converges to ν.

For this purpose, take into account the separability of E and Exercise 12.

Remark 7. The results presented in Exercises 12 and 13 are due to Kolmogorov and Prokhorov.

14. Let $k \geq 1$ be a natural number and let $P_1, P_2, \ldots, P_k$ be convex polyhedra in the space $\mathbf{R}^m$, which all have a common point.

Show that $h_m(P_1 \cup P_2 \cup \ldots \cup P_k) = 1$.

For this purpose, use induction on k.

15. Let $P_1, P_2, \ldots, P_{m+2}$ be convex polyhedra in the space $\mathbf{R}^m$ such that the intersection of any $m+1$ of them is nonempty.

Prove that $P_1 \cap P_2 \cap \ldots \cap P_{m+2} \neq \emptyset$.

For this purpose, keeping in mind Exercise 14 and the Principle of Inclusion and Exclusion (or its dual formula of Exercise 2), infer that

$$h_m(P_1 \cap P_2 \cap \ldots \cap P_{m+2}) = 1,$$

which trivially implies the required result.

16. Taking into account the two preceding exercises, give another proof of Helly's theorem which states that, for a family $\{C_i : i \in I\}$ of compact convex sets in $\mathbf{R}^m$, the following assertions are equivalent:

(a) $\cap\{C_i : i \in I\} \neq \emptyset$;

(b) for each $J \subset I$ with $\mathrm{card}(J) \leq m+1$, all members of the partial family $\{C_j : j \in J\}$ have a common point.

The Erdös–Mordell inequality

Various inequalities of geometric type can be pointed out that are closely connected with different topics of convex, combinatorial and discrete geometry (cf., for instance, [20], [24], [52], [107], [123], [130], [164], [308], [315]). Quite often those inequalities help to solve concrete problems and questions from the above-mentioned fields and related areas of mathematics.

In this Appendix we will be dealing with some inequalities for triangles and convex polygons in the Euclidean plane $\mathbf{R}^2$.

To begin, suppose that a (non-degenerate) triangle $[A, B, C]$ is given in $\mathbf{R}^2$, whose vertices are A, B, and C. As usual, we put

$$a = ||B - C||, \quad b = ||C - A||, \quad c = ||A - B||.$$

Let O be an arbitrary point in $[A, B, C]$, for which we introduce the notation

$$R_A = ||O - A||, \quad R_B = ||O - B||, \quad R_C = ||O - C||.$$

Also, let us define:

r_a = the distance from O to the straight line $l(B, C)$,
r_b = the distance from O to the straight line $l(C, A)$,
r_c = the distance from O to the straight line $l(A, B)$.

P. Erdös conjectured in 1935 that, for every $[A, B, C]$, the relation

$$2(r_a + r_b + r_c) \leq R_A + R_B + R_C$$

holds true and that the equality in this relation is attained only in the case of an equilateral triangle $[A, B, C]$ and its circumcenter O (see [83]). Two years later, L. J. Mordell confirmed this hypothesis (see [264]). Afterwards, many publications were devoted to the Erdös–Mordell inequality and different proofs of it were suggested by various authors. Here we would like to mention only

DOI: 10.1201/9781003458708-A4

several ones (see [15], [42], [71], [96], [159], [162], [273]). Moreover, a generalization of the Erdös–Mordell inequality was obtained for a convex polygon P in $\mathbf{R}^2$ and for a point O inside of P (see, e.g., [71], [127], [262]).

We will present below the generalized version of the Erdös–Mordell inequality. Our argument closely follows the proof given in [71].

The key role is played by the following auxiliary proposition.

Lemma 1. *Let $n \geq 3$ be a natural number, let $x_1, x_2, ..., x_n$ be any strictly positive real numbers, and let $\alpha_1, \alpha_2, ..., \alpha_n$ be real numbers such that*

$$0 < \alpha_k < \pi \quad (1 \leq k \leq n), \qquad \alpha_1 + \alpha_2 + ... + \alpha_n = \pi.$$

Then the inequality

$$\cos(\pi/n)\Sigma\{x_k^2 : 1 \leq k \leq n\} \geq \Sigma\{x_k x_{k+1}\cos(\alpha_k) : 1 \leq k \leq n\}$$

is valid (where it is stipulated $x_{n+1} = x_1$).

Proof. In the plane $\mathbf{R}^2$ consider some rays $l_1, l_2, \ldots, l_n$ such that the origin $O = (0,0)$ of $\mathbf{R}^2$ is their common endpoint and, for any natural number $k \in [1, n-1]$, the measure of the angle between l_k and l_{k+1} is equal to α_k. Let, for each natural number $k \in [1, n]$, the point $P_k = (a_k, b_k) \in \mathbf{R}^2$ belong to the ray l_k and satisfy the relation $||P_k|| = x_k$, i.e., one has $x_k^2 = a_k^2 + b_k^2$.

From the cosine theorem applied to the triangle $[P_k, O, P_{k+1}]$, where a natural index k ranges over $[1, n-1]$, we infer

$$x_k x_{k+1}\cos(\alpha_k) = (1/2)(a_k^2 + b_k^2 + a_{k+1}^2 + b_{k+1}^2 - (a_k - a_{k+1})^2 - (b_k - b_{k+1})^2)$$

$$= a_k a_{k+1} + b_k b_{k+1}.$$

Applying the same cosine theorem to the triangle $[P_n, O, P_1]$ and keeping in mind the relation

$$\alpha_n = \pi - \alpha_1 - \alpha_2 - ... - \alpha_{n-1},$$

we may write

$$x_1 x_n\cos(\alpha_n) = -(1/2)(a_1^2 + b_1^2 + a_n^2 + b_n^2 - (a_1 - a_n)^2 - (b_1 - b_n)^2)$$

$$= -(a_1 a_n + b_1 b_n).$$

Therefore, we get

$$\cos(\pi/n)\Sigma\{x_k^2 : 1 \leq k \leq n\} - \Sigma\{x_k x_{k+1}\cos(\alpha_k) : 1 \leq k \leq n\} =$$

$$\cos(\pi/n)\Sigma\{a_k^2 + b_k^2 : 1 \leq k \leq n\} - \Sigma\{a_k a_{k+1} + b_k b_{k+1} : 1 \leq k \leq n-1\}$$

$$+ a_1 a_n + b_1 b_n.$$

Further, it will be convenient to introduce the following notation:

$$\sigma_1 = \cos(\pi/n)\Sigma\{a_k^2 : 1 \leq k \leq n\} - \Sigma\{a_k a_{k+1} : 1 \leq k \leq n-1\} + a_1 a_n;$$

$$\sigma_2 = \cos(\pi/n)\Sigma\{b_k^2 : 1 \le k \le n\} - \Sigma\{b_k b_{k+1} : 1 \le k \le n-1\} + b_1 b_n.$$

Actually, it turns out that σ_1 is identical with

$$\Sigma_{1\le k\le n-2}\frac{(\sin((k+1)\pi/n)a_k - \sin(k\pi/n)a_{k+1} + \sin(\pi/n)a_n)^2}{2\sin(k\pi/n)\sin((k+1)\pi/n)}$$

and σ_2 is identical with

$$\Sigma_{1\le k\le n-2}\frac{(\sin((k+1)\pi/n)b_k - \sin(k\pi/n)b_{k+1} + \sin(\pi/n)b_n)^2}{2\sin(k\pi/n)\sin((k+1)\pi/n)}.$$

Let us verify the validity of the above assertion for σ_1. For this purpose, treating $a_1, a_2, \ldots, a_n$ as independent variables, we will compare the coefficients at a_k^2, at $a_k a_{k+1}$, and at $a_1 a_n$ in the expression

$$\sigma_1 = \cos(\pi/n)\Sigma\{a_k^2 : 1 \le k \le n\} - \Sigma\{a_k a_{k+1} : 1 \le k \le n-1\} + a_1 a_n$$

and in the expression

$$\Sigma_{1\le k\le n-2}\frac{(\sin((k+1)\pi/n)a_k - \sin(k\pi/n)a_{k+1} + \sin(\pi/n)a_n)^2}{2\sin(k\pi/n)\sin((k+1)\pi/n)}. \qquad (*)$$

We will do it step by step.

Clearly, for all natural numbers $k \in [1, n]$, the coefficient at a_k^2 in σ_1 is $\cos(\pi/n)$. If $2 \le k \le n-2$, then in (*) the coefficient at a_k^2 is equal to

$$(\sin((k+1)\pi/n) + \sin((k-1)\pi/n))/2\sin(k\pi/n) = \cos(\pi/n).$$

In σ_1 the coefficient at a_{n-1}^2 is again $\cos(\pi/n)$. In (*) the coefficient at a_{n-1}^2 is equal to

$$\sin((n-2)\pi/n)/2\sin((n-1)\pi/n) = \cos(\pi/n).$$

In σ_1 the coefficient at a_1^2 is again $\cos(\pi/n)$. In (*) the coefficient at a_1^2 is equal to

$$\sin(2\pi/n)/2\sin(\pi/n) = \cos(\pi/n).$$

In σ_1 the coefficient at a_n^2 is again $\cos(\pi/n)$. In (*) the coefficient at a_n^2 is equal to

$$\Sigma_{1\le k\le n-2}\frac{\sin^2(\pi/n)}{2\sin(k\pi/n)\sin((k+1)\pi/n)} =$$

$$\Sigma_{1\le k\le n-2}\frac{\sin(\pi/n)}{2}(\cot(k\pi/n) - \cot((k+1)\pi/n) =$$

$$\frac{\sin(\pi/n)}{2}(\cot(\pi/n) - \cot((n-1)\pi/n)) = \cos(\pi/n).$$

If $1 \le k \le n-2$, then in σ_1 the coefficient at $a_k a_{k+1}$ is -1. In (*) the coefficient at the same $a_k a_{k+1}$ is equal to

$$\frac{-2\sin((k+1)\pi/n)\sin(k\pi/n)}{2\sin((k+1)\pi/n)\sin(k\pi/n)} = -1.$$

In σ_1 the coefficient at $a_{n-1}a_n$ is again -1. In (*) the coefficient at $a_{n-1}a_n$ is equal to

$$\frac{-2\sin((n-2)\pi/n)\sin(\pi/n)}{2\sin((n-2)\pi/n)\sin((n-1)\pi/n)} = -1.$$

In σ_1 the coefficient at a_1a_n is 1. In (*) the coefficient at a_1a_n is equal to

$$\frac{2\sin(2\pi/n)\sin(\pi/n)}{2\sin(\pi/n)\sin(2\pi/n)} = 1.$$

Finally, if $2 \le k \le n-2$, then in σ_1 the coefficient at a_ka_n is 0. In (*) the coefficient at the same a_ka_n is equal to

$$\Sigma_{2\le k\le n-2}\frac{2\sin((k+1)\pi/n)\sin(\pi/n)}{2\sin(k\pi/n)\sin((k+1)\pi/n)} -$$

$$\Sigma_{1\le k\le n-3}\frac{2\sin(k\pi/n)\sin(\pi/n)}{2\sin(k\pi/n)\sin((k+1)\pi/n)} = 0.$$

Summarizing all the above, we see that σ_1 and (*) are identical.

Using the completely similar argument, we infer that σ_2 and the expression

$$\Sigma_{1\le k\le n-2}\frac{(\sin((k+1)\pi/n)b_k - \sin(k\pi/n)b_{k+1} + \sin(\pi/n)b_n)^2}{2\sin(k\pi/n)\sin((k+1)\pi/n)} \qquad (**)$$

are also identical.

Since both expressions (*) and (**) are non-negative, we conclude that the value $\sigma_1+\sigma_2$ is always non-negative, and the proof of Lemma 1 is complete. □

Lemma 2. *Let $[A, B, C]$ be a non-degenerate triangle in the plane $\mathbf{R}^2$ and let d denote the length of the bisector of $\angle C$.*

Then the inequality $d \le (ab)^{1/2}\cos(\angle C/2)$ holds true.

Proof. In view of the additivity property of area, we obviously may write

$$(1/2)ab\sin(\angle C) = (1/2)ad\sin(\angle C/2) + (1/2)bd\sin(\angle C/2)$$

and, therefore,

$$d = \frac{2ab\cos(\angle C/2)}{a+b} \le \frac{2ab\cos(\angle C/2)}{2(ab)^{1/2}} = (ab)^{1/2}\cos(\angle C/2),$$

which ends the proof of Lemma 2. □

Theorem 1. *Let P be a convex polygon in the plane $\mathbf{R}^2$ whose vertices are $z_1, z_2, \ldots, z_n$. Assume that the enumeration of these vertices is such that the corresponding line segments*

$$[z_1, z_2],\quad [z_2, z_3],\quad \ldots,\quad [z_{n-1}, z_n],\quad [z_n, z_1]$$

give all the sides of P. Suppose also that z is an internal point of P and let

$$d_k = \text{dist}(z, l(z_k, z_{k+1})) \qquad (1 \leq k \leq n),$$

where again it is stipulated $z_{n+1} = z_1$.

Then the following generalized Erdös–Mordell inequality for P is valid:

$$\cos(\pi/n)\Sigma\{||z - z_k|| : 1 \leq k \leq n\} \geq \Sigma\{d_k : 1 \leq k \leq n\}.$$

In addition, the equality takes place in the above formula if and only if P is a regular n-gon and z coincides with the circumcenter (incenter) of P.

Proof. It is convenient to introduce the notation:

$$x_k = ||z - z_k||^{1/2} \qquad (1 \leq k \leq n),$$

$$\alpha_k = \text{ the measure of the angle at } z \text{ in } [z_k, z, z_{k+1}] \qquad (1 \leq k \leq n).$$

For any natural number $k \in [1, n]$, apply Lemma 2 to the triangle $[z_k, z, z_{k+1}]$ and to the length d'_k of its bisector passing through z (of course, we again adopt here $z_{n+1} = z_1$). We thus get the inequalities

$$d'_k \leq (||z - z_k|| \cdot ||z - z_{k+1}||)^{1/2}\cos(\alpha_k/2) \qquad (1 \leq k \leq n).$$

Keeping in mind the relation

$$\Sigma\{\alpha_k/2 : 1 \leq k \leq n\} = \pi,$$

we may write, by virtue of Lemma 1,

$$\cos(\pi/n)\Sigma\{x_k^2 : 1 \leq k \leq n\} \geq \Sigma\{x_k x_{k+1}\cos(\alpha_k/2) : 1 \leq k \leq n\}$$

$$\geq \Sigma\{d'_k : 1 \leq k \leq n\}$$

and, consequently,

$$\cos(\pi/n)\Sigma\{||z - z_k|| : 1 \leq k \leq n\} \geq \Sigma\{d'_k : 1 \leq k \leq n\}.$$

Finally, taking into account that $d_k \leq d'_k$ for each natural number $k \in [1, n]$, we obtain the required generalized Erdös–Mordell inequality.

We leave to the reader specify the case when this inequality becomes the equality (see Exercise 4). □

Putting $n = 3$ in Theorem 1, we immediately get the original Erdös–Mordell inequality

$$R_A + R_B + R_C \geq 2(r_a + r_b + r_c).$$

In Exercise 1 of this chapter another proof of the original Erdös–Mordell inequality is presented, which does not rely on Theorem 1.

Remark 1. Let $S = [z_1, z_2, z_3, z_4]$ be a tetrahedron in the space $\mathbf{R}^3$ with vertices z_1, z_2, z_3, z_4, and let z be an internal point of S. It is natural to conjecture that, similarly to the Erdös–Mordell inequality, its three-dimensional analog

$$||z - z_1|| + ||z - z_2|| + ||z - z_3|| + ||z - z_4|| \geq 3(d_1 + d_2 + d_3 + d_4)$$

holds true, where the numbers d_1, d_2, d_3 and d_4 are defined as follows:

d_1 = the distance from z to the plane containing $[z_2, z_3, z_4]$,
d_2 = the distance from z to the plane containing $[z_1, z_3, z_4]$,
d_3 = the distance from z to the plane containing $[z_1, z_2, z_4]$,
d_4 = the distance from z to the plane containing $[z_1, z_2, z_3]$.

However, the last inequality is not valid in general (in this connection, see Exercise 5 of the present Appendix). It was only proved that, preserving the above notation, the weaker inequality

$$||z - z_1|| + ||z - z_2|| + ||z - z_3|| + ||z - z_4|| \geq 2^{3/2}(d_1 + d_2 + d_3 + d_4)$$

is always satisfied and the coefficient $2^{3/2}$ is precise in a certain sense (for more details, see [163], [164]).

The next statement is an easy consequence of the Erdös–Mordell inequality.

Theorem 2. *Let $[A, B, C]$ be a triangle in $\mathbf{R}^2$, let r denote the radius of the inscribed circle of $[A, B, C]$, and let O be an arbitrary point in $[A, B, C]$. As earlier, put*

$$R_A = ||O - A||, \quad R_B = ||O - B||, \quad R_C = ||O - C||.$$

Then the following relation holds true:

$$R_A + R_B + R_C \geq 6r.$$

Moreover, the equality is attained in the above relation only in the case when $[A, B, C]$ is an equilateral triangle and O coincides with the circumcenter of $[A, B, C]$.

Proof. Let h_a, h_b, h_c denote the lengths of the altitudes of $[A, B, C]$ which pass, respectively, through A, B, and C. Obviously, we may write

$$R_A + r_a \geq h_a, \quad R_B + r_b \geq h_b, \quad R_C + r_c \geq h_c,$$

where r_a, r_b, r_c are defined in the beginning of this chapter. Using the Erdös–Mordell inequality, we readily get

$$(3/2)(R_A + R_B + R_C) \geq h_a + h_b + h_c.$$

Further, for any strictly positive real numbers x, y and z the inequality

$$(x + y + z)(1/x + 1/y + 1/z) \geq 9$$

is valid or, equivalently,

$$x + y + z \geq \frac{9}{1/x + 1/y + 1/z}.$$

Putting $x = h_a$, $y = h_b$, and $z = h_c$, we infer

$$(3/2)(R_A + R_B + R_C) \geq h_a + h_b + h_c \geq \frac{9}{1/h_a + 1/h_b + 1/h_c}.$$

It remains to recall the well-known formula

$$1/h_a + 1/h_b + 1/h_c = 1/r$$

for the triangle $[A, B, C]$, which completes the proof of Theorem 2. □

Remark 2. Let $S = [z_1, z_2, z_3, z_4]$ be a tetrahedron in the space $\mathbf{R}^3$ with vertices z_1, z_2, z_3, z_4, and let z be an internal point of S. Preserving the notation of Remark 1 and denoting by r the radius of the inscribed sphere of S, one may conjecture that the similar inequality

$$||z - z_1|| + ||z - z_2|| + ||z - z_3|| + ||z - z_4|| \geq 12r$$

is always satisfied and the equality in it is realizable only in the case of a regular tetrahedron S with its incenter z. However, taking into account the circumstance described in Remark 1, we cannot argue analogously to the proof of Theorem 2. Indeed, much more delicate reasoning is needed for establishing the validity of the last inequality (in this context, see Exercise 22 of the present chapter).

EXERCISES

1. Let a triangle $[A, B, C]$ be given in the plane $\mathbf{R}^2$, whose vertices are A, B, and C. As usual, put

$$a = ||B - C||, \quad b = ||C - A||, \quad c = ||A - B||.$$

Let O be an arbitrary internal point of $[A, B, C]$ for which the notation introduced at the beginning of this Appendix is preserved:

$R_A = ||O - A||$, $R_B = ||O - B||$, $R_C = ||O - C||$,

r_a = the distance from O to the line $l(B, C)$,

r_b = the distance from O to the line $l(C, A)$,

r_c = the distance from O to the line $l(A, B)$.

Under this notation, give a direct proof of the Erdös–Mordell inequality

$$2(r_a + r_b + r_c) \leq R_A + R_B + R_C$$

and show that the equality is attained only in the case of an equilateral triangle $[A, B, C]$ in which the point O coincides with its circumcenter (incenter).

For this purpose, argue as follows. Take the two points z and z' such that

$$z \in l(A, C) \ \&\ ||O - z|| = r_b, \qquad z' \in l(B, C) \ \&\ ||O - z'|| = r_a,$$

and check that

$$||z - z'|| = R_C \sin(\angle C).$$

Then consider the orthogonal projection of the line segment $[z, z']$ on the straight line $l(A, B)$ and show that the length of this projection is

$$r_a \sin(\angle B) + r_b \sin(\angle A),$$

whence it follows that

$$R_C \sin(\angle C) \geq r_a \sin(\angle B) + r_b \sin(\angle A)$$

or, equivalently,

$$R_C \geq r_a(\sin(\angle B)/\sin(\angle C)) + r_b(\sin(\angle A)/\sin(\angle C)).$$

By using the analogous argument, get two similar relations

$$R_A \geq r_b(\sin(\angle C)/\sin(\angle A)) + r_c(\sin(\angle B)/\sin(\angle A)),$$

$$R_B \geq r_c(\sin(\angle A)/\sin(\angle B)) + r_a(\sin(\angle C)/\sin(\angle B)).$$

Finally, apply the trivial relation

$$x/y + y/x \geq 2 \qquad (x > 0, \ y > 0),$$

and obtain the required inequality $2(r_a + r_b + r_c) \leq R_A + R_B + R_C$.

2. Preserving the notation of Exercise 1, demonstrate the validity of the inequality $8r_a r_b r_c \leq R_A R_B R_C$.

For this purpose, start with the relations

$$R_C \geq r_a(\sin(\angle B)/\sin(\angle C)) + r_b(\sin(\angle A)/\sin(\angle C)),$$

$$R_A \geq r_b(\sin(\angle C)/\sin(\angle A)) + r_c(\sin(\angle B)/\sin(\angle A)),$$

$$R_B \geq r_c(\sin(\angle A)/\sin(\angle B)) + r_a(\sin(\angle C)/\sin(\angle B))$$

of Exercise 1 and take into account that

$$r_a(\sin(\angle B)/\sin(\angle C)) + r_b(\sin(\angle A)/\sin(\angle C)) \geq$$

$$2(r_a r_b \sin(\angle B)\sin(\angle A))^{1/2}/\sin(\angle C),$$

$$r_b(\sin(\angle C)/\sin(\angle A)) + r_c(\sin(\angle B)/\sin(\angle A)) \geq$$
$$2(r_b r_c \sin(\angle C)\sin(\angle B))^{1/2}/\sin(\angle A),$$
$$r_c(\sin(\angle A)/\sin(\angle B)) + r_a(\sin(\angle C)/\sin(\angle B)) \geq$$
$$2(r_c r_a \sin(\angle A)\sin(\angle C))^{1/2}/\sin(\angle B),$$

whence the desired result readily follows.

3*. Preserving again the notation of Exercises 1 and 2, prove the validity of the inequality

$$1/R_A + 1/R_B + 1/R_C \leq (1/2)(1/r_a + 1/r_b + 1/r_c).$$

For this purpose, assume that the point O coincides with the origin $(0, 0)$ of the plane $\mathbf{R}^2$ and consider the polar transform ϕ of $\mathbf{R}^2$ with center O and coefficient 1. Associate with the vertices A, B, and C the straight lines $\phi(A)$, $\phi(B)$, $\phi(C)$, and denote by $[A', B', C']$ the triangle bounded by these three lines.

Check that:

(i) the point O is inside of $[A', B', C']$;

(ii) $||A' - O|| = 1/r_a$, $||B' - O|| = 1/r_b$, $||C' - O|| = 1/r_c$;

(iii) the three equalities

$$\text{dist}(O, l(A', B')) = 1/R_C, \ \text{dist}(O, l(B', C')) = 1/R_A, \ \text{dist}(O, l(C', A')) = 1/R_B$$

are fulfilled.

Finally, apply to the triangle $[A', B', C']$ and its interior point O the Erdös–Mordell inequality.

Using the notation of Theorem 1 and applying the same method of polar transform, deduce the generalized inequality

$$\Sigma\{1/||z - z_k|| : 1 \leq k \leq n\} \leq \cos(\pi/n)\Sigma\{1/d_k : 1 \leq k \leq n\}$$

for a convex n-gon $P = \text{conv}(\{z_1, z_2, ..., z_n\}) \subset \mathbf{R}^2$ and for its interior point z.

Also, specify the case when this inequality is reduced to the equality.

Remark 3. The results of Exercises 1, 2 and 3 admit various analogs for many other means of the values R_A, R_B, R_C and of the values r_a, r_b, r_c. In this connection, see especially [315] and the references therein. A certain version of the Erdös–Mordell inequality was proved in [274] for a spherical triangle $\triangle$ on $\mathbf{S}_2$ and for an interior point of $\triangle$.

4. Show that in the generalized Erdös-Mordell inequality (i.e., Theorem 1) the equality takes place if and only if a convex n-gon P is regular and a point z coincides with the circumcenter (incenter) of P.

For this purpose, preserving the notation of the proof of Theorem 1, first verify that in the case of the equality it must be

$$d'_k = d_k \qquad (1 \le k \le n),$$

whence it follows that z is equidistant from all vertices of P, i.e., z coincides with the center of the circumscribed circle of P. Then deduce that P is a regular polygon (cf. Lemma 1 of this Appendix).

5. Let $S = [z_1, z_2, z_3, z_4]$ be a tetrahedron in the space $\mathbf{R}^3$ with vertices z_1, z_2, z_3, z_4, and let the values d_1, d_2, d_3, d_4 be defined as in Remark 1. In addition to this, suppose that:

(1) the edges $[z_1, z_2]$, $[z_1, z_3]$, $[z_1, z_4]$ are pairwise perpendicular;

(2) $||z_1 - z_2|| = ||z_1 - z_3|| = 1$;

(3) $||z_1 - z_4|| = \varepsilon$, where $\varepsilon > 0$ is sufficiently small.

Let z be an internal point of S closely near to the mid-point of the edge $[z_2, z_3]$.

Verify that the fraction

$$(||z - z_1|| + ||z - z_2|| + ||z - z_3|| + ||z - z_4||)/(d_1 + d_2 + d_3 + d_4)$$

is closely near to $2^{3/2}$, so the direct analog of the Erdös–Mordell inequality written as

$$||z - z_1|| + ||z - z_2|| + ||z - z_3|| + ||z - z_4|| \ge 3(d_1 + d_2 + d_3 + d_4)$$

does not hold in general.

6*. Let $p \ge 3$ be a natural number and let $x_1, x_2, \ldots, x_p$ be some vectors in the Euclidean plane $\mathbf{R}^2$ such that

$$||x_1|| \le 1, \quad ||x_2|| \le 1, \quad \ldots, \quad ||x_p|| \le 1,$$

$$0 \in \mathrm{conv}(\{x_1, x_2, \ldots, x_p\}),$$

where 0 stands for the origin of $\mathbf{R}^2$.

Show that $||x_1 + x_2 + \ldots + x_p|| \le p - 1$ and this inequality is precise in a certain sense. Moreover, if

$$||x_1|| = ||x_2|| = \ \ldots \ = ||x_p|| = 1,$$

then $||x_1 + x_2 + \ldots + x_p|| \le p - 2$.

For this purpose, use the fact that there are three vectors x_i, x_j and x_k among the given ones such that $0 \in \text{conv}(\{x_i, x_j, x_k\})$, and try to reduce the general situation to the case where at least two of these three vectors are collinear.

Also, formulate and prove the analogous result for $p \geq m+1$ vectors in the Euclidean space $\mathbf{R}^m$, where $m > 2$.

7*. Let $S = [x_1, x_2, x_3, x_4]$ be a tetrahedron in the space $\mathbf{R}^3$, whose volume is a fixed real number $v > 0$, and let S satisfy the relations

$$||x_1 - x_2|| = a, \qquad ||x_2 - x_3|| = b, \qquad ||x_3 - x_1|| = c,$$

where a, b, and c are some strictly positive real constants.

Show that the minimum value of the surface area of S is attained when the facets

$$[x_1, x_2, x_4], \qquad [x_2, x_3, x_4], \qquad [x_3, x_1, x_4]$$

form pairwise congruent dihedral angles with the facet $[x_1, x_2, x_3]$.

For this purpose, argue as follows. Denote the measures of the above-mentioned dihedral angles by α, β and γ, respectively. Also, denote by s_1, s_2, s_3 the areas of $[x_1, x_2, x_4]$, $[x_2, x_3, x_4]$, $[x_3, x_1, x_4]$, respectively. Obviously, one may write

$$s_1\cos(\alpha) + s_2\cos(\beta) + s_3\cos(\gamma) = s,$$

where s is the area of the facet $[x_1, x_2, x_3]$. Further, since the volume of S is constant $(= v)$, the length of the altitude of S passing through x_4 is also constant denoted by h. It is clear that

$$2s_1 = ah/\sin(\alpha), \qquad 2s_2 = bh/\sin(\beta), \qquad 2s_3 = ch/\sin(\gamma).$$

So, the problem is reduced to finding the minimum value of the function

$$ah/\sin(\alpha) + bh/\sin(\beta) + ch/\sin(\gamma)$$

of three variables α, β, γ, under the condition that

$$ah\cos(\alpha)/\sin(\alpha) + bh\cos(\beta)/\sin(\beta) + ch\cos(\gamma)/\sin(\gamma) = 2s.$$

By using the Lagrange method of auxiliary multipliers, write

$$ah/\sin(\alpha) + bh/\sin(\beta) + ch/\sin(\gamma) -$$

$$t(ah\cos(\alpha)/\sin(\alpha) + bh\cos(\beta)/\sin(\beta) + ch\cos(\gamma)/\sin(\gamma)),$$

where $t \in \mathbf{R}$, and take the partial derivatives with respect to α, β, and γ. After differentiation, come to the equalities

$$t = \cos(\alpha) = \cos(\beta) = \cos(\gamma),$$

which are equivalent to $\alpha = \beta = \gamma$, as required.

Generalize the obtained result to the case of an m-dimensional simplex in the space $\mathbf{R}^m$ $(m \geq 2)$ by using the analogous argument.

8. Prove the first isoperimetric inequality for an m-dimensional simplex S in the space $\mathbf{R}^m$, where $m \geq 2$.

Namely, show that if the total hypersurface area of S is fixed, then the volume $\mathrm{vol}(S)$ of S attains its maximum value if and only if S is a regular simplex.

For this purpose, apply the result of Exercise 7.

9*. Let $S = [z_1, z_2, z_3, z_4]$ be a tetrahedron in the space $\mathbf{R}^3$ whose vertices are z_1, z_2, z_3, z_4, and suppose that the facet $[z_1, z_2, z_3]$ is not an isosceles triangle; more precisely, assume that $||z_1 - z_3|| \neq ||z_2 - z_3||$.

Let Γ be the plane in $\mathbf{R}^3$ passing through the mid-point of the line segment $[z_1, z_2]$ and perpendicular to $[z_1, z_2]$, and let z_3' and z_4' denote, respectively, the orthogonal projections of z_3 and z_4 on Γ.

Check that the tetrahedron $[z_1, z_2, z_3', z_4']$ has the same volume as S, but the perimeter of S is strictly greater than the perimeter of $[z_1, z_2, z_3', z_4']$.

Generalize the obtained result to the case of an m-dimensional simplex

$$S = [z_1, z_2, ..., z_m, z_{m+1}] \subset \mathbf{R}^m \qquad (m \geq 2)$$

by using the analogous argument.

Remark 4. The described replacing the tetrahedron $[z_1, z_2, z_3, z_4]$ by the tetrahedron $[z_1, z_2, z_3', z_4']$ is a special case of the so-called symmetrization operation of convex bodies, which is very useful in many geometric topics (see, e.g., [20], [24], [123], [130], [249], [360]).

10. Prove the second isoperimetric inequality for an m-dimensional simplex S in the space $\mathbf{R}^m$, where $m \geq 2$.

Namely, show that if the perimeter of S is fixed, then the volume of S attains its maximum value if and only if S is a regular simplex.

For this purpose, apply the result of Exercise 9.

11. Let $m \geq 2$ and let K be a convex polyhedral angle in the space $\mathbf{R}^m$, whose extreme boundary rays are $l_1, l_2, \ldots, l_m$. Fix an internal point z of K. Further, let Γ be an affine hyperplane in $\mathbf{R}^m$ passing through z and having the property that the volume of the non-degenerate simplex bounded by Γ and all facets of K is minimal.

Demonstrate that the point z coincides with the centroid (i.e., geometric barycenter) of the $(m-1)$-dimensional simplex in Γ, whose vertices are produced by the intersections of $l_1, l_2, ..., l_m$ with Γ.

For this purpose, consider first the case $m = 2$ and then use induction on m.

12*. Let $m \geq 2$ and let K be a convex polyhedral angle in the space $\mathbf{R}^m$, whose extreme boundary rays are $l_1, l_2, \ldots, l_m$. Suppose also that B is a smooth strictly convex body inscribed in K. Let Γ be an affine hyperplane in $\mathbf{R}^m$ tangent to B and having the property that the volume of the simplex bounded by Γ and all facets of K is minimal (here the vertex of K and the body B lie in one half-space determined by Γ).

Verify that the unique common point of Γ and B coincides with the barycenter (centroid) of the $(m-1)$-dimensional simplex in Γ, whose vertices are produced by the intersections of $l_1, l_2, \ldots, l_m$ with Γ.

For this purpose, use the result of Exercise 11 and induction on m.

Conclude that if B is inscribed in a simplex S of minimal volume, then all common points of B and S are, respectively, the barycenters (centroids) of the facets of S.

13. Let r_1, r_2, r_3 be strictly positive real numbers and let $[z_1, z_2, z_3]$ be a triangle in the plane $\mathbf{R}^2$ such that, for some internal point z of this triangle, the following conditions are fulfilled:

(a) the distance between z and $l(z_2, z_3)$ is r_1;

(b) the distance between z and $l(z_3, z_1)$ is r_2;

(c) the distance between z and $l(z_1, z_2)$ is r_3.

Show that if the area of $[z_1, z_2, z_3]$ takes minimum value, then z coincides with the circumcenter of $[z_1, z_2, z_3]$.

14. Let r_1, r_2 and r_3 be three strictly positive real numbers and let $[A, B, C]$ be a triangle in the plane $\mathbf{R}^2$, which satisfies the relations

$$\mathrm{dist}(O, l(B, C)) = r_1, \quad \mathrm{dist}(O, l(C, A)) = r_2, \quad \mathrm{dist}(O, l(A, B)) = r_3,$$

where O is the circumcenter of $[A, B, C]$.

Verify that the radius R of the circumcircle of $[A, B, C]$ is a strictly positive root of the following cubic equation:

$$x^3 - (r_1^2 + r_2^2 + r_3^2)x - 2r_1r_2r_3 = 0.$$

In particular, consider the case when $r_1 = 1$, $r_2 = 2$, and $r_3 = 3$. Then

$$R^3 - 14R - 12 = 0.$$

Check that the polynomial $x^3 - 14x - 12$ is irreducible over the field $\mathbf{Q}$ of all rational numbers and conclude that in this case the triangle $[A, B, C]$ cannot be constructed by using the straight edge and compass.

15. Let r be a strictly positive real number. Consider the family of all m-dimensional simplexes in the space $\mathbf{R}^m$ for which the radius of the inscribed sphere is equal to r.

Show that among all such simplexes the least volume has a regular simplex (and only it). This means that the inequality

$$\mathrm{vol}(S) \geq \frac{(m+1)^{(m+1)/2} m^{m/2} r^m}{m!}$$

is valid for every simplex S from the above-mentioned family (and the equality is attained only for a regular simplex).

For this purpose, first infer from the result of Exercise 12 that if the volume of a simplex

$$T = [z_1, z_2, ..., z_{m+1}]$$

from the family takes minimum value, then the conjunction of the following two assertions is true:

(a) for any natural number $k \in [1, m+1]$, the perpendicular dropped from the incenter of T to the facet $[z_1, ..., z_{k-1}, z_{k+1}, ..., z_{m+1}]$ passes through the centroid of this facet;

(b) all altitudes of T have a common point (i.e., T is an orthocentric simplex).

Then argue by induction on m and establish that the same T is a regular simplex.

16. Let $m \geq 2$ be a natural number, let $R_1, R_2, \ldots, R_{m+1}$ be fixed strictly positive real numbers, and let $S = [z_1, z_2, ..., z_{m+1}]$ be an m-dimensional simplex in $\mathbf{R}^m$ with vertices $z_1, z_2, \ldots, z_{m+1}$ such that

$$||z - z_k|| = R_k \qquad (1 \leq k \leq m+1)$$

for some internal point z of S.

Verify that the volume of S attains its maximum value if and only if z is a common point of all altitudes of S (in other words, $\mathrm{vol}(S)$ attains its maximum value if and only if S is an orthocentric simplex).

17. In this exercise we assume that the space $\mathbf{R}^m$ is canonically embedded into the space $\mathbf{R}^{m+1}$, i.e., $\mathbf{R}^{m+1} = \mathbf{R}^m \times \mathbf{R}$.

Let S be an m-dimensional orthocentric simplex in $\mathbf{R}^m$ with its orthocenter O and let l denote the straight line in $\mathbf{R}^{m+1}$ passing through O and perpendicular to $\mathbf{R}^m$.

Show that, for any point $z \in l \backslash \{O\}$, the $(m+1)$-dimensional polyhedron

$$S' = \mathrm{conv}(S \cup \{z\})$$

is an orthocentric simplex in the space $\mathbf{R}^{m+1}$.

For this purpose, use induction on m.

18. Check that:

(a) all facets of an m-dimensional orthocentric simplex in the Euclidean space $\mathbf{R}^m$ ($m \geq 3$) are orthocentric simplexes in the corresponding affine hyperplanes of $\mathbf{R}^m$;

(b) if the orthocenter of an orthocentric simplex $S \subset \mathbf{R}^m$ lies in the interior of S, then the set of all vertices of S forms a strong at-subset of $\mathbf{R}^m$ (for the definition of strong at-sets in a real pre-Hilbert space, see Chapter 3);

(c) if a strong at-set $\{z_1, z_2, ..., z_{m+1}\} \subset \mathbf{R}^m$ does not lie in an affine hyperplane of $\mathbf{R}^m$ and the m-dimensional simplex $S = [z_1, z_2, ..., z_{m+1}]$ is orthocentric, then the orthocenter of S belongs to the interior of S.

In addition, for $m > 3$ prove that an m-dimensional simplex in $\mathbf{R}^m$ is orthocentric if and only if all its facets are orthocentric $(m-1)$-dimensional simplexes (in fact, if $m > 3$ and an m-simplex S has at least three orthocentric facets, then S itself is orthocentric).

19*. Let a_1, a_2, . . . , a_{m+1} be any strictly positive real numbers and let $\{e_1, e_2, ..., e_{m+1}\}$ denote the canonical orthonormal basis of the space $\mathbf{R}^{m+1}$, where $m \geq 2$. Consider the vectors $a_1^{1/2}e_1$, $a_2^{1/2}e_2$, . . . , $a_{m+1}^{1/2}e_{m+1}$.

Verify that:

(a) these vectors form a strong at-set in the affine hyperplane of $\mathbf{R}^{m+1}$ which contains all of them;

(b) these vectors are the vertices of a certain m-dimensional orthocentric simplex S;

(c) the orthocenter O of S is determined in $\mathbf{R}^{m+1}$ by the formula

$$O = (a/a_1^{1/2},\ a/a_2^{1/2},\ \ldots,\ a/a_{m+1}^{1/2}),$$

where $1/a = 1/a_1 + 1/a_2 + ... + 1/a_{m+1}$;

(d) for each natural number $i \in [1, m+1]$, the equality

$$||O - a_i^{1/2}e_i||^2 = a_i - a$$

holds true.

Conversely, demonstrate that every m-dimensional orthocentric simplex in $\mathbf{R}^m$, whose vertices form a strong at-set, is isometric to some simplex S described above (for appropriate values a_1, a_2, . . . , a_{m+1}).

To establish the latter fact, keep in mind the two preceding exercises and use induction on m.

20. Let $z_1, z_2, \ldots, z_k$ be some fixed points in the space $\mathbf{R}^m$, where $m \geq 2$, and let z_0 be a point in the same space such that

$$\Sigma\{||z_i - z_0|| : 1 \leq i \leq k\} \leq \Sigma\{||z_i - z|| : 1 \leq i \leq k\}$$

for every point $z \in \mathbf{R}^m$.

Supposing that z_0 differs from all the points $z_1, z_2, \ldots, z_k$, show that

$$\Sigma\{(z_i - z_0)/||z_i - z_0|| : 1 \leq i \leq k\} = 0,$$

where 0 denotes the origin of $\mathbf{R}^m$.

For this purpose, take the partial derivatives of the expression

$$\phi(z) = \Sigma\{||z_i - z|| : 1 \leq i \leq k\},$$

where z ranges over the whole of $\mathbf{R}^m$.

21*. Let $t > 0$ be a real constant, let $S = [z_1, z_2, ..., z_{m+1}]$ be an m-dimensional simplex in the space $\mathbf{R}^m$, where $m \geq 2$, and let

$$||z_0 - z_1|| + ||z_0 - z_2|| + ... + ||z_0 - z_{m+1}|| = t \qquad (*)$$

for some internal point z_0 of S.

Prove that the volume of S attains its maximum if and only if the following two conditions are satisfied:

(a) S is an orthocentric simplex and its orthocenter coincides with z_0;

(b) for all points z which lie in the interior of S, one has

$$||z - z_1|| + ||z - z_2|| + ... + ||z - z_{m+1}|| \geq t.$$

Finally, applying the results of Exercises 19 and 20, show that vol(S) is maximal, under the same condition (*), if and only if S is a regular simplex whose incenter coincides with z_0. In particular, the inequality

$$\text{vol}(S) \leq \frac{(m+1)^{(m+1)/2}(t/(m+1))^m}{m^{m/2}m!}$$

holds true for every simplex S satisfying condition (*).

22*. Let $S = [z_1, z_2, ..., z_{m+1}]$ be an m-dimensional simplex in the space $\mathbf{R}^m$, where $m \geq 2$, and let z_0 be an internal point of S. Denote

$$R_1 = ||z_0 - z_1||, \quad R_2 = ||z_0 - z_2||, \quad \ldots, \quad R_{m+1} = ||z_0 - z_{m+1}||,$$

$$r \;=\; \text{the radius of the iscribed sphere of } S.$$

Demonstrate that the relation

$$R_1 + R_2 + ... + R_{m+1} \geq (m+1)mr$$

is valid, and the equality is realizable in this relation if and only if S is a regular simplex and z_0 coincides with the incenter of S.

For this purpose, combine the results of Exercise 15 and Exercise 21.

23. Let $k \geq 2$ be a natural number and let $a_1, a_2, \ldots, a_k$ be strictly positive real numbers such that

$$\sum\{1/(1+a_i) : 1 \leq i \leq k\} = 1.$$

Show the validity of the inequality

$$\prod\{a_i : 1 \leq i \leq k\} \geq (k-1)^k.$$

Also, supposing that $k \geq 3$, verify that the equality is attained here if and only if $a_i = k-1$ for all natural numbers $i \in [1, k]$.

Argue as follows. First, observe that if $k = 2$, then $a_1 a_2 = 1$ in view of

$$1/(1+a_1) + 1/(1+a_2) = 1.$$

Further, assuming that $k \geq 3$ and denoting $p = \prod\{a_i : 1 \leq i \leq k\}$, check that if, for some natural index $i \in [1, k]$, the term a_i tends either to zero or to infinity, then p tends to infinity. Afterwards, take into account the relation

$$c^2 < ab,$$

where $a > 0$, $b > 0$, $c > 0$ and

$$c \neq a, \quad c \neq b, \quad 1/(1+a) + 1/(1+b) = 2/(1+c).$$

Keeping in mind all these circumstances, obtain the required result.

24*. Let $S = [z_1, z_2, ..., z_{m+1}]$ be an m-dimensional simplex in the space $\mathbf{R}^m$, where $m \geq 2$, and let z be an internal point of S. For each natural index $i \in [1, k]$, let z'_i be the common point of the line $l(z_i, z)$ and of the facet of S opposite z_i. Denote

$$R_i = ||z_i - z||, \quad d_i = ||z'_i - z|| \quad (1 \leq i \leq m+1).$$

Prove that the following two inequalities are fulfilled:

(a) $R_1 R_2 ... R_{m+1} \geq m^{m+1} d_1 d_2 ... d_{m+1}$;

(b) $(R_1 + R_2 + ... + R_{m+1})(1/d_1 + 1/d_2 + ... + 1/d_{m+1}) \geq (m+1)^2 m.$
Moreover, show that the equality in (a) holds true if and only if z coincides with the centroid of S, while the equality in (b) holds true if and only if z coincides with the centroid of S and all medians of S are pairwise congruent.

For this purpose, introduce the notation

$$a_i = R_i/d_i \qquad (1 \leq i \leq m+1)$$

and apply to the real numbers $a_1, a_2, \ldots, a_{m+1}$ the inequality of Exercise 23.

Remark 5. It is clear that the relation (a) of Exercise 24 generalizes the inequality of Exercise 2. In other words, the inequality of Exercise 2 is immediately obtained from (a) by putting $m = 2$.

25. Demonstrate that $m+1$ is the maximum number of those rays in the space $\mathbf{R}^m$ $(m \geq 1)$ whose common endpoint coincides with the origin of $\mathbf{R}^m$ and which have the property that the angle produced by any two distinct rays from them is strictly greater than $\pi/2$.

For this purpose, use induction on m.

Also, prove that $2m$ is the maximum number of those rays in the space $\mathbf{R}^m$ $(m \geq 1)$ whose common endpoint coincides with the origin of $\mathbf{R}^m$ and which have the property that the angle produced by any two distinct rays from them is greater than or equal to $\pi/2$.

26*. Preserving for a convex polygon $P \subset \mathbf{R}^2$ and for its internal point z the notation of Theorem 1 and of its proof, show that

$$(\cos(\pi/n))^n \prod\{||z - z_k|| : 1 \leq k \leq n\} \geq \prod\{d'_k : 1 \leq k \leq n\}.$$

Also, specify the case when this inequality becomes the equality.

Argue as follows. Start with the inequalities

$$d'_k \leq (||z - z_k|| \cdot ||z - z_{k+1}||)^{1/2} \cos(\alpha_k/2),$$

where $1 \leq k \leq n$ and it is again stipulated $z_{n+1} = z_1$. So, one can write

$$(d'_k)^2 \leq ||z - z_k|| \cdot ||z - z_{k+1}|| (\cos(\alpha_k/2))^2,$$

$$2\ln(d'_k) \leq \ln(||z - z_k||) + \ln(||z - z_{k+1}||) + 2\ln(\cos(\alpha_k/2)).$$

Keeping in mind the relations

$$0 < \alpha_k/2 < \pi/2 \qquad (1 \leq k \leq n),$$

$$\Sigma\{\alpha_k/2 : 1 \leq k \leq n\} = \pi,$$

the upper convexity of the function

$$x \to \ln(\cos(x)) \qquad (0 \leq x < \pi/2),$$

and Jensen's inequality (i.e., Theorem 3 of Appendix 2), infer that

$$\Sigma\{\ln(\cos(\alpha_k/2)) : 1 \leq k \leq n\} \leq n\ln(\cos(\pi/n)),$$

and obtain the desired result by summing the above inequalities.

Also, deduce that the equality takes place here if and only if the given polygon P is regular and z coincides with the incenter of P.

Remark 6. The inequality of Exercise 26 was first proved by L. Fejes Tóth. Obviously, the result presented in Exercise 26 substantially generalizes the result given in Exercise 2 for the special case $n = 3$.

APPENDIX 5

Some facts from graph theory

Among various mathematical structures, the structure of a graph is closely related to geometry and elementary topology. Indeed, the graphs regarded as a certain kind of diagrams serve for visualization of diverse objects (patterns, images, designs, etc.), which can be met in purely theoretical studies and in many practical situations as well.

The area of applications of graph theory is enormously vast. Different types of graphs occur in geometry itself, topology, algebra, mathematical logic and set theory, mathematical analysis, optimization theory, general system theory, and so on.

Naturally, there are a lot of text-books and monographs devoted to graph theory and to its theoretical or applied aspects. We would like to especially refer to the following two classical books: [135] and [275]. However, our preference is very individual and the reader may study the subject by using many other well-written manuals.

The present manuscript is oriented to a concrete circle of geometric topics and, for our purposes, we need some simplest auxiliary notions and facts from graph theory. So, it seems reasonable to remind them here.

As a rule, the graphs considered in the sequel will be assumed to be finite, non-directed (undirected, non-oriented) and without loops. But, sometimes, we will also be dealing with infinite (even uncountable) graphs and with directed (oriented) graphs. Actually, the latter ones can be identified with ordinary binary relations.

Let us begin with recalling basic ideas and concepts of this beautiful and helpful theory (cf. the terminology adopted in [135] and [275]).

A graph is any pair of the form (V, E) where V is a ground set and E is a subset of the family $\mathcal{F}_2(V)$ of all two-element subsets of V.

Notice that quite often, especially in set-theoretical works and publications, the family $\mathcal{F}_2(V)$ is denoted by $[V]^2$.

DOI: 10.1201/9781003458708-A5

The set V is called the set of vertices (nodes) of (V, E) and the set E is called the set of edges (arcs) of (V, E).

The cardinal number $\mathrm{card}(V)$ is called the cardinality of (V, E). As we have already mentioned, in our further considerations we will primarily deal with finite graphs, i.e., in most cases we will assume that $\mathrm{card}(V) < \mathrm{card}(\mathbf{N})$, where

$$\mathbf{N} = \{0,\ 1,\ 2,\ \dots,\ n,\ \dots\}$$

denotes the set of all natural numbers (the symbol ω is also used for denoting the same set).

Let $v \in V$ and $e \in E$. As usual, we say that v and e are incident if $v \in e$.

Two vertices $v \in V$ and $u \in V$ are adjacent (or neighboring) if $\{v, u\} \in E$.

Similarly, two distinct edges from E are adjacent (or neighboring) if they are incident to a certain vertex of (V, E).

For any vertex $v \in V$, the set

$$U(v) = \{u \in V : \{u, v\} \in E\}$$

is called the neighborhood of v. The cardinal number $\mathrm{card}(U(v))$ is called the degree (sometimes, valence) of v in (V, E) and is denoted by $\deg(v)$.

A graph (V, E) is called locally finite if the neighborhoods of all its vertices are finite.

A graph (V, E) is trivial (or null) if $E = \emptyset$.

A graph (V, E) is full (or complete) if E coincides with the family $\mathcal{F}_2(V)$ of all two-element subsets of V.

The notion of an isomorphism between two graphs can be introduced in the standard manner, in accordance with the general concept of an isomorphism between two mathematical structures of a given type or of a given signature (cf. [44], [59], [151], [241], [296]). In our special case, the definition of an isomorphism is as follows.

Two graphs (V_1, E_1) and (V_2, E_2) are isomorphic if there exists a bijection

$$f : V_1 \to V_2$$

such that

$$\{v, u\} \in E_1 \Leftrightarrow \{f(v), f(u)\} \in E_2$$

for any two vertices $v \in V_1$ and $u \in V_1$.

Now, let us recall several important inner (intrinsic) objects and corresponding numerical characteristics in graph theory.

Let (V, E) be a graph. A graph (V', E') is called a subgraph of (V, E) if

$$V' = U, \quad E' = \mathcal{F}_2(U) \cap E,$$

where U is some subset of V.

If (V, E) is a graph, then the pair (V', E'), where

$$V' = V, \quad E' = \mathcal{F}_2(V) \setminus E,$$

is called the complementary graph of (V, E).

A chain (or path) in a graph (V, E) is any finite sequence $(v_1, v_2, ..., v_k)$ of its vertices, where $k \geq 2$ and $\{v_i, v_{i+1}\} \in E$ for all natural indices $i \in [1, k-1]$.

The vertices v_1 and v_k are usually called the endpoints of this chain.

A chain $(v_1, v_2, ..., v_k)$ in (V, E) is simple if $(v_1, v_2, ..., v_k)$ is injective. In this case, the number $k - 1$ is called the length of the chain.

Sometimes, it is convenient to introduce chains of length zero; they may be identified with the singletons of the form $\{v\}$, where $v \in V$, or simply with the vertices of (V, E).

A path $(v_1, v_2, ..., v_k)$ in (V, E) is called a cyclic path (or, briefly, a cycle) if $k \geq 4$ and $v_k = v_1$.

A cycle $(v_1, v_2, ..., v_k)$ in (V, E) is called simple (or circuit) if the relation $i \neq j \ \& \ v_i = v_j$ implies the disjunction

$$(i = 1 \ \& \ j = k) \ \vee \ (i = k \ \& \ j = 1).$$

In this case, $k - 1$ is called the length of the simple cycle $(v_1, v_2, ..., v_k)$. Sometimes it is convenient to treat the singletons of the form $\{v\}$, where $v \in V$, as simple cycles of length zero.

A graph (V, E) is called acyclic if there are no simple cycles in it.

A graph (V, E) is connected if any two distinct vertices u and w of (V, E) are linked by some chain, i.e., there exists a chain $(v_1, v_2, ..., v_k)$ in (V, E) such that $v_1 = u$ and $v_k = w$.

Obviously, it can be assumed in this definition (without loss of generality) that $(v_1, v_2, ..., v_k)$ is a simple chain.

It is easy to see that a nonempty graph (V, E) is not connected if and only if the set V admits a partition $\{V_1, V_2\}$ possessing the following property:

no edge of (V, E) is incident to a vertex of V_1 and to a vertex of V_2 simultaneously.

An important subclass of the class of all nonempty connected graphs is constituted by the trees.

A nonempty graph (V, E) is called a tree if it is connected and acyclic.

Every tree can be represented in a very convenient form which reflects the general and fruitful idea of hierarchy between various objects.

Let (V, E) be a tree and let v_0 be any of its vertices. We define by induction (more exactly, by recursion) the sequence

$$V_0, \ V_1, \ . \ . \ . \ , \ V_k, \ . \ . \ .$$

of subsets of V.

First of all, we put $V_0 = \{v_0\}$ and call v_0 the root of the given tree (V, E).

Suppose now that, for a natural number k, the sets $V_0, \ V_1, \ . \ . \ . \ , \ V_k$ have already been determined. Then we define

$$V_{k+1} = \{v \in V \setminus \cup\{V_i : 0 \leq i \leq k\} : (\exists w \in V_k)(\{v, w\} \in E)\}.$$

It directly follows from this definition that all the sets V_k ($k \in \mathbf{N}$) are pairwise disjoint. They are usually called the levels of the tree (V, E) whose root is v_0.

More precisely, for each natural number k, we say that the set V_k is the k-th level of the tree (V, E) whose root is v_0.

It is clear that only two cases are possible.

Case 1. There exists an empty level of (V, E).

In this case, only finitely many levels of (V, E) are nonempty, and the height of (V, E) is defined as the least natural number k for which $V_{k+1} = \emptyset$.

Case 2. All levels of (V, E) are nonempty.

In this case, the height of (V, E) is equal (by definition) to the least infinite ordinal number ω ($= \text{card}(\mathbf{N})$).

Since any tree (V, E) is connected, we readily deduce that

$$V = \cup\{V_k : k \in \mathbf{N}\},$$

i.e., each vertex v of (V, E) belongs to a certain level V_k, where $k = k(v)$. This level is uniquely determined by v, and $k(v)$ coincides with the length of the shortest chain connecting v with the root v_0. Notice, by the way, that no two distinct vertices from V_k are adjacent.

In various set-theoretical reasonings and constructions, the notion of a tree needs a substantial extension in order to involve those tree-like structures which have heights strictly greater than ω (in this connection, see, for instance, [151], [158], [237]). In most cases such structures are uncountable.

Let us stress once more that in our further considerations we primarily will be dealing with finite and countably infinite trees.

It turns out that every nonempty connected graph (V, E) contains a tree which is maximal with respect to the inclusion relation and, in some sense, serves as a skeleton of (V, E). The proof of this result is heavily based on the Axiom of Choice (**AC**) or on the logically equivalent maximality principle widely known as Kuratowski–Zorn lemma (see [44], [59], [82], [150], [241], [259], [275]).

Theorem 1. *Let (V, E) be an arbitrary nonempty connected graph. Then there exists a tree of the form (V, E'), where $E' \subset E$.*

Proof. Consider the family $\mathcal{T}$ of all those graphs T which satisfy the following two conditions:

(1) T is a subgraph of (V, E);

(2) T is a tree.

Equip this family $\mathcal{T}$ with the canonical partial ordering, i.e., with the standard inclusion relation.

Notice that if $\{T_i : i \in I\}$ is an arbitrary linearly ordered subfamily of $\mathcal{T}$, then the union $\cup\{T_i : i \in I\}$ of this subfamily is a tree contained in (V, E), so also belongs to $\mathcal{T}$.

By virtue of the Kuratowski–Zorn lemma, there exists a maximal tree

$$T' = (V', E') \in \mathcal{T}.$$

It remains to check that $V' = V$. Suppose otherwise, i.e., $V \setminus V' \neq \emptyset$. Since (V, E) is a connected graph, there exists a simple chain in (V, E), one endpoint of which belongs to V' and the other endpoint belongs to $V \setminus V'$. We may assume, without loss of generality, that this chain has minimal length. Then it can easily be seen that the above-mentioned chain is trivial, i.e., it consists only of one edge $e = \{v', v\} \in E$, where $v' \in V'$ and $v \in V \setminus V'$. So we may consider the graph

$$(V'', E'') = (V' \cup \{v\}, E' \cup \{e\}).$$

Now, a straightforward verification shows that (V'', E'') is a tree in (V, E) and (V'', E'') properly contains (V', E'), which contradicts the maximality of (V', E').

The contradiction obtained completes the proof of Theorem 1. □

Of course, a maximal tree in a given nonempty connected graph (V, E) is not uniquely determined in general. In other words, a nonempty connected graph (V, E) may have various skeleton-trees (they are called the spanning trees for (V, E)).

Remark 1. Actually, Theorem 1 turns out to be equivalent to the Axiom of Choice (in this connection, see Exercise 2). Notice also that this theorem has some interesting and unexpected applications. For instance, it is essentially exploited in a proof of the deep and important fact of the general theory of groups, which states that any subgroup of a free group is also free (see, e.g., [268]).

The following simple proposition is a very particular case of the famous Euler formula which will be discussed later in this Appendix (recall that the much more general Euler–Poincaré formula was considered and proved in Appendix 3).

Theorem 2. *Let* (V, E) *be an arbitrary finite tree.*

Then the equality $\mathrm{card}(V) = \mathrm{card}(E) + 1$ *holds true.*

Proof. We use the method of induction on $\mathrm{card}(V)$.

If $\mathrm{card}(V) = 1$, then there is nothing to prove.

Suppose now that $\mathrm{card}(V) \geq 2$ and that the assertion of this theorem has already been established for all those trees whose cardinalities are strictly less than $\mathrm{card}(V)$.

Let us show that there exists a vertex $v \in V$ incident to exactly one edge from E (such a vertex is called a leaf of (V, E)). Indeed, supposing otherwise and keeping in mind the finiteness of V, we readily get a nontrivial simple cycle in (V, E), which contradicts the definition of a tree. So we can pick a leaf v in (V, E). Consider the reduced graph

$$(V', E') = (V \setminus \{v\}, E \cap \mathcal{F}_2(V \setminus \{v\})).$$

Clearly, V' is not empty, (V', E') is also a tree, and

$$\text{card}(E) = \text{card}(E') + 1.$$

Applying the inductive assumption to (V', E'), we may write

$$\text{card}(V') = \text{card}(E') + 1,$$

whence it immediately follows that

$$\text{card}(V) = \text{card}(V') + 1 = (\text{card}(E') + 1) + 1 = \text{card}(E) + 1.$$

This proves (by induction) the assertion of Theorem 2. □

Theorems 1 and 2 directly imply that a nonempty finite connected graph (V, E) is a tree if and only if $\text{card}(V) = \text{card}(E) + 1$.

The next statement is concerned with a situation where a finite graph *a priori* cannot be connected.

Theorem 3. *If a finite graph* (V, E) *is such that*

$$\text{card}(V) = n, \qquad \text{card}(E) < n - 1,$$

then this graph is not connected.

Proof. Suppose to the contrary that (V, E) is connected. Let (V, E') be a spanning tree for (V, E). According to Theorem 2, we may write

$$\text{card}(E) \geq \text{card}(E') = \text{card}(V) - 1 = n - 1,$$

which contradicts the given inequality $\text{card}(E) < n - 1$.

Theorem 3 has thus been proved. □

Recall that a graph (V, E) is realized in the plane $\mathbf{R}^2$ if V is a subset of $\mathbf{R}^2$ and E is a set of simple arcs (i.e., homeomorphic images of $[0, 1]$) whose endpoints belong to V. It is also required that any two distinct arcs $e \in E$ and $e' \in E$ either have no common points, or their intersection $e \cap e'$ is a singleton whose unique element belongs to V.

More generally, a graph (V, E) is called planar if there exists an isomorphic copy of (V, E) which is realized in $\mathbf{R}^2$. It can readily be checked that (V, E) is planar if and only if (V, E) admits a realization in the two-dimensional unit sphere

$$\mathbf{S}_2 = \{(x, y, z) \in \mathbf{R}^3 : x^2 + y^2 + z^2 = 1\}$$

of the Euclidean space $\mathbf{R}^3$.

Remark 2. The definition of a graph realized in $\mathbf{R}^2$ requires that all edges of the graph should be simple arcs. However, such arcs may be very "wild" from the geometrical or topological viewpoint. Indeed, it is well known that:

(1) there exists a continuous nowhere differentiable function

$$f : [0, 1] \to \mathbf{R},$$

so its graph $\mathrm{Gr}(f) \subset \mathbf{R}^2$ is a simple arc in $\mathbf{R}^2$ with very bad analytical properties;

(2) there exists a simple arc $L \subset \mathbf{R}^2$ such that $\lambda_2(L) > 0$, where λ_2 denotes the two-dimensional Lebesgue measure on $\mathbf{R}^2$; so this L is not a thin subset of $\mathbf{R}^2$.

For more details about such somewhat pathological situations, we refer the reader to any advanced course of Real Analysis (see also [108]). To avoid pathologies of the above-mentioned types (1) and (2), it is reasonable to assume that the edges of a graph realized in $\mathbf{R}^2$ are polygonal curves consisting of finitely many (or countably many) line segments. Moreover, if a given finite graph is planar, then we will suppose that all the edges of its realization (V, E) in $\mathbf{R}^2$ are polygonal curves with finitely many sides. In this situation it becomes quite evident that (V, E) produces a decomposition of $\mathbf{R}^2$ into finitely many domains (regions). Actually, the fact just mentioned can rigorously be proved with the aid of the Jordan theorem for simple polygons lying in $\mathbf{R}^2$. Notice that, in the case of simple polygons in $\mathbf{R}^2$, the Jordan theorem admits a completely elementary proof (see, for instance, [4], [33]). The case of arbitrary curves lying in $\mathbf{R}^2$ and homeomorphic to the unit circle $\mathbf{S}_1 \subset \mathbf{R}^2$ needs a more delicate argument (cf. [5], [193], [240], [268]).

Planar graphs permanently appear in various questions of combinatorial geometry and topology. Let us remind the celebrated Euler formula for nonempty finite connected planar graphs. This formula is one of the most beautiful results in all mathematics and has numerous applications (some of them are discussed in the present book).

Theorem 4. *Let (V, E) be a nonempty finite connected graph realized in $\mathbf{R}^2$ (or in $\mathbf{S}_2$), let $v = \mathrm{card}(V)$, and let $e = \mathrm{card}(E)$.*

Then the formula of Euler

$$v - e + f = 2$$

is valid, where f denotes the total number of regions (domains) into which $\mathbf{R}^2$ (or $\mathbf{S}_2$) is dissected by the edges of this graph.

Proof. Suppose first that (V, E) is a finite tree. In this case, we obviously have $f = 1$. According to Theorem 2, we may write $v = e + 1$. Therefore,

$$v + f = (e + 1) + 1 = e + 2,$$

$$v - e + f = 2,$$

which shows that Euler's formula holds true for all finite trees.

Suppose now that our graph (V, E) is not a tree. Then (V, E) contains a nontrivial simple cycle. We may assume that the length of this cycle is minimal. Removing from the above-mentioned cycle only one edge (without touching upon its vertices), we simultaneously reduce some two neighboring regions to exactly one region, i.e., the expression $v - e + f$ does not change its value after making the indicated elementary operation. At the same time, the obtained reduced graph again will be connected (which is not difficult to check). Repeating this process finitely many times, we necessarily come to a tree. But, as has already been shown, in the case of a nonempty tree the expression $v - e + f$ is equal to 2. In view of the invariance of this expression under the described edge-removing operations, we conclude the validity of Euler's formula for all nonempty finite connected planar graphs.

Theorem 4 has thus been proved (in this connection, see also Appendix 3 where a generalization of Euler's formula for multi-dimensional convex polyhedra in $\mathbf{R}^m$ is presented). □

Remark 3. A straightforward corollary of the above theorem states that the number f of all those regions (domains) which are produced by an arbitrary realization of a nonempty finite connected graph (V, E) in $\mathbf{R}^2$ or in $\mathbf{S}_2$ does not depend on this realization. In other words, the value f is an invariant for all possible realizations of (V, E) in $\mathbf{R}^2$ or in $\mathbf{S}_2$.

As another corollary of Theorem 4, we have the following helpful statement.

Theorem 5. *Let (V, E) be a finite planar graph with $\mathrm{card}(V) \geq 3$. Then the inequality*

$$e \leq 3(v - 2)$$

holds true for the number $v = \mathrm{card}(V)$ of vertices of this graph and for the number $e = \mathrm{card}(E)$ of its edges.

Proof. We may assume, without loss of generality, that the given graph (V, E) is realized in the sphere $\mathbf{S}^2$. It is not difficult to see that, by adding some new edges, we can obtain a finite graph (V, E') also realized in $\mathbf{S}_2$ and such that:

(a) $E \subset E'$;

(b) the graph (V, E') is connected;

(c) all regions in $\mathbf{S}^2$ determined by (V, E') are triangles in the generalized sense (more precisely, the boundary of every region is a simple cycle of length 3).

Now, let us make some easy calculations for the new graph (V, E'). Obviously, we may write $2e' = 3f'$, where $e' = \mathrm{card}(E')$ and f' is the number of regions in $\mathbf{S}^2$ produced by (V, E'). Further, by virtue of Euler's formula, we have $v + f' = e' + 2$. Consequently, we readily obtain

$$v + 2e'/3 = e' + 2, \quad v - 2 = e'/3, \quad 3(v - 2) = e' \geq e,$$

which completes the proof of Theorem 5. □

We would like to mention (without any proof) one application of the Euler formula in computational geometry closely related to discrete and combinatorial geometry.

Suppose that a finite set Z with $\mathrm{card}(Z) = n$ is given in the plane $\mathbf{R}^2$ and it is required to find a maximally simple construction of the convex hull of Z (denoted, as usual, by $\mathrm{conv}(Z)$). Clearly, any construction of $\mathrm{conv}(Z)$ consists of a finite number of elementary geometric steps which do not depend on the basic parameter n.

Using Euler's formula, it can be shown that the algorithmic complexity of an optimal construction of $\mathrm{conv}(Z)$ does not exceed $O(n\ln(n))$.

Moreover, an appropriate application of Euler's formula also yields that the same estimate remains valid for the problem of finding $\mathrm{conv}(Z)$, where Z is an n-element subset of the space $\mathbf{R}^3$.

In this context, it should be underlined that, for a dimension $m > 3$ and for finite subsets Z of $\mathbf{R}^m$, the situation becomes much more complicated.

Further details, concerning various constructions of convex hulls of finite point sets, can be found e.g. in the works [3], [110], [112], [115], [149], [284], [285], [345].

EXERCISES

1. Let (V, E) be a finite graph with $\mathrm{card}(V) = k$ and let $(n_1, n_2, ..., n_k)$ denote the injective sequence of all degrees of the vertices of (V, E).

Verify that

$$n_1 + n_2 + ... + n_k = 2\mathrm{card}(E).$$

Conclude from this equality that the number of odd degrees is always even.

Sometimes, this equality is called (in a jocose manner) the handshaking lemma.

Further, suppose that $\mathrm{card}(V) \geq 2$.

Demonstrate that there are at least two distinct vertices of (V, E) which have equal degrees.

Also, check that the last assertion fails to be true even for infinite locally finite trees.

Remark 4. There are some necessary and sufficient conditions obtained by Erdös and Gallai, under which a finite sequence $(n_1, n_2, ..., n_k)$ of natural numbers turns out to be the sequence of all degrees of a graph with exactly k vertices (see, for instance, [57]).

2*. For a nonempty finite connected graph (V, E), prove Theorem 1 of this chapter without appealing to the Axiom of Choice, i.e., within **ZF** set theory.

For this purpose, use induction on card(E) (actually, formulate a certain algorithm of constructing a spanning tree of (V, E)).

Further, demonstrate that the following two assertions are equivalent within the same **ZF** theory:

(a) the Axiom of Choice (**AC**);

(b) every nonempty connected graph contains at least one spanning tree.

The implication (a) $\Rightarrow$ (b) has already been established (see the proof of Theorem 1).

To check the validity of the converse implication (b) $\Rightarrow$ (a), suppose that (b) holds true. Take an arbitrary family $\{X_i : i \in I\}$ of nonempty sets and assume, without loss of generality, that all members of this family are pairwise disjoint. Let

$$a, \quad b_i \ (i \in I), \quad c_i \ (i \in I)$$

be pairwise different elements not belonging to $\cup\{X_i : i \in I\}$ (show the existence of such elements within **ZF** theory). Then put

$$V = \cup\{X_i : i \in I\} \cup \{a\} \cup \{b_i : i \in I\} \cup \{c_i : i \in I\},$$

$$E = \{\{a, b_i\} : i \in I\} \cup$$

$$(\cup\{\{\{b_i, x\} : x \in X_i\} : i \in I\}) \cup (\cup\{\{\{c_i, x\} : x \in X_i\} : i \in I\}),$$

and consider the connected graph (V, E). According to (b), there is a spanning tree (V, E') of (V, E). Verify that, for every $i \in I$, there exists a unique element $x_i \in X_i$ such that $\{x_i, b_i\} \in E'$ and $\{x_i, c_i\} \in E'$. This yields the selector $\{x_i : i \in I\}$ of the family $\{X_i : i \in I\}$, i.e., the Axiom of Choice holds true.

3*. Let (V, E) be a finite complete graph with card$(V) \geq 2$. Take any edge $e = \{v, u\} \in E$ and replace it either by the ordered pair (v, u) or by the ordered pair (u, v). Proceeding in this manner for all edges of (V, E), one finally comes to a certain binary relation $S \subset V \times V$ which is connected in the sense that

$$(\forall u \in V)(\forall v \in V)(u \neq v \Rightarrow (S(u, v) \ \vee \ S(v, u))).$$

Demonstrate that there exists a bijection

$$\phi : \{1, 2, ..., \text{card}(V)\} \to V$$

having the following property: for any index $i \in \{1, 2, ..., \text{card}(V) - 1\}$, the pair $(\phi(i), \phi(i + 1))$ belongs to S.

In order to establish this fact (which is known as Redei's theorem), argue by induction on card(V).

4. Show that if everyone in some group of people is a friend of at least half the people in this group, then all members of the group can take places around a table in such a way that everyone would be seated between two of his friends.

Reformulate the above fact in terms of graph theory.

5. Prove that any graph (V, E) admits a unique representation in the form of the union of a family of nonempty pairwise disjoint connected subgraphs.

These subgraphs are usually called the connected components of (V, E). Thus, a nonempty graph (V, E) is connected if and only if it coincides with its unique connected component.

Check that if a graph (V, E) is locally finite, then all its connected components are at most countable.

Suppose that (E, V) is a finite graph and card$(V) = n$. Suppose also that there are precisely k connected components of this graph.

Demonstrate that the inequality

$$\text{card}(E) \leq (n - k)(n - k + 1)/2$$

is valid and deduce from this fact that if

$$\text{card}(E) > (n - 1)(n - 2)/2,$$

then (V, E) is a connected graph.

Compare Theorem 3 with the result of this exercise.

6. Let (V, E) be a graph. An injective family of nonempty sets $\{X_v : v \in V\}$ is called a set-theoretical representation of (V, E) if the equivalence

$$\{v, u\} \in E \Leftrightarrow X_v \cap X_u \neq \emptyset$$

holds true for any two distinct vertices $v \in V$ and $u \in V$.

Prove that every graph admits at least one set-theoretical representation (Marczewski's theorem).

Remark 5. In connection with Exercise 6, let us notice that set-theoretical representations of abstract graphs are useful in various questions of graph theory and its applications. In this context, those representations of abstract graphs should be mentioned especially, which consist of point sets lying in Euclidean spaces and having sufficiently nice geometric structure (for example, line segments, convex sets, polyhedra, etc.).

7*. Let (V, E) be an arbitrary finite graph.

Show that the following two assertions are equivalent:

(a) (V, E) is representable in the form of the union of a family of simple cycles which pairwise have no common edges;

(b) for every vertex v of (V, E), the number $\deg(v)$ is even.

Show also that this result admits a certain extension to the case of an infinite locally finite graph (V, E). More precisely, an infinite injective sequence

$$(\ldots, z_{-n}, \ldots, z_{-2}, z_{-1}, z_0, z_1, z_2, \ldots, z_n, \ldots)$$

of vertices of (V, E), where $\{z_i, z_{i+1}\} \in E$ for all integers i, is called a generalized simple cycle in (V, E) (because one may imagine that the left and right infinite branches of this sequence meet each other at infinity).

Under this convention, demonstrate that if (V, E) is an arbitrary locally finite graph, then the following assertions are equivalent

(i) (V, E) is representable in the form of the union of a family of simple cycles and generalized simple cycles, which pairwise have no common edges;

(ii) for every vertex v of (V, E), the number $\deg(v)$ is even.

In order to prove the equivalence (i) $\Leftrightarrow$ (ii), use the partition of (V, E) into its connected components and take into account the circumstance that all isolated vertices of (V, E) are regarded as simple cycles of length zero.

8. Verify that if (V, E) is a tree, then for any two distinct vertices u and v of (V, E) there exists a unique simple chain whose endpoints are u and v.

Keeping in mind this fact, define the distance $d(u, v)$ between u and v as the length of the unique simple chain connecting u and v (the introduced function d trivially satisfies all axioms for metric spaces).

Conversely, show that if a nonempty graph (V, E) is such that any two vertices of (V, E) can be connected by a unique simple chain, then (V, E) is a tree.

9*. Suppose that the following information about a finite tree (V, E) is available:

(a) it is known that V contains precisely n leaves $v_1, v_2, \dots, v_n$;

(b) there are known all the distances

$$d(v_i, v_j) \quad (1 \leq i \leq n,\ 1 \leq j \leq n,\ i \neq j).$$

Assuming that there exists at least one tree satisfying the conditions (a) and (b), give a construction of such a tree.

For this purpose, consider one of the leaves of (V, E) as a root of (V, E) and argue by induction on the number of levels of (V, E).

10*. Let (V, E) be a connected graph, all vertices of which have countable degrees.

Verify that (V, E) itself is at most countable.

Conclude that if a connected graph is uncountable, then there exists at least one vertex of this graph, whose degree is also uncountable (in particular, there are no uncountable locally finite connected graphs).

Let (V, E) be an infinite locally finite connected graph and let a vertex $v \in V$ be chosen arbitrarily.

Prove that there exists an infinite simple chain in (V, E) whose endpoint coincides with v. In other words, demonstrate that there exists an infinite injective sequence $(v_0, v_1, ..., v_n, ...)$ of vertices of (V, E) such that $v_0 = v$ and

$$\{v_i, v_{i+1}\} \in E \quad (i = 0, 1, ..., n, ...).$$

This result is known as König's lemma (cf. [137], [150], [241], [275]).

Verify that, in **ZF** theory, the following two assertions are equivalent:

(a) König's lemma;

(b) any countable family of nonempty finite sets has a selector.

Conclude that König's lemma is logically equivalent to a certain weak form of the Axiom of Choice.

Remark 6. König's lemma found many successful applications in various areas of contemporary mathematics (infinite combinatorics, game theory, general topology, measure theory, etc.).

11*. Let (V, E) be an arbitrary countable graph.

Show that the following two assertions are equivalent:

(a) (V, E) is planar;

(b) every finite subgraph of (V, E) is planar.

For this purpose, apply König's lemma (see the previous exercise).

Conclude from the equivalence of (a) and (b) that every countable acyclic graph is planar.

Remark 7. In connection with (b) of Exercise 11, let us recall that an efficient criterion for recognizing the planarity (or non-planarity) of any given finite graph was first stated by Kuratowski [239]. This criterion is formulated in terms of the two famous Kuratowski graphs K_5 and $K_{3,3}$ and admits an appropriate algorithmic procedure for establishing the planarity or non-planarity of abstract finite graphs (see, e.g., [135], [251], [318]). Recall that K_5 is a complete (full) graph with five vertices, and $K_{3,3} = (V, E)$ is described as follows:

$$V = \{u_1, u_2, u_3, v_1, v_2, v_3\},$$

where all $u_1, u_2, u_3, v_1, v_2, v_3$ are pairwise distinct and

$$E = \{\{u_i, v_j\} : i \in \{1, 2, 3\},\ j \in \{1, 2, 3\}\}.$$

Actually, $K_{3,3}$ is a special case of a bichromatic (bipartite) graph (see Exercise 17 below).

12. Any finite graph (V, E) can be represented in the plane $\mathbf{R}^2$, but some of its edges may have common points distinct from all vertices of the graph.

By definition, the crossing number of (V, E) (denoted by $c(V, E)$) is the smallest number of such common points when one considers all possible representations of (V, E) in $\mathbf{R}^2$. So, a finite graph (V, E) is planar if and only if $c(V, E) = 0$.

Check that the crossing numbers of both Kuratowski graphs K_5 and $K_{3,3}$ are equal to 1.

13. Any graph of the form $(\{v_0, v_1, v_2, v_3\}, \{\{v_0, v_1\}, \{v_0, v_2\}, \{v_0, v_3\}\})$ is called a triode.

Demonstrate that if a graph (V, E) contains an uncountable disjoint family of triodes, then (V, E) cannot be realized in the plane $\mathbf{R}^2$ (in the sphere $\mathbf{S}_2$), i.e., (V, E) is not planar.

Conclude from this fact that there exists an uncountable non-planar graph all countable subgraphs of which are planar (cf. Exercise 11).

Remark 8. It was proved in [199] that a graph (V, E) is planar if and only if the conjunction of these three conditions is fulfilled:

(i) $\mathrm{card}(V) \le \mathbf{c}$, where $\mathbf{c}$ denotes the cardinality of the continuum;

(ii) the set of all those vertices of (V, E) whose degrees are greater than or equal to 3 is at most countable;

(iii) every finite subgraph of (V, E) is planar.

14. Let (V, E) be a nonempty finite planar graph with precisely k connected components.

Show that the equality

$$v - e + f = k + 1$$

is valid, where, as usual, $v = \text{card}(V)$, $e = \text{card}(E)$, and f denotes the number of regions in $\mathbf{R}^2$ produced by (V, E).

This is the generalized Euler formula for nonempty finite planar graphs.

15. Verify that a graph is planar if and only if it is isomorphic to some graph which is realized in the unit sphere $\mathbf{S}_2 \subset \mathbf{R}^3$.

Taking this circumstance into account, demonstrate that the generalized Euler formula given in Exercise 14 is also valid for any nonempty finite graph (V, E) realized in the sphere $\mathbf{S}_2$ and having exactly k connected components.

16*. Let (V, E) be a nonempty planar graph, all vertices of which have degrees greater than or equal to 3.

By starting with generalized Euler's formula, verify the validity of the following relation:

$$e + 6 \leq 3v \leq 2e \leq 6v - 12.$$

Deduce from this relation the inequality

$$3v_3 + 2v_4 + v_5 \geq 12,$$

where v_i $(i = 3, 4, 5)$ denotes the number of all those vertices of (V, E) whose degrees are equal to i.

Conclude from the last inequality that the first Kuratowski graph K_5 is not planar.

By using a similar argument, show that the second Kuratowski graph $K_{3,3}$ is not planar.

Finally, demonstrate that if (V, E) is any finite graph with $v \geq 3$, then

$$c(V, E) \geq e - 3v + 6 > e - 3v.$$

For this purpose, use Theorem 5 of the present chapter.

Remark 9. In connection with Exercise 16, see also Chapter 10.

17*. A nonempty graph (V, E) is called bichromatic (2-chromatic, bipartite) if there exists a coloring of all its vertices by exactly two colors (red and blue, say) in such a manner that the endpoints of any edge of (V, E) carry distinct colors.

In other words, (V, E) is bichromatic if and only if V admits a partition $\{V_1, V_2\}$ such that any $e \in E$ is simultaneously incident to a vertex from V_1 and to a vertex from V_2.

Assuming the Axiom of Choice (**AC**), demonstrate that the following two assertions are equivalent:

(a) (V, E) is bichromatic;

(b) $\mathrm{card}(V) \geq 2$ and the length of any simple cycle in (V, E) is even.

In particular, every tree containing at least two vertices is 2-chromatic.

Remark 10. It should be noticed, in connection with Exercise 17, that the implication (b) $\Rightarrow$ (a) cannot be proved without the aid of uncountable forms of the Axiom of Choice, because the validity of this implication implies (within a certain weak fragment of set theory) the existence of a Lebesgue nonmeasurable subset of the real line $\mathbf{R}$. For more details, see Exercise 16 from Chapter 11.

18. Let $k > 0$ be a natural number. A graph (V, E) is called k-colorable if there exists a surjective mapping

$$\phi : V \to \{1, 2, ..., k\}$$

such that $\phi(v) \neq \phi(u)$ for every edge $e = \{v, u\} \in E$.

The chromatic number of (V, E) is the least natural number k (if it exists) such that (V, E) is k-colorable.

A graph (V, E) is called k-chromatic if its chromatic number equals k. In particular, for $k = 2$, we get the definition of a bichromatic graph (see Exercise 17 above).

Suppose that (V, E) is a finite graph such that the degrees of all its vertices do not exceed k.

Show that the chromatic number of (V, E) does not exceed $k + 1$.

For this purpose, argue by induction on $\mathrm{card}(V)$.

Also, verify that this estimate is precise in a certain sense.

Remark 11. A much stronger result was obtained by Hajnal and Szemerédi [133] and is now known as the Hajnal–Szemerédi theorem. Their original argument was rather complicated. A more simple proof of their theorem can be found, e.g., in [231]. Notice also that the result of Exercise 18 remains valid for infinite graphs, too, but the proof is essentially based on the Axiom of Choice. Actually, by starting with the result for finite graphs, it becomes possible to give a nice proof of the generalized result with the aid of Tychonov's celebrated theorem on topological products of compact spaces.

19. Let (V, E) and (V', E') be two graphs and let $h : V \to V'$ be a mapping. This mapping is called a morphism (homomorphism) acting from (V, E) into (V', E') if, for any vertices $v \in V$ and $u \in V$, the implication

$$\{v, u\} \in E \Rightarrow \{h(v), h(u)\} \in E'$$

holds true. Moreover, a morphism h is called strong if, for every edge

$$e' = \{v', u'\} \in E',$$

there exist two vertices $v \in V$ and $u \in V$ such that

$$\{v, u\} \in E, \quad h(v) = v', \quad h(u) = u'.$$

For any natural number $k > 0$, let C_k denote a full (complete) graph with precisely k vertices. Hence the number of all edges of C_k is maximal and is equal to $k(k-1)/2$.

Demonstrate that:

(a) a graph (V, E) is k-colorable if and only if there exists a morphism from (V, E) onto C_k;

(b) a graph (V, E) has chromatic number k if and only if k is the least natural number for which there exists a strong morphism from (V, E) onto C_k.

20*. Let $G = (V, E)$ be a finite graph. Mycielski's extension G^* of G is defined as follows. Let $\{v_1, v_2, ..., v_n\}$ be an injective enumeration of all vertices of G, i.e., $V = \{v_1, v_2, ..., v_n\}$.

Consider any set $U = \{u_1, u_2, ..., u_n\}$ having the same number of elements and such that $U \cap V = \emptyset$. Let w be any element not belonging to $U \cup V$. Denote

$$V' = U \cup V \cup \{w\}.$$

Further, equip V' with the graph structure by adding to E some new edges. Firstly, all two-element sets

$$\{u_1, w\}, \quad \{u_2, w\}, \quad \ldots, \quad \{u_n, w\}$$

should be added to E. Secondly, if $\{v_i, v_j\} \in E$, where $\{i, j\} \subset \{1, 2, ..., n\}$, then also add two new edges $\{u_i, v_j\}$ and $\{u_j, v_i\}$.

No other new edges are added to E (except those which were just described). Thus, one comes to the graph $G^* = (V', E')$ which is called Mycielski's extension of G.

Demonstrate that:

(a) if G is triangle-free (i.e., G does not contain any simple cycles of length 3), then G^* is triangle-free, too;

(b) the chromatic number of G^* is strictly greater than the chromatic number of G.

Argue in the following manner.

Actually, (a) can easily be verified, so the main difficulty is concentrated in checking the validity of (b). For this purpose, denote by k the chromatic number of G (it does exist, because of finiteness of G). Since G^* contains an isomorphic copy of G, the chromatic number of G^* is greater than or equal to k. So it must be shown that the equality cannot be realized. Suppose otherwise, i.e., suppose that there exists a k-coloring

$$\phi : U \cup V \cup \{w\} \to \{1, 2, ..., k-1, k\}$$

of G^* having the property that every edge in G^* carries exactly two colors. Since w is incident to all vertices from U, the number of colors carried by U is strictly less than k, and one may assume that $\phi(w) = k$ and only colors from $\{1, 2, ..., k-1\}$ are used for U. Now, define a coloring

$$\psi : V \to \{1, 2, ..., k-1\}$$

of the initial graph G by putting:

$\psi(v_i) = \phi(v_i)$ if $\phi(v_i) \neq k$, and $\psi(v_i) = \phi(u_i)$ if $\phi(v_i) = k$.

Then the coloring ψ is such that every edge of G carries exactly two colors, which contradicts the definition of k. Thus, the chromatic number of G^* is strictly greater than k.

Finally, deduce from (a) and (b) that there exist finite triangle-free graphs whose chromatic numbers are arbitrarily large (Mycielski's theorem).

21. Let (V, E) be a finite graph with $\text{card}(V) = n$ and suppose that this graph is triangle-free.

Demonstrate the validity of Turán's inequality $\text{card}(E) \leq \lfloor n^2/4 \rfloor$, where the symbol $\lfloor n^2/4 \rfloor$ denotes the greatest integer not exceeding $n^2/4$.

For this purpose, consider separately the case of an odd n and the case of an even n, and use induction on n in both these cases.

In addition, give examples which show that the above inequality is precise, i.e., for certain finite bichromatic graphs it becomes the equality.

22. Let (V, E) be a graph. Recall that, for any vertex $v \in V$, the symbol $U(v)$ denotes the neighborhood of v.

A partition $\{A, B\}$ of V is called unfriendly if it satisfies the following relation:

$$(\forall v \in A)(\text{card}(U(v) \cap A) \leq \text{card}(U(v) \cap B)) \ \&$$

$$(\forall v \in B)(\text{card}(U(v) \cap B) \leq \text{card}(U(v) \cap A)).$$

Show that any finite graph (V, E) with $\text{card}(V) \geq 2$ admits an unfriendly partition of V.

For this purpose, consider a partition $\{A, B\}$ of V such that the number of all those edges whose endpoints belong to different members of $\{A, B\}$ is maximal.

Remark 12. Notice that the result of Exercise 22 remains valid for locally finite graphs, too. More precisely, any nonempty locally finite graph admits at least one unfriendly partition. The proof is substantially based on the Axiom of Choice (**AC**).

23*. Let n be a natural number, $X = \{x_1, x_2, ..., x_n\}$ be an n-element set, and let $\{X_1, X_2, ..., X_m\}$ be a family of nonempty proper subsets of X.

Suppose that every two-element subset of X is contained in exactly one set X_i, where $i \in \{1, 2, ..., m\}$.

For each index $j \in \{1, 2, ..., n\}$, denote by k_j the number of all those sets X_i which contain x_j.

Dually, for each index $i \in \{1, 2, ..., m\}$, denote by l_i the cardinality of X_i.

Show that:

(a) $\sum\{k_j : 1 \leq j \leq n\} = \sum\{l_i : 1 \leq i \leq m\}$;

(b) if $x_j \notin X_i$, then $l_i \leq k_j$;

(c) $n \leq m$.

The proof of (a) and (b) is not difficult. In order to demonstrate (c), suppose for a moment that $n > m$ and argue in the following manner.

Firstly, it can be assumed without loss of generality that

$$\text{card}(X_i) \geq 2 \qquad (i = 1, 2, ..., m)$$

and that k_n is a smallest number among all the numbers k_j $(1 \leq j \leq n)$.

Secondly, check that $k_n < m$ and that one also can assume the validity of these two conditions:

(*) if $k_n < i \leq m$, then $x_n \notin X_i$;

(**) if $i \leq k_n$, $j \leq k_n$ and $i \neq j$, then $x_j \notin X_i$.

Further, denoting $\theta = k_n$ and taking into account (b), (*) and (**), write

$$x_\theta \notin X_1, \quad x_1 \notin X_2, \quad x_2 \notin X_3, \; .. \; , \quad x_{\theta-1} \notin X_\theta,$$

$$l_1 \leq k_\theta, \quad l_2 \leq k_1, \quad l_3 \leq k_2, \; \ldots \; , \quad l_\theta \leq k_{\theta-1},$$

$$l_i \leq k_n \leq k_i \qquad (i = \theta+1, \theta+2, ..., m),$$

$$\sum\{l_i : 1 \leq i \leq m\} \leq \sum\{k_i : 1 \leq i \leq m\} < \sum\{k_j : 1 \leq j \leq n\}.$$

Clearly, the last relation contradicts (a). The obtained contradiction yields (c).

In addition, specify the two particular cases when the inequality $n \leq m$ is reduced to the equality $n = m$.

Remark 13. The result of Exercise 23 is due to de Bruijn and Erdös (see [50]).

24*. Let P be a finite set of points in the plane $\mathbf{R}^2$, satisfying the following condition:

For any two distinct points p_1 and p_2 from P, there exists a point p_3 from $P \setminus \{p_1, p_2\}$ which belongs to the line $l(p_1, p_2)$ determined by p_1 and p_2.

Prove that all points of P are collinear (Sylvester's problem).

For this purpose, suppose to the contrary that P is not collinear and consider a non-degenerate triangle whose vertices belong to P and whose altitude has the least length. Verify that the existence of such a triangle contradicts the above condition.

Infer from the presented result that if Q is a finite non-collinear subset of $\mathbf{R}^2$ with $\text{card}(Q) = n$, then there are at least n distinct straight lines determined by couples of points of Q.

For this purpose, argue by induction on n (or directly refer to Exercise 23).

Remark 14. There are many works devoted to Sylvester's problem and related questions (see, for instance, [41], [64], [116], [140], [167], [266]).

25. Let $\mathcal{L}$ be a finite family of pairwise non-parallel straight lines in $\mathbf{R}^2$ satisfying the following condition:

If any two distinct lines l_1 and l_2 from $\mathcal{L}$ have a common point p, then there exists a line l_3 from $\mathcal{L} \setminus \{l_1, l_2\}$ also passing through p.

Demonstrate that all lines from $\mathcal{L}$ have a common point.

For this purpose, using an appropriate polar transform of $\mathbf{R}^2$, reduce this exercise to the previous one.

At the same time, demonstrate that the required result follows also from the inequality

$$3v_3 + 2v_4 + v_5 \geq 12$$

of Exercise 16, and give another purely combinatorial solution to Sylvester's problem.

26. Let (V, E) be a nonempty finite planar connected graph such that every region produced by some concrete realization of (V, E) in the plane $\mathbf{R}^2$ has at least four sides.

Verify that the inequality $e \leq 2(v - 2)$ holds true.

For this purpose, observe that $4f \leq 2e$ and use the Euler formula $f = e + 2 - v$.

Deduce from this result that if (V, E) is an arbitrary nonempty 2-chromatic planar graph containing at least three vertices, then $e \leq 2(v - 2)$.

27. Let α belong to the open interval $]0, 2[$ and let $a > 0$ be a real constant.

Prove that there exist a natural number n_0 and a real constant $b > 0$ having the following property:

If (V, E) is a finite graph such that $\text{card}(V) = n \geq n_0$ and $\text{card}(E) \leq an^\alpha$, then there is a null subgraph $(V', \emptyset)$ of (V, E) for which the inequality

$$\text{card}(V') \geq bn^{(2-\alpha)/2}$$

takes place.

28. Let n be a nonzero natural number, let X be a subset of $\mathbf{R}^2$ with $\text{card}(X) = n$, and let $\triangle$ be a fixed non-degenerate triangle in $\mathbf{R}^2$.

Show that there exists a subset Z of X satisfying the following two conditions:

(a) $\text{card}(Z) \geq (5 + (25 + 24n)^{1/2})/12$;

(b) no three points of Z are vertices of a triangle similar to $\triangle$.

Try to obtain an analogous result in the case of $\mathbf{R}^m$, where $m \geq 3$.

Remark 15. In connection with Exercise 28, cf. Example 4 from Chapter 5.

Bibliography

[1] P. K. Agarwal, F. Hurtado, G. T. Toussaint, J. Trias, *On polyhedra induced by point sets in space*, Discrete Applied Mathematics, v. 156, issue 1, 2008, pp. 42–54.

[2] M. Aigner, G. M. Ziegler, *Proofs from The Book*, Springer-Verlag, New York-Heidelberg, 2004.

[3] S. G. Akl, G. T. Toussaint, *A fast convex hull algorithm*, Information Processing Letters, v. 7, 1978, pp. 219–222.

[4] A. D. Alexandrov, *Convex Polyhedra*, Gostechizdat, Moscow, 1950 (in Russian).

[5] P. S. Alexandroff, *Combinatorial Topology*, Dover, New York, 1998 (translation from the Russian).

[6] N. Alon, *Splitting necklaces*, Advances in Mathematics, v. 63, 1987, pp. 247–253.

[7] N. Alon, *Discrete Mathematics: Methods and Challenges*, Proc. Intern. Congress of Math., Beijing, August 20–28, 2002, v. I, pp. 119–136.

[8] N. Alon, D. West, *The Borsuk–Ulam theorem and bisection of necklaces*, Proc. Amer. Math. Soc., v. 98, 1986, pp. 623–628.

[9] J. Anusiak, *On set-theoretically independent collections of balls*, Colloquium Mathematicum, v. 13, 1965, pp. 223–233.

[10] E. Artin, *Über die Zerlegung definiter Funktionen in Quadrate*, Abh. Math. Sem. Univ. Hamburg, v. 5, 1927, pp. 85–99.

[11] F. Bagemihl, *On indecomposable polyhedra*, Amer. Math. Monthly, v. 55, n. 7, 1948, pp. 411–413.

[12] R. Ball, *The plank problem for symmetric bodies*, Invent. Math., v. 104, n. 3, 1991, pp. 535–543.

[13] S. Banach, A. Tarski, *Sur la décomposition des ensembles de points en parties respectivement congruentes*, Fund. Math., v. 6, 1924, pp. 244–277.

[14] Th. Bang, *A solution of the plank problem*, Proc. Amer. Math. Soc., v. 2, 1951, pp. 990–993.

[15] L. Bankoff, *An elementary proof of the Erdös–Mordell theorem*, Amer. Math. Monthly, v. 65, 1958, pp. 521.

[16] Ei. Bannai, Et. Bannai, D. Stanton, *An upper bound for the cardinality of an s-distance set in real Euclidean space*, II, Combinatorica, v. 3, 1983, pp. 147–152.

[17] J. Beck, *On the lattice property of the plane and some problems of Dirac, Motzkin, and Erdös in combinatorial geometry*, Combinatorica, v. 3, 1983, pp. 281–297.

[18] J. L. Bell, D. H. Fremlin, *A geometric form of the Axiom of Choice*, Fund. Math., v. 77, n. 2, 1972, pp. 167–170.

[19] R. V. Benson, *Euclidean Geometry and Convexity*, New York, 1966.

[20] M. Berger, *Geometry*, v. I, v. II, Springer-Verlag, Berlin-Heidelberg, 1987.

[21] J. Berkes, *Einfacher Beweis und Verallgemeinerung einer Dreiecksungleichung*, Elemente der Mathematik, v. 22, n. 6, 1967, pp. 135–136.

[22] K. Bezdek, *Tarski's plank problem revisited*, in: Geometry - Intuitive, Discrete, and Convex, Springer, Berlin, 2013, pp. 45–64.

[23] C. E. Blair, *The Baire category theorem implies the principle of dependent choices*, Bull. Acad. Polon. Sci., Ser. Math., v. 25, 1977, pp. 933-934.

[24] W. Blaschke, *Kreis und Kugel*, Leipzig, 1916.

[25] A. Blass, *Existence of bases implies the Axiom of Choice*, Contemporary Mathematics, v. 31, 1984, pp. 31-33.

[26] A. Blokhuis, *Few-distance sets*, CWI Tract., v. 7, 1984, pp. 1–70.

[27] L. Blumental, *Theory and Applications of Distance Geometry*, Oxford University Press, Oxford, 1953.

[28] V. Bogachev, *Measure Theory*, Springer-Verlag, Berlin-Heidelberg, 2007.

[29] J. Böhm, E. Hertel, *Polyedergeometrie in n-dimensionalen Räumen konstanter Krümmung*, Birkhäuser, Basel-Boston-Stuttgart, 1981.

[30] B. Bollobás, *The Art of Mathematics*, Cambridge University Press, Cambridge, 2006.

[31] V. G. Boltyanskii, *Hilbert's Third Problem*, Izd. Nauka, Moscow, 1977 (in Russian).

[32] V. Boltyanskii, *Combinatorial Geometry*, in the Series: "Algebra, Topology, Geometry", v. 19, 1981, pp. 209 - 274 (in Russian).

[33] V. G. Boltyanskii, V. A. Efremovich, *Intuitive Topology*, Izd. Nauka, Moscow, 1982 (in Russian).

[34] V. G. Boltyanskii, I. Ts. Gokhberg, *Theorems and Problems in Combinatorial Geometry*, Izd. Nauka, Moscow, 1965 (in Russian).

[35] V. Boltyanski, H. Martini, *Illumination of direct vector sums of convex bodies*, Studia Scientiarum Mathematicarum Hungarica, v. 44, n. 3, 2007, pp. 367–376.

[36] V. Boltyanskii, H. Martini, P. Soltan, *Excursions into Combinatorial Geometry*, Springer-Verlag, Berlin, 1997.

[37] V. Boltyanskii, A. Soifer, *Geometric Etudes in Combinatorial Mathematics*, Colorado Springs, 1991.

[38] V. Boltyanskii, P. Soltan, *Combinatorial Geometry of Various Classes of Convex Sets*, Acad. Nauk Mold. SSR, "Stinitsa", Kishinev, 1978 (in Russian).

[39] A. Bondarenko, *On Borsuk's conjecture for two-distance sets*, Discrete and Computational Geometry, v. 51, 2014, pp. 509–515.

[40] K. Borsuk, *Drei Sätze über die n-dimensionale euklidische Sphäre*, Fund. Math., v. 20, 1933, pp. 177–190.

[41] P. Borwein, W. O. J. Moser, *A survey of Sylvester's problem and its generalizations*, Aequationes Mathematicae, v. 40, 1990, pp. 111–135.

[42] O. Botten, R. Z. Djordjević, R. R. Janić, D. S. Mitrinović, P. M. Vasić, *Geometric Inequalities*, Lters-Noordhoff Publishing, Groningen, 1969.

[43] K. Bouhjar, J. J. Dijkstra, R. D. Mauldin, *No n-point set is σ-compact*, Proc. Amer. Math. Soc., v. 129, n. 2, pp. 621–622.

[44] N. Bourbaki, *Theory of Sets*, Hermann et Cie, Paris, 1968 (in French).

[45] N. Bourbaki, *Topological Vector Spaces*, Springer, Berlin-Heidelberg, 2002.

[46] J. Brandts, S. Korotov, M. Krizek, J. Solc, *On non-obtuse simplicial partitions*, SIAM Review, v. 51, n. 2, 2009, pp. 317–335.

[47] R. Brooks, C. Smith, A. Stone, W. Tutte, *The dissection of rectangles into squares*, Duke Mathematical Journal, v. 7, 1940, pp. 312–340.

[48] R. A. Brualdi, *Introductory Combinatorics*, Prentice-Hall, N.J., 2010.

[49] A. Bruckner, *Differentiation of Real Functions*, Springer-Verlag, Berlin, 1978.

[50] N. G. de Bruijn, P. Erdös, *On a combinatorial problem*, Indag. Math., v. 10, 1948, pp. 421–423.

[51] N. G. de Bruijn, P. Erdös, *A colour problem for infinite graphs and a problem in the theory of relations*, Nederl. Akad. Wetensch. Proc., Ser. A, v. 54, n. 5, 1951, pp. 371–373.

[52] V. V. Buldygin, A. B. Kharazishvili, *Geometric Aspects of Probability Theory and Mathematical Statistics*, Kluwer Academic Publishers, Dordrecht, 2000.

[53] P. J. Cameron, *Combinatorics: Topics, Techniques, Algorithms*, Cambridge University Press, Cambridge, 1994.

[54] A. Chandoul, *The Pell equation* $x^2 - Dy^2 = \pm k^2$, Advances in Pure Mathematics, v. 1, 2011, pp. 16–22.

[55] M. N. Charatonik, W. J. Charatonik, *On Mazurkiewicz sets*, Comment. Math. Univ. Carolinae, v. 41, n. 4, 2000, pp. 817–819.

[56] B. Chazelle, *Triangulating a simple polygon in linear time*, Discrete and Computational Geometry, v. 6, 1991, pp. 485–524.

[57] S. A. Choudum, *A simple proof of the Erdös-Gallai theorem on graph sequences*, Bulletin of the Australian Mathematical Society, v. 33, n. 1, 1986, pp. 67–70.

[58] J. Cobb, *Nice decompositions of* $\mathbf{R}^n$ *entirely into nice sets are mostly impossible*, Geometriae Dedicata, v. 62, n. 1, 1996, pp. pp. 107–114.

[59] P. M. Cohn, *Universal Algebra*, Harper and Row, Publishers, New York, 1965.

[60] R. Connelly *Comments on generalized Heron polynomials and Robbins' conjectures*, Discrete Mathematics, v. 309, issue 12, 2009, pp. 4192–4196.

[61] J. H. Conway and H. T. Croft, *Covering a sphere with congruent great-circle arcs*, Proc. Cambridge Phil. Soc., v. 60, 1964, pp. 787–800.

[62] T. H. Cormen, C. E. Leiserson, R. L. Rivest, C. Stein, *Introduction to Algorithms*, MIT Press, 2009.

[63] D. A. Coulson, *A* 15*-coloring of* 3*-space omitting distance* 1, Discrete Math., v. 256, 2002, pp. 83–90.

[64] H. S. M. Coxeter, *A problem of collinear points*, Amer. Math. Monthly, v. 55, n. 1, 1948, pp. 26–28.

[65] H. S. M. Coxeter, S. L. Greitzer, *Geometry Revisited*, The Mathematical Association of America, Washington, 1967.

[66] H. T. Croft, 9-*point and* 7-*point configurations in* 3-*space*, Proc. London Math. Soc., v. 12, n. 3, 1962, pp. 400–424.

[67] L. Danzer, B. Grünbaum, *Über zwei Probleme bezüglich konvexer Körper von P. Erdös und von V. L. Klee*, Math. Zeitschrift, v. 79, 1962, pp. 95–99.

[68] L. Danzer, B. Grünbaum, V. Klee, *Helly's theorem and its relatives*, Convexity, Symp. Pure Math., v. VII, Amer. Math. Soc., Providence, R.I., 1963, pp. 101–179.

[69] B. N. Delaunay, *Sur la sphere vide*, in: Proceedings of the International Mathematical Congress held in Toronto, August 11-16, 1928, pp. 695–700.

[70] C. N. Delzell, *A continuous, constructive solution to Hilbert's 17th problem*, Inventiones Mathematicae, v. 76, issue 3, 1984, pp. 365–384.

[71] M. Dinca, *A simple proof of the Erdös–Mordell inequality for polygons in N-dimensional spaces*, Gazeta Matematica, n. 4, 2009, pp. 172–177 (in Romanian).

[72] V. L. Dol'nikov, *Transversals of families of sets and a relationship between Helly and Borsuk theorems*, Matem. Sbornik, v. 184, n. 5, 1993, pp. 111–132 (in Russian).

[73] A. J. W. Duijvestijn, *A simple perfect square of lowest order*, Journal Comb. Theory, Ser B, v. 25, 1978, pp. 240–243.

[74] B. Dushnik, E. W. Miller, *Partially ordered sets*, Amer. Journ. Math., v. 63, n. 3, 1941, pp. 600–610.

[75] H. Edelsbrunner, *Algorithms in Combinatorial Geometry*, EATCS Monographs in Computer Science, 10, Springer-Verlag, Berlin-New York, 1987.

[76] H. G. Eggleston, *Problems in Euclidean Space: Applications of Convexity*, London, 1957.

[77] H. G. Eggleston, *Convexity*, Cambridge University Press, Cambridge, 1958.

[78] H. G. Eggleston, *A triangle inequality*, Math. Gaz., v. 42, 1958, pp. 54–55.

[79] S. J. Einhorn, I. J. Schoenberg, *On Euclidean sets having only two distances between points*, Indagationes Mathematicae, v. 28, 1966, pp. 479–504.

[80] V. A. Emelichev, M. M. Kovalev, and M. K. Kravtsov, *Polyhedra, Graphs and Optimization*, Izd. Nauka, Moscow, 1981 (in Russian).

[81] E. Engeler, *Metamathematik der Elementarmathematik*, Springer-Verlag, Berlin, 1983.

[82] R. Engelking, *General Topology*, PWN, Warszawa, 1985.

[83] P. Erdös, *Problem 3740*, Amer. Math. Monthly, v. 42, 1935, p. 396.

[84] P. Erdös, *Problem 4065*, Amer. Math. Monthly, v. 51, 1944, pp. 169–171.

[85] P. Erdös, *Some unsolved problems*, Publ. Math. Inst. Hung. Acad. Sci., Ser. A, v. 6, 1961, pp. 221–254.

[86] P. Erdös, *Problems and results in combinatorial geometry. Discrete geometry and convexity*, Ann. New York Acad. Sci., v. 440, 1985, pp. 1–11.

[87] P. Erdös, N. G. de Bruijn, *A colour problem for infinite graphs and a problem in the theory of relations*, Indag. Math., v. 13, 1951, pp. 371–373.

[88] P. Erdös, Z. Füredi, *The greatest angle among n points in the d-dimensional Euclidean space*, Annals of Discrete Mathematics, v. 17, 1983, pp. 275–283.

[89] P. Erdös, A. Hajnal, A. Máté, R. Rado, *Combinatorial Set Theory: Partition Relations for Cardinals*, North-Holland Publ. Co., Amsterdam, 1984.

[90] P. Erdös, R. Rado, *A partition calculus in set theory*, Bull. Amer. Math. Soc., v. 62, 1956, pp. 427–489.

[91] P. Erdös, G. Szekeres, *A combinatorial problem in geometry*, Compositio Math., v. 2, 1935, pp. 463–470.

[92] K. Fan, *A generalization of Tucker's combinatorial lemma with topological applications*, Ann. of Math., v. 56, n. 2, 1952, pp. 431–437.

[93] F. Faruk, Abi-Khuzam, *A trigonometric inequality and its geometric applications*, Mathematical Inequalities and Applications, v. 3, 2003, pp. 437–442.

[94] H. Federer, *Geometric Measure Theory*, Springer-Verlag, Berlin-New York, 1969.

[95] W. Fenchel, *On the Bang's solution of the plank problem*, Math. Tidsskr., Ser. B, v. 14, 1951, pp. 49–56.

[96] A. Florian, *Zu einem Satz von P. Erdös*, Elem. Math., v. 13, 1958, pp. 55–58.

[97] L.R. Ford, S. M. Johnson, *A tournament problem*, American Mathematical Monthly, v. 66, n. 5, 1959, p. 387.

[98] P. Frankl, R. M. Wilson, *Intersection theorems with geometric consequences*, Combinatorica, v. 1, 1981, pp. 357–368.

[99] R. M. Freund, M. J. Todd, *A constructive proof of Tucker's combinatorial lemma*, Journal of Combinatorial Theory, Series A, v. 30, 1981, pp. 321–325.

[100] H. Furstenberg, *Recurrence in Ergodic Theory and Combinatorial Number Theory*, Princeton University Press, Princeton, 1981.

[101] H. Furstenberg, Y. Katznelson, D. Ornstein, *The ergodic theoretical proof of Szemerédi's theorem*, Bull. Amer. Math. Soc., N.S., v. 7, n. 3, 1982, pp. 527–552.

[102] A. Gabrichidze, *One class of symmetric sets*, Bull. Acad. Sci. GSSR, v. 114, n. 1, 1984, pp. 41–43 (in Russian).

[103] A. A. Gaifullin, *Generalization of Sabitov's theorem to polyhedra of arbitrary dimensions*, Discrete Comput. Geometry, v. 52, n. 2, 2014, pp. 195–220.

[104] D. Gale, *Neighboring vertices on a convex polyhedron*, in: Linear Inequalities and Related Systems (H. W. Kuhn and A. W. Tucker, eds.), Annals of Math. Studies, Princeton University Press, v. 38, 1956, pp. 255–263.

[105] V. Galperin, G. Galperin, *Illumination of the space by searchlights*, Journal "Kvant", n. 11, 1981, pp. 28–30 (in Russian).

[106] R. J. Gardner, *Geometric Tomography*, Cambridge University Press, Cambridge, 1995.

[107] R. J. Gardner, *The Brunn–Minkowski inequality*, Bull. Amer. Math. Soc., v. 39, n. 3, 2002, pp. 355–405.

[108] B. R. Gelbaum and J. M. H. Olmsted, *Counterexamples in Analysis*, Holden-Day, San Francisco, 1964.

[109] L. Gerber, *The orthocentric simplex as an extreme simplex*, Pacific Journ. of Mathematics, v. 56, n. 1, 1975, pp. 97–111.

[110] J. E. Goodman, J. O'Rourke (editors), *Handbook of Discrete and Computational Geometry*, second edition, CRC Press, New York, 2004.

[111] M. T. Goodrich, R. Tamassia, *Algorithm Design: Foundation, Analysis, and Internet Examples*, Wiley and Sons, Inc., 2001.

[112] R. L. Graham, *An efficient algorithm for determining the convex hull of a finite planar set*, Information Processing Letters, v. 1, 1972, pp. 132–133.

[113] R. L. Graham, D. E. Knuth, O. Patashnik, *Concrete Mathematics*, Addison-Wesley Publishing Company, New York, 1994.

[114] R. L. Graham, B. L. Rothschild, J. H. Spencer, *Ramsey Theory*, John Wiley and Sons, New York, 1990.

[115] R. L. Graham, F. F. Yao, *Finding the convex hull of a simple polygon*, Journal of Algorithms, v. 4, 1983, pp. 324–331.

[116] B. Green, T. Tao, *On sets defining few ordinary lines*, Discrete and Computational Geometry, v. 50, n. 2, 2013, pp. 409–468.

[117] A. D. N. J. de Grey, *The chromatic number of the plane is at least* 5, Geombinatorics, v. 28, 2018, pp. 5–18.

[118] J. L. Gross, *Combinatorial Methods with Computer Applications*, Chapman and Hall/CRC, Boca Raton, 2008.

[119] P. M. Gruber, *Über den Durchschnitt einer abnehmenden Folge von Parallelepipeden*, Elemente der Mathematik, v. 32, n. 1, 1977, pp. 13–15.

[120] P. M. Gruber, *Aspects of convexity and its applications*, Expositiones Mathematicae, v. 2, 1984, pp. 47–83.

[121] P. M. Gruber, C. G. Lekkerkerker, *Geometry of Numbers*, North-Holland Publ. Co., Amsterdam, 1987.

[122] P. Gruber, J. Wills, *Convexity and its Applications*, Birkhauser, Basel, 1983.

[123] P. M. Gruber, J. Wills (editors), *Handbook of Convex Geometry*, v. 1, v. 2, North-Holland Publ. Comp., Amsterdam, 1990–1991.

[124] B. Grünbaum, *Venn diagrams and independent families of sets*, Mathematics Magazine, v. 48, 1975, pp. 12–23.

[125] B. Grünbaum, *Hamiltonian polygons and polyhedra*, Geombinatorics, v. 3, 1994, pp. 83–89.

[126] B. Grünbaum, *Convex Polytopes*, Springer-Verlag, Berlin-Heidelberg, 2003.

[127] S. Gueron, I. Shafrir, *A weighted Erdös–Mordell inequality for polygons*, Amer. Math. Monthly, v. 112, 2005, pp. 257–263.

[128] H. Guggenheimer, E. Lutwak, *Characterization of n-dimensional parallelotope*, Amer. Math. Monthly, v. 83, n. 6, 1976, pp. 475–478.

[129] H. Hadwiger, *Eulers Charakteristik und kombinatorische Geometrie*, J. Reine Angew. Math., v. 194, 1955, pp. 101–110.

[130] H. Hadwiger, *Vorlesungen über Inhalt, Oberfläche und Isoperimetrie*, Springer-Verlag, Berlin, 1957.

[131] H. Hadwiger, H. Debrunner, *Combinatorial Geometry in the Plane*, Izd. Nauka, Moscow, 1964 (in Russian; translation from the German, with comments by I. M. Iaglom).

[132] H. Hadwiger, H. Debrunner, V. Klee, *Combinatorial Geometry in the Plane*, New York, 1964.

[133] A. Hajnal, E. Szemerédi, *Proof of a conjecture of P. Erdös*, Combinatorial Theory and its Applications, II, North-Holland Publ. Co., 1970, pp. 601–623.

[134] M. Hall, *Combinatorial Theory*, Wiley and Sons, New York, 1998.

[135] F. Harary, *Graph Theory*, Addison-Wesley, Reading, MA, 1969.

[136] M. Henle, *A Combinatorial Introduction to Topology*, Dover, New York, 1994.

[137] H. Herrlich, *Axiom of Choice*, Springer-Verlag, Berlin, 2006.

[138] E. Hertel, *Some generalizations of equidecomposability theory of polyhedra*, Forschungsergebnisse, Universitat Jena, N/82/49, 1982 (in Russian).

[139] E. Hertel, *Zerlegungsähnlichkeit von Polygonen*, Elemente der Mathematik, v. 41, 1986, pp. 139–143.

[140] F. Herzog, L. M. Kelly, *A generalization of a theorem of Sylvester*, Proc. Amer. Math. Soc., v. 11, 1960, pp. 327–331.

[141] E. Hewitt, K. A. Ross, *Abstract Harmonic Analysis*, v. 1, Springer-Verlag, Berlin, 1963.

[142] D. Hilbert, *Mathematische Probleme*, Nachrichten von der Königlichen Gesellschaft der Wissenschaften zu Göttingen, 1900, pp. 253–297.

[143] D. Hilbert, *Foundations of Geometry*, Moscow-Leningrad, 1948 (in Russian; translation from the German, with comments by P. K. Rashevskii).

[144] C. D. Hodgson, I. Rivin, W. D. Smith, *A characterization of convex hyperbolic polyhedra and of convex polyhedra inscribed in the sphere*, Bull. Amer. Math. Soc., v. 27, n. 2, 1992, pp. 246–251.

[145] F. Hurtado, G. T. Toussaint, and J. Trias, *On polyhedra induced by point sets in space*, in: Proceedings of 15th Canadian Conference on Computational Geometry, Dalhousie University, Halifax, Nova Scotia, Canada, August 11-13, 2003, pp. 107–110.

[146] Y. J. Ionin, *Isosceles sets*, The Electronic Journal of Combinatorics, v. 16, R 141, 2009, pp. 1–24.

[147] Y. J. Ionin, *Strongly isosceles sets*, Mathematical Education, Ser. 3, issue 15, 2011, pp. 154–175 (in Russian).

[148] N. Jacobson, *Lectures in Abstract Algebra*, Part III, Springer, New York, 1975.

[149] R. A. Jarvis, *On the identification of the convex hull of a finite set of points in the plane*, Information Processing Letters, v. 2, 1973, pp. 18–21.

[150] T. J. Jech, *The Axiom of Choice*, North-Holland Publishing Company, Amsterdam, 1973.

[151] T. J. Jech, *Set Theory*, Springer-Verlag, Berlin, 2003.

[152] Z. Jiang, A. Polyanskii, *Proof of Lászlό Fejes Toth's zone conjecture*, Geom. and Funct. Analysis, v. 27, issue 6, 2017, pp. 1367–1377.

[153] M. Jonsson, J. Wastlund, *Partitions of* $\mathbf{R}^3$ *into curves*, Mathematica Scandinavica, v. 83, n. 2, 1998, pp. 192–204.

[154] J. Kahn, G. Kalai, *A counterexample to Borsuk's conjecture*, Bull. Amer. Math. Soc. (N.S.), v. 29, n. 1, 1993, pp. 60–62.

[155] H. Kaiser, *Mehrfach-orthogonale Simplexe in Räumen konstanter Krümmung*, Beitrage zur Algebra und Geometrie, v. 19, 1985, pp. 131–143.

[156] H. Kaiser, *On a classification of n-dimensional simplices*, Bull. Acad. Sci. GSSR, v. 125, n. 1, 1987, pp. 29–31 (in Russian).

[157] G. Kalai, *Some old and new problems in combinatorial geometry I: Around Borsuk's problem*, Surveys in Combinatorics, Cambridge University Press, Cambridge, 2015, pp. 147–174.

[158] A. Kanamori, *The Higher Infinite*, Springer-Verlag, Berlin, 2003.

[159] L. Karlitz, *Some inequalities for a triangle*, Amer. Math. Monthly, v. 71, 1964, pp. 881–885.

[160] E. A. Kasimatis, *Dissections of regular polygons into triangles of equal areas*, Discrete and Computational Geometry, v. 4, n. 1, 1989, pp. 375–381.

[161] M. Katz, J. Reimann, *An Introduction to Ramsey Theory: Fast Functions, Infinity, and Metamathematics*, AMS, New York, 2018.

[162] D. K. Kazarinoff, *A simple proof of the Erdös–Mordell inequality for triangles*, Michigan Math. Journal, v. 4, 1957, pp. 97–98.

[163] N. D. Kazarinoff, *D. K. Kazarinoff's inequality for tetrahedra*, Michigan Math. Journal, v. 4, 1957, pp. 99–104.

[164] N. D. Kazarinoff, *Geometric Inequalities*, New York-Toronto, 1961.

[165] J. L. Kelley, *The Tychonoff product theorem implies the Axiom of Choice*, Fund. Math., v. 37, 1950, pp. 75–76.

[166] L. M. Kelly, *E 735*, The Amer. Math. Monthly, v. 54, 1947, pp. 227–229.

[167] L. M. Kelly, W. O. J. Moser, *On the number of ordinary lines determined by n points*, Canad. Journ. of Mathematics, v. 10, 1958, pp. 210–219.

[168] A. B. Kharazishvili, *On a property of convex hulls*, Bull. Acad. Sci. GSSR, v. 71, n. 1, 1973, pp. 25–27 (in Russian).

[169] A. B. Kharazishvili, *To the problem of illumination*, Bull. Acad. Sci. GSSR, v. 71, n. 2, 1973, pp. 289–291 (in Russian).

[170] A. B. Kharazishvili, *Characteristic properties of parallelepipeds*, Bull. Acad. Sci. GSSR, v. 72, n. 1, 1973, pp. 17–19 (in Russian).

[171] A. B. Kharazishvili, *Some combinatorial properties of subsets of the Euclidean spaces*, Thesis, Izd. Tbil. Gos. Univ., Tbilisi, 1974 (in Russian).

[172] A. B. Kharazishvili, *Realization of the coloured graphs in Hilbert spaces*, Bull. Acad. Sci. GSSR, v. 76, n. 1, 1974, pp. 53–56 (in Russian).

[173] A. B. Kharazishvili, *On decompositions of polytopes*, Bull. Acad. Sci. GSSR, v. 77, n. 1, 1975, pp. 37–39 (in Russian).

[174] A. B. Kharazishvili, *On locally conical sets*, Bull. Acad. Sci. GSSR, v. 80, n. 1, 1975, pp. 29–32 (in Russian).

[175] A. B. Kharazishvili, *On intransitivity classes of motion groups of the Euclidean spaces*, DAN SSSR, v. 226, n. 3, 1976, pp. 520–522 (in Russian).

[176] A. B. Kharazishvili, *Fully symmetric configurations*, Bull. Acad. Sci. GSSR, v. 85, n. 2, 1977, pp. 297–300 (in Russian).

[177] A. B. Kharazishvili, *A characterization of rt-sets in Euclidean spaces*, Bull. Acad. Sci. GSSR, v. 86, n. 3, 1977, pp. 537–540 (in Russian).

[178] A. B. Kharazishvili, *Orthogonal simplexes in the four-dimensional space*, Bull. Acad. Sci. GSSR, v. 88, n. 1, 1977, pp. 33–36 (in Russian).

[179] A. B. Kharazishvili, *Equidecomposability of polyhedra with respect to the group of homotheties and translations*, DAN SSSR, v. 236, n. 3, 1977, pp. 552–555 (in Russian).

[180] A. B. Kharazishvili, *Some properties of the centre of symmetry*, Bull. Acad. Sci. GSSR, v. 89, n. 1, 1978, pp. 21–24 (in Russian).

[181] A. B. Kharazishvili, *Affine diameters of convex bodies*, Bull. Acad. Sci. GSSR, v. 90, n. 3, 1978, pp. 541–544 (in Russian).

[182] A. B. Kharazishvili, *Questions in the Theory of Sets and in Measure Theory*, Izd. Tbil. Gos. Univ., Tbilisi, 1978 (in Russian).

[183] A. B. Kharazishvili, *On equidecomposability of three-dimensional polyhedra with respect to the group of parallel translations and central symmetries*, DAN SSSR, v. 243, n. 6, 1978, pp. 1410–1413 (in Russian).

[184] A. B. Kharazishvili, *On the index of an isometric embedding*, Proc. I. Vekua Inst. Appl. Math., v. 5-6, Tbilisi, 1978, pp. 259–264 (in Russian).

[185] A. B. Kharazishvili, *Selected Topics in the Geometry of Euclidean Spaces*, Izd. Tbil. Gos. Univ., Tbilisi, 1978 (in Russian).

[186] A. B. Kharazishvili, *Elements of the Combinatorial Theory of Infinite Sets*, Izd. Tbil. Gos. Univ., Tbilisi, 1981 (in Russian).

[187] A. B. Kharazishvili, *On a property of families of convex sets*, Bull. Acad. Sci. GSSR, v. 102, n. 1, 1981, pp. 25–28 (in Russian).

[188] A. B. Kharazishvili, *To the theory of equidecomposability*, Bull. Acad. Sci. GSSR, v. 103, n. 3, 1981, pp. 537–540 (in Russian).

[189] A. B. Kharazishvili, *To the general theory of volumes*, Bull. Acad. Sci. GSSR, v. 108, n. 3, 1982, pp. 485–488 (in Russian).

[190] A. B. Kharazishvili, *On simple polyhedra*, Reports of Seminar of I. Vekua Inst. Appl. Math., v. 18, Tbilisi, 1984, pp. 34–38 (in Russian).

[191] A. B. Kharazishvili, *On a problem of combinatorial geometry*, Bull. Acad. Sci. GSSR, v. 118, n. 1, 1985, pp. 37–40 (in Russian).

[192] A. B. Kharazishvili, *Partition of the three-dimensional space into congruent circles*, Bull. Acad. Sci. GSSR, v. 119, n. 1, 1985, pp. 57–60 (in Russian).

[193] A. B. Kharazishvili, *Introduction to Combinatorial Geometry*, Tbilisi State University Press, Tbilisi, 1985 (in Russian).

[194] A. B. Kharazishvili, *Quasi-polygons and their uncountable unions*, Bull. Acad. Sci. GSSR, v. 124, n. 3, 1986, pp. 465–468 (in Russian).

[195] A. B. Kharazishvili, *On some properties of Jordan curves*, Bull. Acad. Sci. GSSR, v. 129, n. 1, 1988, pp. 29–32 (in Russian).

[196] A. B. Kharazishvili, *Groups of motions and the uniqueness property of Lebesgue measure*, Bull. Acad. Sci. GSSR, v. 130, n. 1, 1988, pp. 29–32 (in Russian).

[197] A. B. Kharazishvili, *Realizations of graphs in the three-dimensional Euclidean space*, Abstracts of the IX-th Soviet Union Conference in Geometry, Kishinev, 1988, p. 338 (in Russian).

[198] A. B. Kharazishvili, *Applications of Set Theory*, Izd. Tbil. Gos. Univ., Tbilisi, 1989 (in Russian).

[199] A. B. Kharazishvili, *On some topological realizations if infinite graphs*, Bull. Acad. Sci. Georgia, v. 142, 1991, n. 3, pp. 485–488 (in Russian).

[200] A. B. Kharazishvili, *Some partitions consisting of Jordan curves*, Georgian Math. Journ., v. 3, n. 3, 1996, pp. 233–238.

[201] A. B. Kharazishvili, *A note on convexly independent subsets of an infinite set of points*, Georgian Math. Journ., v. 9, n. 2, 2002, pp. 303–307.

[202] A. B. Kharazishvili, *On maximal ot-subsets of the Euclidean plane*, Georgian Math. Journ., v. 10, n. 1, 2003, pp. 127–131.

[203] A. B. Kharazishvili, *On faces of convex polyhedra in* $\mathbf{R}^3$ *with a small number of sides*, Georgian Math. Journ., v. 10, n. 4, 2003, pp. 709–715.

[204] A. B. Kharazishvili, *On homogeneous coverings of Euclidean spaces*, Georgian Math. Journ., v. 11, n. 1, 2004, pp. 105–109.

[205] A. B. Kharazishvili, *On decompositions of a cube into cubes and simplexes*, Georgian Math. Journ., v. 13, n. 2, 2006, pp. 285–290.

[206] A. B. Kharazishvili, *Some combinatorial properties of finite line-systems in the Euclidean plane*, Georgian Math. Journ., v. 14, n. 4, 2007, pp. 681–686.

[207] A. B. Kharazishvili, *On some families of convex functions*, Workshop: Variable Exponent Analysis and Related Topics, 2-5 September, Tbilisi, Georgia, 2008, p. 9.

[208] A. B. Kharazishvili, *On some applications of set-theoretical methods in Euclidean geometry*, Bulletin of TICMI, v. 13, 2009, 12–20.

[209] A. B. Kharazishvili, *Piecewise affine approximations of continuous functions of several variables and Gale polyhedra*, Proc. A. Razmadze Math. Institute, v. 152, 2010, pp. 133–140.

[210] A. B. Kharazishvili, *On non-elementary methods in elementary geometry*, Journal of Geometry, v. 97, numbers 1-2, 2010, p. 125.

[211] A. B. Kharazishvili, *Some discrete geometric structures and associated algorithms*, Proc. A. Razmadze Math. Inst., v. 155, 2011, pp. 142–148.

[212] A. B. Kharazishvili, *A note on algebraic convex curves of constant width*, Georgian Math. Journ., v. 18, n. 4, 2011, pp. 727–733.

[213] A. B. Kharazishvili, *On finite and infinite ot-sets*, Bulletin of TICMI, v. 15, 2011, pp. 27–34.

[214] A. B. Kharazishvili, *Some properties of at-sets and ot-sets in a Hilbert space*, Proc. A. Razmadze Math. Inst., v. 159, 2012, pp. 151–155.

[215] A. B. Kharazishvili, *On some topologic-geometrical properties of external bisectors of a triangle*, Georgian Math. Journ., v. 19, n. 4, 2012, pp. 697–704.

[216] A. B. Kharazishvili, *Some geometric consequences of Ramsey's combinatorial theorem*, Bulletin of TICMI, v. 16, n. 1, 2012, pp. 34–42.

[217] A. B. Kharazishvili, *On three-colorings of the Euclidean plane and associated triangles of a prescribed type*, Journal of Geometry, v. 105, issue 1, 2014, p. 193.

[218] A. B. Kharazishvili, *On iscribed and circumscribed convex polyhedra*, Proc. A. Razmadze Math. Inst., v. 167, 2015, pp. 123–129.

[219] A. B. Kharazishvili, *On the cardinalities of at-sets in a real Hilbert space*, Georgian Math. Journ., v. 22, n. 2, 2015, pp. 259–264.

[220] A. Kharazishvili, *Elements of Combinatorial Geometry*, Part I, The Publishing House of Georgian National Academy of Sciences, Tbilisi, 2016.

[221] A. B. Kharazishvili, *Acute triangles in the context of the illumination problem*, Annuaire de L'Universite de Sofia "St. Kliment Ohridski", Faculty of Mathematics and Informatics, v. 103, 2016, pp. 1–6.

[222] A. B. Kharazishvili, *On $(n+1)$-colorings of the n-space and associated isosceles simplexes*, Journal of Geometry, v. 111, issue 2, 2020, p. ?.

[223] A. Kharazishvili, *Elements of Combinatorial Geometry*, Part II, The Publishing House of Georgian National Academy of Sciences, Tbilisi, 2020.

[224] A. B. Kharazishvili, *On a geometric statement of Ramsey type*, Georgian Math. Journ., v. 29, n. 2, 2022, pp. 229–232.

[225] A. B. Kharazishvili, *On rainbow isosceles n-simplexes*, Georgian Math. Journ., v. 29, n. 4, 2022, pp. 543–549.

[226] A. B. Kharazishvili, T. Sh. Tetunashvili, *Combinatorial properties of families of sets and Euler-Venn diagrams*, Proc. A. Razmadze Math. Inst., v. 146, 2008, pp. 115–119.

[227] A. B. Kharazishvili, T. Sh. Tetunashvili, *On some combinatorial problems concerning geometrical realizations of finite and infinite families of sets*, Georgian Math. Journ., v. 15, n. 4, 2008, pp. 665–675.

[228] A. B. Kharazishvili, T. Sh. Tetunashvili, *On some coverings of the Euclidean plane with pairwise congruent circles*, Amer. Math. Monthly, v. 117, n. 5, 2010, pp. 414–423.

[229] A. Kharazishvili, A. Kirtadze, *On weakly metrically transitive measures and nonmeasurable sets*, Real Analysis Exchange, v. 32, n. 2, 2007, pp. 553–562.

[230] H. Kido, *Classification of isosceles eight-point sets in three-dimensional Euclidean space*, Europ. J. Combin., v. 27, 2006, pp. 329–341.

[231] H. Kierstead, A. Kostochka, *A short proof of the Hajnal–Szemerédi theorem on equitable coloring*, Combinatorics, Probability and Computing, v. 17, n. 2, 2008, pp. 265–270.

[232] V. Klee, *What is a convex set?*, American Mathematical Monthly, v. 78, 1971, pp. 616–631.

[233] F. Klein, *Vergleichende Betrachtungen über neuere geometrische Forschungen*, (1872), Math. Annalen, v. 43, 1893, pp. 63–100.

[234] N. Koblitz, P-adic Numbers, P-adic Analysis, and Zeta-Functions, Springer-Verlag, Berlin-Heidelberg, 1984.

[235] P. Komjáth, V. Totik, *Problems and Theorems in Classical Set Theory*, Springer-Verlag, New York, 2006.

[236] E. Kopczyński, I. Pak, P. Przytycki, *Acute triangulations of polyhedra and* $\mathbf{R}^n$, Combinatorica, v. 32, issue 1, 2012, pp. 85–110.

[237] K. Kunen, *Set Theory*, North-Holland Publishing Company, Amsterdam, 1980.

[238] A. B. Kupavskii, J. Pach, *From Tarski's plank problem to simultaneous approximation*, Amer. Math. Monthly, v. 124, issue 6, 2017, pp. 494–505.

[239] K. Kuratowski, *Sur le problème des courbes gauches en topologie*, Fund. Math., v. 15, 1930, pp. 271–283.

[240] K. Kuratowski, *Topology*, v. 1, v. 2, Academic Press, New York-London, 1966, 1969.

[241] K. Kuratowski, A. Mostowski, *Set Theory*, North-Holland Publ. Co., Amsterdam, 1967.

[242] K. Kuratowski, H. Steinhaus, *Une application géométrique du théoréme de Brouwer sur les points invariants*, Bull. Acad. Polon. Sci., v. 1, 1953, pp. 83–86.

[243] A. S. Kuzichev, *Venn Diagrams*, Izd. Nauka, Moscow, 1968 (in Russian).

[244] J.-Ph. Labbé, G. Rote, G. M. Ziegler, *Area difference bounds for dissections of a square into an odd number of triangles*, arXiv: 1708.02891v3

[245] 6 Jun 2018.

[246] D. G. Larman, *A problem of incidence*, J. London Math. Soc., v. 43, 1968, pp. 407–409.

[247] D. G. Larman, C. A. Rogers, J. J. Seidel, *On two-distance sets in Euclidean space*, Bull. London Math. Soc., v. 9, 1977, pp. 261–267.

[248] H.-Chr. Lenhard, *Verallgemeinerung und Verschärfung der Erdös–Mordellschen Satz für Polygone* Archiv der Mathematik, v. 12, 1961, pp. 311–314.

[249] P. Lisonek, *New maximal two-distance sets*, Journ. Combin. Theory, Ser. A, v. 77, 1997, pp. 318–338.

[250] L. A. Lusternik, *Convex Figures and Polyhedra*, Heath, Boston, 1966 (translation from the Russian).

[251] D. W. Lyons, *An elementary introduction to the Hopf fibration*, Mathematics Magazine, v. 76, n. 2, 2003, pp. 87–98.

[252] Yu. Makarychev, *A short proof of Kuratowski's graph planarity criterion*, Journal of Graph Theory, v. 25, 1997, pp. 129–131.

[253] H. Martini, V. Soltan, *Combinatorial problems on the illumination of convex bodies*, Aequationes Mathematicae, v. 57, 1999, pp. 121–152.

[254] J. Matousek, *Lectures on Discrete Geometry*, Springer-Verlag, New York, 2002.

[255] S. Mazurkiewicz, *Sur un ensemble plan qui a avec chaque droite deux et seulement deux points communs*, C. R. Varsovie, v. 7, 1914, pp. 382–384.

[256] D. G. Mead, *Dissection of the hypercube into simplexes*, Proc. Amer. Math. Soc., v. 76, n. 2, 1979, pp. 302–304.

[257] E. Melchior, *Über Vielseite der projektiven Ebene*, Deutsche Mathematik, v. 5, 1940, pp. 461–475.

[258] Z. A. Melzak, *Problems connected with convexity*, Canad. Math. Bull., v. 8, 1965, pp. 565–573.

[259] Z. A. Melzak, *More problems connected with convexity*, Canad. Math. Bull., v. 11, 1968, pp. 489–494.

[260] E. Mendelson, *Introduction to Mathematical Logic*, 6th edition, CRC Press, Boca Raton, 2015.

[261] H. Minkowski, *Geometrie der Zahlen*, Lfg 1-2, Leipzig-Berlin, 1896–1910.

[262] P. Mironescu, L. Panaitopol, *The existence of a triangle with prescribed angle bisector lengths*, Amer. Math. Monthly, v. 101, n. 1, 1994, pp. 58–60.

[263] D. S. Mitrinović, J. E. Pecarić, *On the Erdös–Mordell inequality for a polygon*, J. College Arts Sci., Chiba Univ., B-19, 1986, pp. 3–6.

[264] P. Monsky, *On dividing a square into triangles*, Amer. Math. Monthly, v. 77, n. 2, 1970, pp. 161–164.

[265] L. J. Mordell, D. F. Barrow, *Solution of Problem 3740*, Amer. Math. Monthly, v. 44, 1937, pp. 252–254.

[266] W. Morris and V. Soltan, *The Erdös–Szekeres problem on points in convex position – a survey*, Bull. Amer. Math. Soc., v. 37, n. 4, 2000, pp. 437–458.

[267] A. Mukhopadhyay, E. Greene, *The ordinary line problem revisited*, Computational Geometry, v. 45, issue 3, 2012, pp. 127–130.

[268] S. Mukhopadhyaya, *New methods in the geometry of a plane arc*, Bull. Calcutta Math. Soc., I, 1909, pp. 31–37.

[269] J. R. Munkres, *Topology*, Pearson, NJ, 2000.

[270] O. R. Musin, *Spherical two-distance sets*, Journ. Combin. Theory, Ser. A, v. 116, 2009, pp. 988–995.

[271] M. B. Nathanson, *Additive Number Theory*, Springer-Verlag, New York, 1996.

[272] O. Nechushtan, *On the space chromatic number*, Discrete Math., v. 256, 2002, pp. 499–507.

[273] K. L. Nyman, F. E. Su, *A Borsuk–Ulam equivalent that directly implies Sperner's lemma*, Amer. Math. Monthly, v. 120, n. 4, 2013, pp. 346–354.

[274] A. Oppenheim, *The Erdös inequality and other inequalities for a triangle*, Amer. Math. Monthly, v. 68, 1961, pp. 226–230.

[275] A. Oppenheim, *Some inequalities for a spherical triangle and an internal point*, Publ. Elektrotehn. fak. Univ. Beogradu, Ser. Math. i Fiz., N 200–209, 1967, pp. 13–16.

[276] O. Ore, *Theory of Graphs*, Amer. Math. Soc., Providence, Rhode Island, 1962.

[277] S. F. Osinkin, *On the existence of a triangle with prescribed bisector lengths*, Forum Geometr., v. 16, 2016, pp. 399–405.

[278] J. Pach (ed.), *New Trends in Discrete and Computational Geometry*, Springer-Verlag, Berlin, 1993.

[279] I. Pak, *Lectures on Discrete and Polyhedral Geometry*, UCLA, 2010.

[280] L. Pakula, *A note on Venn diagrams*, Amer. Math. Monthly, v. 96, 1989, pp. 38–39.

[281] V. Pambuccian, *Existence of special rainbow triangles in weak geometries*, Georgian Mathematical Journal, v. 26, n. 4, 2019, pp. 489–498.

[282] J. Pawlikowski, *The Hahn–Banach theorem implies the Banach–Tarski paradox*, Fund. Math., v. 138, 1991, pp. 21–22.

[283] C. Petty, D. Waterman, *An extreme theorem for n-simplexes*, Monatsh. Math., v. 59, 1955, pp. 320–322.

[284] V. Pokrovskii, *Decompositions of n-dimensional polyhedra and the theorem of Zylev and Debrunner*, Proc. A. Razmadze Math. Inst., v. 85, 1987, pp. 34–36 (in Russian).

[285] F. P. Preparata and S. J. Hong, *Convex hulls of finite sets of points in two and three dimensions*, Communications of the ACM, v. 20, n. 2, 1977, pp. 87–93.

[286] F. P. Preparata and M. I. Shamos, *Computational Geometry: An Introduction*, Springer-Verlag, New York, 1985.

[287] S. Rabinowitz, *A polynomial curve of constant width*, Missouri Journal of Mathematical Sciences, v. 9, 1997, pp. 23–27.

[288] A. M. Raigorodskii, *Borsuk's problem and the chromatic numbers of some metric spaces*, Russian Math. Surveys, v. 56, 2001, pp. 103–139.

[289] A. M. Raigorodskii, *Chromatic Numbers*, Series: Mathematical Education, Moscow, 2003 (in Russian).

[290] A. M. Raigorodskii, *Borsuk's Problem*, Series: Mathematical Education, Moscow, 2006 (in Russian).

[291] A. M. Raigorodskii, *Acute Triangles of Danzer and Grünbaum*, Series: Mathematical Education, Moscow, 2009 (in Russian).

[292] F. P. Ramsey, *On a problem of formal logic*, Proc. London Math. Soc., v. 30, 1930, pp. 264–286.

[293] M. Reid, *Undergraduate Algebraic Geometry*, Cambridge University Press, Cambridge, 1998.

[294] A. Rényi, V. Rényi, J. Surányi, *Sur l'indépendance des domaines simples dans l'espace Euclidien a n dimensions*, Colloq. Mathem., v. 2, 1951, pp. 130–135.

[295] F. Richman, J. Thomas, *Problem 5471*, Amer. Math. Monthly, v. 74, 1967, p. 329.

[296] D. P. Robbins, *Areas of polygons inscribed in a circle*, Amer. Math. Monthly, v. 102, 1995, pp. 523–530.

[297] A. Robinson, *Introduction to Model Theory and to the Metamathematics of Algebra*, North-Holland Publ. Co., Amsterdam, 1963.

[298] C. V. Robinson, *A characterization of the disc*, Bull. Amer. Math. Soc., v. 47, 1941, pp. 818–819.

[299] R. T. Rockafellar, *Convex Analysis*, Princeton University Press, Princeton, 1997.

[300] J. Rosen, *How to tee a hyperplane*, The American Mathematical Monthly, v. 129, issue 8, 2022, pp. 781–784.

[301] M. E. Rudin, *An unshellable triangulation of a tetrahedron*, Bulletin Amer. Math. Soc., v. 64, 1958, pp. 90–91.

[302] I. Kh. Sabitov, *The polyhedron's volume as a function of lengths of its edges*, Fundam. Prikl. Mat., v. 2, n. 1, 1996, pp. 305–307 (in Russian).

[303] I. Kh. Sabitov, *The generalized Heron-Tartaglia formula and some of its consequences*, Sbornik Math., v. 189, 1998, pp. 1533–1561.

[304] T. Sajeniuk, *On some inductive constructions in combinatorial geometry*, Proc. I. Vekua Inst. of Appl. Math., v. 20, 1987, pp. 32–50 (in Russian)

[305] T. Sajeniuk, *On combinatorial properties of subsets of the Euclidean plane*, Proc. A. Razmadze Math. Inst., v. 85, 1987, pp. 37–39 (in Russian).

[306] L. A. Santaló, *Integral Geometry and Geometric Probability*, Addison-Wesley Publishing Company, Inc., Reading, Massachusetts, 1976.

[307] H. H. Schaefer, M. P. Wolff, *Topological Vector Spaces*, Springer, Berlin-Heidelberg, 1999.

[308] J. H. Schmerl, *Some 2-point sets*, Fund. Math., v. 208, n. 1, 2010, pp. 87–91.

[309] R. Schneider, *Convex Bodies: The Brunn-Minkowski Theory*, Cambridge University Press, Cambridge, 1993.

[310] E. Schönhardt *Über die zerlegung von dreieckspolyedern in tetraeder*, Mathematische Annalen, v. 98, 1928, pp. 309–312.

[311] A. Seidenberg, *A new decision method for elementary algebra*, Ann. Math., v. 60, 1954, pp. 365–374.

[312] M. Shamos, *Geometric complexity*, Proceedings of the 7th ACM Symposium on the Theory of Computing, 1975, pp. 224–253.

[313] I. Shashkin, *Euler characteristic*, Izd. Nauka, Moscow, 1984 (in Russian).

[314] S. Shelah, A. Soifer, *Axiom of Choice and chromatic number of the plane*, Journal of Combinatorial Theory, Ser. A, v. 103, 2003, pp. 387–391.

[315] D. O. Shkliarskij, N. N. Chentsov, I. M. Iaglom, *Selected Problems and Theorems in Plane Geometry*, Izd. Nauka, Moscow, 1967 (in Russian).

[316] D. O. Shkliarskij, N. N. Chentsov, I. M. Iaglom, *Geometric Inequalities and Problems on Maxima and Minima*, Izd. Nauka, Moscow, 1970 (in Russian).

[317] W. Sierpiński, *Cardinal and Ordinal Numbers*, PWN, Warszawa, 1958.

[318] A. B. Skopenkov, *n-dimensional cube, polynomials, and solution of Borsuk's problem*, Mathematical Education, Ser. 3, issue 3, 1999, pp. 184–188.

[319] A. B. Skopenkov, *Around Kuratowski's graph planarity criterion*, Mathematical Education, Ser. 3, v. 9, 2005, pp. 116–128 (in Russian).

[320] A. Soifer, *The Mathematical Coloring Book*, Springer-Verlag, New York, 2009.

[321] A. Soifer, S. Shelah, *Axiom of Choice and chromatic number: examples on the plane*, Journal of Combinatorial Theory, Ser. A, v. 105, 2004, pp. 359–364.

[322] R. M. Solovay, *A model of set theory in which every set of reals is Lebesgue measurable*, Ann. Math., Ser. 2, v. 92, 1970, pp. 1–56.

[323] V. Soltan, *Pairs of convex bodies with centrally symmetric intersections of translates*, Discrete Comp. Geometry, v. 33, n. 4, 2005, pp. 605–616.

[324] J. Solymosi, T. Tao, *An incidence theorem in higher dimensions*, Discrete and Computational Geometry, v. 48, issue 2, 2012, pp. 255–280.

[325] R.Sprague, *Beispiel einer Zerlegung des Quadrats in lauter verschiedene Quadrate*, Math. Z., 45 (1939), pp. 607–608.

[326] S. K. Stein, *Mathematics: The Man-Made Universe*, Dover Publications, New York, 1998.

[327] Sh. Stein, *Cutting a polygon into triangles of equal areas*, The Mathematical Intelligencer, v. 26, n. 1, 2004, pp. 17–21.

[328] J. Steinig, *A comparison of two inequalities for the triangle*, Acta Math. Acad. Hungar., v. 16, 1965, pp. 19–22.

[329] E. Steinitz, *Bedingt konvergente Reihen und konvexe Systeme*, J. Reine Angew. Math., v. 143, 1913, pp.128–175; v. 144, 1914, pp. 1–40; v. 146, 1916, pp. 1–52.

[330] E. Steinitz, *Polyeder und Raumeinteilungen*, Enzylk. Math. Wiss., v. 3, 1922, pp. 1–139.

[331] J. Stillwell, *Mathematics and its History*, 3rd edition, Springer-Verlag, Berlin-New York, 2010.

[332] A. Stöhr, *Über Zerlegungen von Rechtecken in inkongruente Quadrate*, Diss. Univ. Berlin, 1938, Schriften des Math. Inst. und des Inst. für angew. Math. der Univ. Berlin, 4 (1939), pp. 119–140.

[333] F. E. Su, *Borsuk–Ulam implies Brouwer: a direct construction*, Amer. Math. Monthly, v. 104, 1997, pp. 855–859.

[334] Sun-Joo-Shin, *The Logical Status of Diagrams*, Cambridge University Press, Cambridge, 1994.

[335] J. J. Sylvester, *Question 11851*, Educational Times, v. 59, 1893, p. 98.

[336] L. Székely, *Crossing numbers and hard Erdös problems in discrete geometry*, Combinatorics, Probability and Computing, v. 6, issue 3, 1997, pp. 353–358.

[337] E. Szemerédi, W. T. Trotter, *Extremal problems in discrete geometry*, Combinatorica, v. 3, n. 3–4, 1983, pp. 381–392.

[338] A. Szulkin, $\mathbf{R}^3$ *is the union of disjoint circles*, Amer. Math. Monthly, vol. 90, 1983, pp 640–641.

[339] A. Tarski, *Uwagi o stopniu równowazności wielokatów*, Parametr, v. 2, 1932, pp. 310–314 (in Polish).

[340] A. Tarski, *Der Wahrheitsbegriff in den formalisierten Sprachen*, Studia Philosophica, v. 1, 1936, pp. 261–405.

[341] A. Tarski, *A Decision Method of Elementary Algebra and Geometry*, University of California Press, Berkeley, 1951.

[342] A. Tarski, *What is elementary geometry?*, in: The Axiomatic Method (editors: Henkin, Suppes, and Tarski), North-Holland Publ. Co., Amsterdam, 1959, pp. 1–15.

[343] J. Thomas, *A dissection problem*, Math. Magazine, v. 41, 1968, pp. 187–190.

[344] R. Thomas, *A combinatorial construction of a non-measurable set*, Amer. Math. Monthly, v. 92, 1985, pp. 421–422.

[345] H. Tietze, *Über das Problem der Nachbargite im Raum*, Monatsh. Math. Phys., v. 16, 1905, pp. 211–216.

[346] G. T. Toussaint, D. Avis, *On convex hull algorithm for polygons and its application to triangulation problems*, Pattern Recognition, v. 15, n. 1, 1982, pp. 23–29.

[347] K. Tschirpke, *On the dissection of simplices into orthoschemes*, Geometriae Dedicata, v. 46, n. 3, 1993, pp. 313–329.

[348] K. Tschirpke, *The dissection of five-dimensional simplices into orthoschemes*, Beiträge Algebra Geom., v. 35, 1994, pp. 1–11.

[349] A. W. Tucker, *Some topological properties of disk and sphere*, Proc. First Canad. Math. Congress, Montreal; University of Toronto Press, Toronto, 1945, pp. 285–309.

[350] D. de Turck, H. Gluck, D. Pomerleano, D. S. Vick, *The four vertex theorem and its converse*, Notices of the Amer. Math. Soc., v. 54, n. 2, 2007, pp. 192–207.

[351] F. A. Valentine, *Convex Sets*, McGraw-Hill, New York, 1964.

[352] E. VanderZee, A. N. Hirani, V. Zharnitsky, D. Guoy, *A dihedral acute triangulation of the cube*, Computational Geometry, v. 43, n. 5, 2010, pp. 445–452.

[353] J. Venn, *On the diagrammatic and mechanical representation of propositions and reasonings*, The London, Edinburgh, and Dublin Philosophical Magazine and Journal of Science, v. 9, 1880, pp. 1–18.

[354] G. Vitali, *Sul problema della misura dei gruppi di punti di una retta*, Tip. Gamberini e Parmeggiani, Bologna, 1905.

[355] A. Yu. Volovikov, *Brouwer, Kakutani, and Borsuk–Ulam theorems*, Matem. Zametki, v. 79, issue 3, 2006, pp. 471–473 (in Russian).

[356] S. Wagon, *The Banach–Tarski Paradox*, Cambridge University Press, Cambridge, 1985.

[357] R. Webster, *Convexity*, Oxford University Press, Oxford, 1994.

[358] B. Weglorz, *Nerves and set-theoretical independence*, Colloquium Mathematicum, v. 13, 1964, pp. 17–19.

[359] J. B. Wilker, *Tiling* $\mathbf{R}^3$ *with circles and disks*, Geometriae Dedicata, v. 32, issue 2, 1989, pp. 203–209.

[360] I. M. Yaglom, *How to Dissect a Square?*, Izd. Nauka, Moscow, 1968 (in Russian).

[361] I. M. Yaglom, V. G. Boltyanskii, *Convex Figures*, Holt, Rinehart and Winston, New York, 1961 (translation from the Russian).

[362] J. Zaks, *Neighborly families of congruent convex polytopes*, Amer. Math. Monthly, v. 94, 1987, pp. 151–155.

[363] E. Zermelo, *Beweis dass jede Menge wohlgeordnet werden kann*, Math. Ann., v. 59, 1904, pp. 514–516.

[364] A. Zhukov, I. Akulich, *Is a triangle uniquely determined?*, Kvant, n. 1, 2003, pp. 29–31 (in Russian).

[365] G. M. Ziegler, *Lectures on Polytopes*, Springer-Verlag, New York, 1995.

[366] Ch. Zong, *Strange Phenomena in Convex and Discrete Geometry*, Springer-Verlag, Berlin-Heidelberg, 2012.

Index